The Minerals, Metals & Materials Series

Hojong Kim · Bradford Wesstrom
Shafiq Alam · Takanari Ouchi
Gisele Azimi · Neale R. Neelameggham
Shijie Wang · Xiaofei Guan
Editors

Rare Metal Technology 2018

TMS

Springer

Editors
Hojong Kim
The Pennsylvania State University
University Park, PA
USA

Gisele Azimi
University of Toronto
Toronto, ON
Canada

Bradford Wesstrom
Freeport-McMoRan
El Paso, TX
USA

Neale R. Neelameggham
IND LLC
South Jordan, UT
USA

Shafiq Alam
University of Saskatchewan
Saskatoon, SK
Canada

Shijie Wang
Rio Tinto Kennecott Utah Copper Corp
South Jordan, UT
USA

Takanari Ouchi
Massachusetts Institute of Technology
Cambridge, MA
USA

Xiaofei Guan
Watertown, MA
USA

ISSN 2367-1181 ISSN 2367-1696 (electronic)
The Minerals, Metals & Materials Series
ISBN 978-3-319-72349-5 ISBN 978-3-319-72350-1 (eBook)
https://doi.org/10.1007/978-3-319-72350-1

Library of Congress Control Number: 2017960779

Printed on acid-free paper

This Springer imprint is published by Springer Nature
The registered company is Springer International Publishing AG
The registered company address is: Gewerbestrasse 11, 6330 Cham, Switzerland

Preface

Rare Metal Technology 2018 is the proceedings of the symposium on Rare Metal Extraction and Processing sponsored by the Hydrometallurgy and Electrometallurgy Committee of the TMS Extraction and Processing Division. The symposium has been organized to encompass the extraction of rare metals as well as rare extraction processing techniques used in metal production and mineral processing. This is the fifth symposium since 2014, which will be held in Phoenix, Arizona, USA.

This symposium intends to cover research and developments in the extraction and processing of less common, rare metals that are not covered by other TMS symposia. These elements include antimony, bismuth, barium, beryllium, boron, calcium, chromium, gallium, germanium, hafnium, indium, manganese, molybdenum, platinum group metals, rare-earth metals, rhenium, scandium, selenium, sodium, strontium, tantalum, tellurium, and tungsten. These are rare metals of low tonnage sales compared to high tonnage metals such as iron, copper, nickel, lead, tin, zinc, or light metals such as aluminum, magnesium, or titanium and electronic metalloid silicon. Rare processing includes biometallurgy, hydrometallurgy, and electrometallurgy, as well as extraction of values from electric arc furnace (EAF) dusts, and less common waste streams not discussed in recycling symposia. Rare high-temperature processes included microwave heating, solar-thermal reaction synthesis, molten salt electrochemical processes, cold crucible synthesis of the rare metals, and the design of extraction equipment used in these processes as well as laboratory and pilot plant studies.

This volume covers extraction and processing techniques of various platinum group metals, rare earth elements as well as other less common metals such as lead, antimony, molybdenum, tungsten, chromium, titanium, and vanadium, including electrochemical processing, aqueous processing, biological separation, and microwave heating. The symposium is organized into the following sessions: (1) rare earth elements, (2) platinum group metals, (3) base and rare metals (Ni, Cu, Zn, Pb, Sb, Mn, Li, Cs, and Ga), and (4) Ti, V, Mo, and W.

We acknowledge the efforts of the symposium organizers and proceedings editors: Hojong Kim, Bradford Wesstrom, Shafiq Alam, Takanari Ouchi, Gisele Azimi, Neale R. Neelameggham, Xiaofei Guan, and Shijie Wang. The support from

TMS staff members Carol Matty and Patricia Warren is greatly appreciated in assembling and publishing the proceedings. We sincerely thank all the authors, speakers, and participants and look forward to continued collaboration in the advancement of science and technology in the area of rare metal extraction and processing.

Hojong Kim
Lead Organizer

Contents

Part II Rare Earth Elements II and Platinum Group Metals

Part III Base and Rare Metals

Part V Poster Session

About the Editors

Hojong Kim is an Assistant Professor of Materials Science and Engineering, and Norris B. McFarlane Faculty Professor at the Pennsylvania State University. Dr. Kim received his B.S from Seoul National University in South Korea in 2000 and Ph.D. degree from Massachusetts Institute of Technology (MIT) in 2004 both in Materials Science and Engineering. His doctoral research sought to identify the corrosion mechanisms of constructional alloys in high-temperature and high-pressure steam environments under Prof. Ronald Latanision in the Uhlig Corrosion Laboratory at MIT. After graduate research, he worked as a Senior Research Scientist at Samsung-Corning Precision Glass Co. Ltd. and as a project lead to improve the process yield for TFT-LCD glass melting processes. After five years of industrial experience, he returned to MIT as a post-doctoral associate to contribute to the growing need for sustainable technology. He conducted research on high-temperature electrochemical processes, including molten oxide electrolysis for carbon-free steel production with focus on developing inert anode materials in molten slags as well as molten salt liquid metal batteries for large-scale energy storage.

His current research interests embrace the development of environment-friendly electrochemical processes for resource extraction/recycling, corrosion-resistant materials, as well as energy storage systems. He was awarded The Minerals, Metals & Materials Society (TMS) Young Leaders Professional

Development Award in 2013 from the Extraction and Processing Division. In 2015, he was chosen to receive a Doctoral New Investigator Award from the American Chemical Society Petroleum Research Fund. He is currently leading efforts to separate alkali/alkaline-earth fission products from molten salt electrolytes used for recycling used nuclear fuel. He is the lead organizer of the Rare Metal Extraction and Processing Symposium at the TMS 2018 Annual Meeting and is the Chair of the Hydrometallurgy and Electrometallurgy Committee of the Extraction and Processing Division.

Bradford Wesstrom is the Senior Process Engineer for Freeport McMoRan at the El Paso Refinery. In 1989, he received his B.S. degree in Metallurgical Engineering from the University of Texas at El Paso. He has worked within the copper industry for over 34 years, covering areas of smelting, casting, electrorefining, electrowinning, acid purification, metal hydrolysis, metal salts production, roasting, precious metals recovery by pyrometallurgy and hydrometallurgy, pressure leaching, cementation, mechanical vapor recompression for water recovery, copper rod production, and water treatment. He has published on the above areas and given numerous talks at conferences, including TMS, ASTM, STDA, IPMI, IMM, and CANMET.

Shafiq Alam is Associate Professor at the University of Saskatchewan, Canada. In 1998, he received his Ph.D. degree in Chemical Engineering from Saga University, Japan. From 1999 to 2001, he was appointed as a Postdoctoral Research Fellow at the University of British Columbia and the University of Toronto, Canada.

Dr. Alam has extensive experience in industrial operations, management, engineering, design, consulting, teaching, research, and professional services. Before joining the University of Saskatchewan in 2014, he was an Assistant/Associate Professor at Memorial University of Newfoundland for about 7 years. Prior to starting his career in academia, he worked with many different companies, such as Shell, Process Research ORTECH Inc., Fluor Canada Ltd., and the National Institute of Advanced Industrial Science and Technology (AIST), Japan. He is highly experienced in the area of mineral processing and extractive metallurgy, and he possesses 2 patents and has over 150 publications. He is the co-editor of six books and an associate editor of the *International Journal of Mining, Materials and Metallurgical Engineering* (IJMMME). He is the winner of the 2014 TMS Extraction and Processing Division's Technology Award.

He is a registered professional engineer and has worked on projects with many different mining companies including Falconbridge, INCO (Vale), Barrick, Hatch, Phelps Dodge, Rambler, and Anaconda. He is an Executive Committee Member of the Hydrometallurgy Section of the Canadian Institute of Mining, Metallurgy and Petroleum (CIM). During 2015–2017, he served as the Chair on the Hydrometallurgy and Electrometallurgy Committee of the Extraction and Processing Division (EPD) of The Minerals, Metals & Materials Society (TMS). He is a co-organizer of many symposia at the international conferences through CIM and TMS. He is one of the founding organizers of the Rare Metal Extraction and Processing Symposium at TMS.

Takanari Ouchi is a Research Associate in the Institute of Industrial Science at the University of Tokyo. He received his Ph.D. in Nanoscience and Nanoengineering from Waseda University in 2011. He developed electrochemical deposition processes to fabricate metal nanostructures with both well-controlled crystallinity and uniformity at the single nanometer scale and demonstrated the applicability of these processes to fabrication of bit-patterned magnetic recording media for future hard disk drives. After completing his doctoral degree, Dr. Ouchi joined the Massachusetts Institute of Technology (MIT), where he developed liquid metal batteries, which is in principle a bidirectional electrolysis (electrorefining) cell, to apply for the grid-scale energy storage. As a research scientist, he led the systematic investigation of electrochemical properties of liquid metal electrodes in molten salt electrolytes and developed novel lithium, calcium, and sodium liquid metal batteries. He authored 17 peer-reviewed papers and conference proceedings and presented 50 talks at conferences. He has constantly contributed to create the vibrant field of metal extraction by working as a Member of Hydrometallurgy and Electrometallurgy Committee at TMS, organizing technical symposiums at TMS and soliciting papers as a guest editor of *JOM*. He has earned several awards and honors, such as TMS Extraction and Processing Division (EPD) Young Leaders Professional Development Award in 2015 based on his reputation in the electrochemical metal extraction processing.

Gisele Azimi is an Assistant Professor cross-appointed between the Departments of Chemical Engineering and Applied Chemistry, and Materials Science and Engineering at the University of Toronto. She is an expert in the fields of materials extraction, processing, and recovery using hydrometallurgical, electrochemical, and supercritical processes. Her research program is mainly focused on the extraction and recovery of strategic materials, mainly rare earth elements, precious metals, lithium, and iron. This includes molten salt electrolysis and supercritical fluid extraction of rare earth elements as well as recovery of these elements from secondary sources. Another thrust of her research program is focused on the design and fabrication of advanced materials and surfaces with controlled properties. She has received a number of awards including the TMS Light Metals/Extraction and Processing Subject Award, CIM Gordon Ritcey Outstanding Award, and TATP Teaching Excellence Award.

Neale R. Neelameggham is 'The Guru' at IND LLC, involved in international consulting in the fields of metals and associated chemicals (boron, magnesium, titanium, and lithium and rare earth elements), thiometallurgy, energy technologies, soil biochemical reactor design, etc. He was a Visiting Expert at Beihang University of Aeronautics and Astronautics, Beijing, China. He was a Plenary Speaker at the Light Metal Symposium in South Africa—on low carbon dioxide emission processes for magnesium. Dr. Neelameggham has more than 38 years of expertise in magnesium production and was involved in process development of its start-up company NL Magnesium through to the present US Magnesium LLC, UT until 2011. Neelameggham and Brian Davis authored the ICE-JNME award-winning (2016) paper "21st Century Global Anthropogenic Warming Convective Model," which notes that constrained air mass warming is independent of the energy conversion source—fossil or renewable energy. He is presently developing Agricoal™ and agricoalture to improve arid soils. He holds 16 patents and patent applications and has published several technical papers. He has served in the Magnesium Committee of the Light Metals Division (LMD) of TMS since its inception in 2000, chaired it in 2005, and in 2007, he was made a permanent co-organizer for the Magnesium Symposium. He has been a member of the Reactive Metals Committee, Recycling Committee, Titanium Committee, and Program Committee Representative of LMD and LMD council. He was the inaugural chair, when in 2008, LMD and EPD (Extraction & Processing Division) created the Energy Committee, and has been a co-editor of the energy technology symposium proceedings through the present. He received the LMD Distinguished Service Award in 2010. While he was the chair of Hydro- and Electrometallurgy Committee, he initiated the rare metal technology symposium in 2014. He is co-editor for the 2018 proceedings for the symposia on Magnesium Technology, Energy Technology, Rare Metal Technology, and Solar Cell Silicon.

Shijie Wang is a Principal Advisor at the Rio Tinto Kennecott Utah Copper in Salt Lake City. Dr. Wang has been active in extractive metallurgy and has 30 years of experience and expertise in metallurgical process development, existing operation optimization, and troubleshooting. His work interests include extractive and process metallurgy, recycling, waste treatment, metal recovery, operational efficiency, and profitability. He holds three U.S. patents and has published more than 40 journal papers including non-ferrous metals', precious metals', rare metals', and rare earth metals' resourcefulness and recoveries. He received a B.S. degree in Mineral Process from China, and an M.S. degree and a Ph.D. degree in Metallurgical Engineering from the University of Nevada, Reno. He has been very active in TMS since 1991 and is the Former Chair of the Hydrometallurgy and Electrometallurgy Committee of TMS from 2011 to 2013. He has been the lead and co-organizer of six symposia at the international conferences through TMS. He received the TMS Extraction and Processing Division (EPD) Distinguished Service Award in 2017.

Xiaofei Guan is an Assistant Professor in School of Physical Science and Technology at ShanghaiTech University. His primary research interest is in understanding chemical and electrochemical processes for materials synthesis and energy conversion with the goal of improving efficiency and reducing environment impact. He received his B.S. degree in Applied Physics from Nankai University in 2009 and Ph.D. degree in Materials Science and Engineering from Boston University in 2013. His Ph.D. research was on magnesium recycling and electrolytic production of energy-intensive metals from oxides. In 2014, he joined Harvard University as a Postdoctoral Fellow and led research on energy conversion and storage devices including solid oxide fuel cells, protonic ceramic fuel cells, and hydride-air batteries and later worked on synthesis of iron sulfides particularly for solar energy application, which is a joint project between materials, microbiology, and electrochemistry.

Dr. Guan received the Outstanding Ph.D. Dissertation Award in Materials Science and Engineering from Boston University in 2014 and the Young Leaders Professional Development Award from the TMS Extraction and Processing Division in 2015. He also serves as an Advisor and Guest Editor in the Recycling and Environmental Technology Committee for *JOM*.

Part I
Rare Earth Element I

The Demonstration Pilot Plant Results for the Search Minerals Direction Extraction Process for Rare Earth Recovery

David Dreisinger, Greg Andrews, Niels Verbaan, Mike Johnson and Ernesto Bourricaudy

Abstract Search Minerals has developed a Direct Extraction Process for recovery of rare earths from the Foxtrot and Deep Fox deposits in Labrador Canada. The process includes crushing, acid treatment, water leaching, impurity removal, precipitation of an intermediate rare earth product and finally refining of the intermediate product to produce a high purity mixed rare earth oxide ($\sim$99% total rare each oxides). The Direct Extraction Process has been validated in an extensive demonstration pilot plant program at SGS Minerals. A total of 3 tonnes of material was processed to produce final mixed rare earth product. The results of the demonstration pilot plant program are reported.

Keywords Rare earths · Extraction · Mixed rare earth oxide · Foxtrot Deep Fox

Introduction

Search Minerals Inc. (Search) is exploring and developing a number of deposits for rare earth element (REE) recovery in Labrador, Canada. The Port Hope Simpson (PHS) District is approximately 70 km by 8 km in size, and is highly prospective for heavy and light rare earth elements.

There are a total of more than 20 prospects that have been identified in the PHS District. The Foxtrot deposit sits within the Port Hope Simpson REE District. The infrastructure available at Foxtrot is excellent; a deep water port, air strip, road and power infrastructure are preexisting at Port Hope SimpsonSt. Lewis (Fig. 2). The

D. Dreisinger (✉) · G. Andrews
Search Minerals, #211, 901 West 3rd Street, V7P 3P9 North Vancouver, BC, Canada
e-mail: david_dreisinger@yahoo.com

N. Verbaan · M. Johnson · E. Bourricaudy
SGS Minerals Services, 185 Concession Street, PO Box 4300, K0L 2H0 Lakefield, ON, Canada

H. Kim et al. (eds.), *Rare Metal Technology 2018*, The Minerals, Metals & Materials Series, https://doi.org/10.1007/978-3-319-72350-1_1

three communities of Port Hope Simpson, St. Lewis and Mary's Harbour are in close proximity to the site.

The Foxtrot deposit has a NI43-101 resource of 7,392,000 t (indicated) at 1.09% TREO 1,958,000 t (inferred) at 1.17% TREO (Table 1). A Preliminary Economic Assessment [1] concluded that the deposit would support a 14 year mine life at 1000 tpd. The first 8 years would use open pit mining and the last 6 years underground mining. The revenue estimates were dominated by Nd(39%), Dy (29%), Pr(14%) and Tb(8%).

In addition to the Foxtrot Deposit, Search Minerals are also exploring the Deep Fox deposit in the PHS District. The Deep Fox deposit will be drilled in 2017. The surface analysis of the Deep Fox deposit has a greater width and higher grades as shown in Table 2.

Process Development and Pilot Plant Results

The Search Minerals Direct Extraction Process was developed to treat the Foxtrot mineralization.

The process involves crushing to fine particle size, application of modest amounts of H_2SO_4 to the ore at ~200 °C followed by a water leach to produce a weakly acidic product leach solution. The acid/feed mixture and calcine have favourable materials handling characteristics and the water leach slurry is easily filtered. It is believed that this gives the Search Minerals process the advantage of a much more operable acid bake circuit than other "typical" REE acid bake circuits.

After oxidation and pH adjustment with $MgCO_3$ slurry to precipitate the bulk of the iron and thorium from the solution, and ion exchange removal of uranium, sodium carbonate is added to precipitate a mixed carbonate rare earth product. The mixed carbonate is then re-leached with sulfuric acid to produce a strong rare earth salt solution. The pH is again increased with MgO to reject residual amounts of thorium and other impurities. Zinc is removed by sulfide precipitation. The rare earths are then precipitated with oxalic acid addition to produce a high quality mixed rare earth oxalate for calcination. This calcine product may be acid leached at a refinery to produce a high strength solution for rare earth separation by multi-stage solvent extraction processing or one of a number of emerging separation and refining processes.

The initial development of the process was based on bench scale and small continuous testing. In order to build confidence in the process and develop further engineering designs and cost estimates, Search Minerals embarked on an extensive pilot plant program to treat approximately 3 tonnes of Foxtrot sample.

Three pilot plant campaigns were conducted to test the Search Minerals Process treating Foxtrot ore (0.90% TREE) from the Acid Bake circuit to the Rare Earth Element precipitation circuit. The "as piloted" flowsheet is depicted in Fig. 1, with the pilot campaign battery limits indicated.

Table 1 Mineral resource estimate for the Foxtrot project [1]

Classification	Cut-off NSR	Tonnage (Tonnes)	Dy (ppm)	Nd (ppm)	Pr (ppm)	LREE (%)	LREO (%)	HREE (%)	HREO (%)	TREE (%)	TREO (%)
Open pit											
Indicated	$165	4,129,000	177	1393	372	0.69	0.83	0.17	0.2	0.86	1.03
Inferred	$165	228,000	179	1378	368	0.68	0.82	0.17	0.2	0.85	1.02
Underground											
Indicated	$260	3,263,000	209	1602	429	0.78	0.94	0.19	0.23	0.97	1.17
Inferred	$260	1,730,000	201	1602	430	0.8	0.96	0.19	0.23	0.99	1.19
Total indicated		7,392,000	191	1485	397	0.73	0.88	0.18	0.21	0.91	1.09
Total inferred		1,958,000	199	1576	423	0.79	0.94	0.18	0.22	0.97	1.17

Table 2 Foxtrot versus Deep Fox Comparison

Assays	Quantity (t)	Dy (ppm)	Nd (ppm)	Pr (ppm)
Foxtrot				
Indicated	7,392,000	191	1485	397
Inferred	1,938,000	199	1576	423
Deep Fox	Length			
Channel 14-01	5.96 m	286	2156	567
Channel 15-07	9.29 m	270	1911	504
Channel 15-06	10.17 m	260	2049	525
Channel 15-11	11.40 m	253	1732	461
Channel 14-01	17.50 m	241	1893	507
Surface expression	Width	Length		
Foxtrot	Mostly 5–14 m	~400 m		
Deep Fox	Mostly 7–30 m	~400 m		

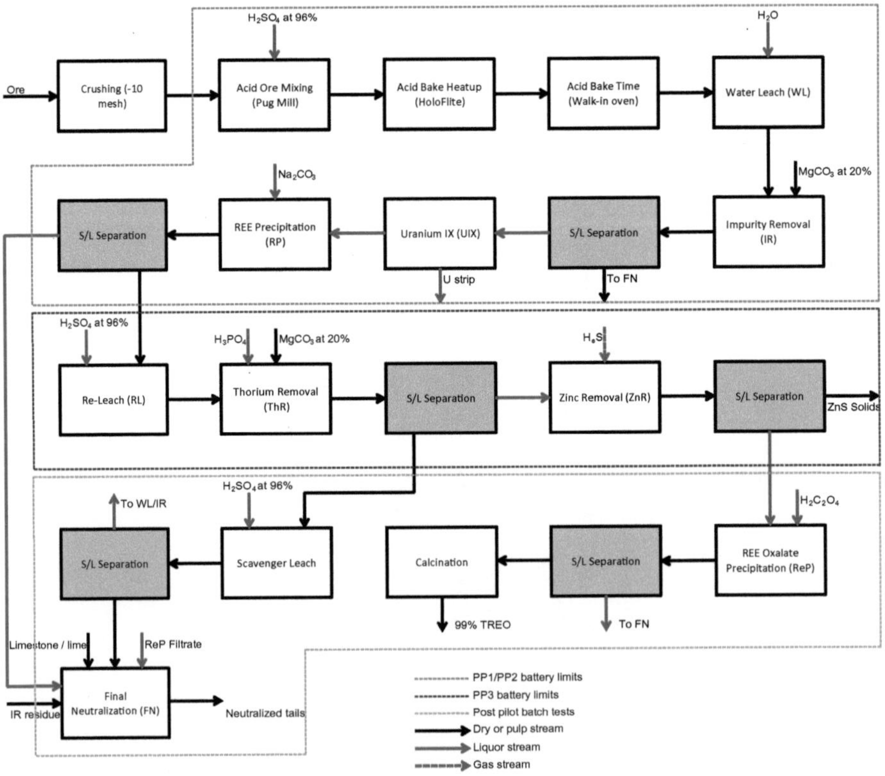

Fig. 1 Conceptual process flow diagram—as piloted

The overall flow sheet was split into two different pilot circuits, consisting of:

- Upstream circuits: Acid Bake (AB), Water Leach (WL), Impurity Removal (IR), Uranium IX (UIX), and REE Precipitation (RP). Two separate campaigns (PP1 & PP2) were carried out for a total run time of 12 days. Product from RP was used in the downstream circuit;
- Downstream circuits: Re-leach (RL), Thorium Removal (ThR), and Zinc Removal (ZnR). One single campaign (PP3) of 4 days was carried out.

Upon completion of the pilot campaigns, the final rare earth oxalate precipitation, calcination, scavenger leach, and final neutralization were operated batch-wise.

Praseodymium, neodymium and dysprosium leach extractions were 86, 85 and 68% respectively. After IR, the recoveries to solutions were 84% Pr, 83% Nd and 68% Dy, as also shown in Table 3.

Overall praseodymium, neodymium and dysprosium recoveries from ore to final rare earth oxide product (99% TREO, composition shown in Table 4) were calculated to be 83.5, 82.7 and 65.7%, respectively.

Based on post-pilot plant bench testwork, a simplified and improved flowsheet capable of treating Foxtrot ore was developed and this is depicted in Fig. 2.

Table 3 Rare earth and thorium/uranium recoveries to solution after iron removal

Recovery to solution (%)																	
La	Ce	Pr	Nd	Sm	Eu	Gd	Tb	Dy	Ho	Y	Er	Tm	Yb	Lu	Sc	Th	U
84	84	84	83	80	76	75	70	68	65	68	64	60	56	49	8	14	46

Table 4 Final product (REO) composition

La	%	12.9	Ag	g/t	<200
Ce	%	29.9	Al	g/t	<53
Pr	%	3.84	As	g/t	<200
Nd	%	15.5	Ba	g/t	<2
Sm	%	2.25	Be	g/t	<2
Eu	%	0.15	Bi	g/t	30
Gd	%	2.56	Ca	g/t	1144
Tb	%	0.34	Cd	g/t	<3
Dy	%	1.84	Co	g/t	<5
Ho	%	0.35	Cr	g/t	<68
Er	%	0.89	Cu	g/t	<100
Tm	%	0.11	Fe	g/t	<70
Yb	%	0.55	K	g/t	<42
Lu	%	0.07	Li	g/t	<400
Y	%	9.26	Mg	g/t	377

(continued)

Table 4 (continued)

Sc	g/t	<70	Mn	g/t	<77
U	g/t	<0.5	Mo	g/t	<40
Th	g/t	5.5	Na	g/t	<200
C(t)	%	0.01	Ni	g/t	50
S	%	0.19	P	g/t	<44
F	%	0.042	Pb	g/t	<400
Zr	%	<0.01	Sb	g/t	<30
Nb	%	<0.01	Se	g/t	<30
Ta	%	<0.01	Sn	g/t	<40
Si	%	0.12	Sr	g/t	<20
LOI	%	0.17	Ti	g/t	<60
TREE	%	80.5	Tl	g/t	<40
TREO (direct)[a]	%	96.8	V	g/t	<56
TREO (100-Imp)	%	99.0	Zn	g/t	<40

[a]TREO = La_2O_3, CeO_2, Pr_6O_{11}, Nd_2O_3, Sm_2O_3, Eu_2O_3, Gd_2O_3, Tb_4O_7, Dy_2O_3, Ho_2O_3, Er_2O_3, Tm_2O_3, Yb_2O_3, Lu_2O_3 and Y_2O_3

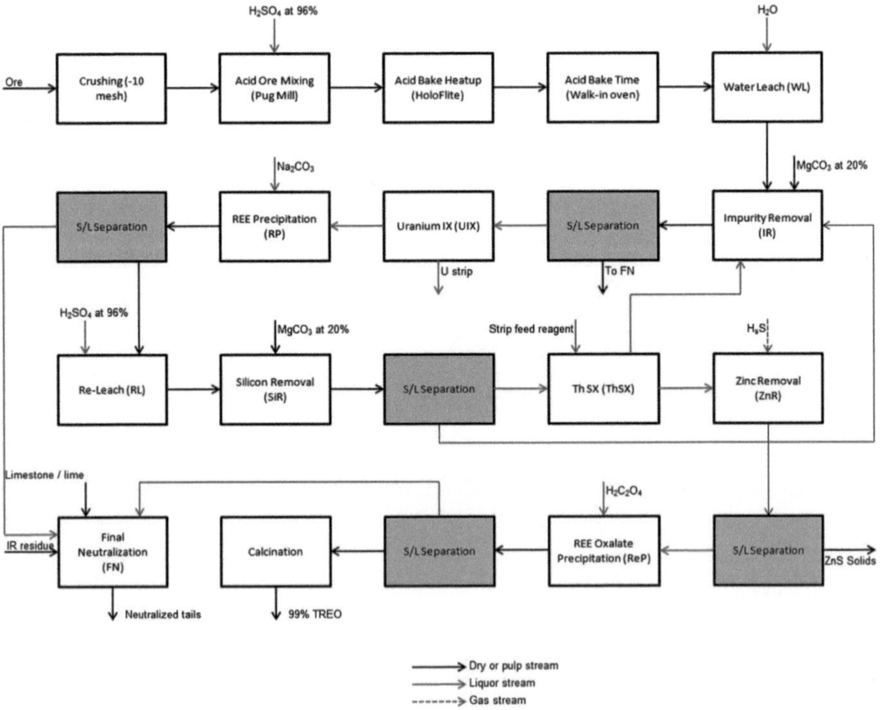

Fig. 2 Conceptual optimized process flow diagram

Pilot Operation

Crushed ore (−10 mesh) was treated in a three step acid bake process consisting of acid/ore mixing (at a sulphuric acid dosage of 144 kg/t) in a pug mill, followed by heating the acid/ore mixture in a HoloFlite to approximately 190 °C. The heated material was then maintained at temperature in a static walk-in oven. While this circuit operated well for periods of time, significant wear and tear of the HoloFlite screws interfered with its capability to transfer material adequately.

The calcined material was removed from the walk-in oven and processed in a water leach tank train to selectively dissolve REE. Due to the coarse size of the solids, challenges were encountered around advancing the solids from tank to tank. Using a pump to transfer pulp between tanks solved much of this problem. Measured extractions were high, with neodymium extraction ranging between 82 and 89% and dysprosium extraction ranging between 63 and 73%. Water leach pulp was treated with hydrogen peroxide and magnesium carbonate in an impurity removal circuit, which precipitated 92% of the dissolved thorium as well as close to 100% of the iron. Rare earth losses were below 2% under optimized conditions, leading to an overall extraction of 83% Nd and 66% dysprosium. The combined water leach/impurity removal pulp filtered well on a filter press, producing a PLS of around 0.75 g/L TREE.

Uranium was successfully extracted using two ion exchange columns (lead and lag) filled with a strong base anion resin (the pilot plant used Purolite A660). Uranium extraction was virtually quantitative (>98%) while REE co-extraction was negligible (<0.1%). The circuit did not run long enough to reach maximum loading capacity on the column and no elution testwork was carried out.

Raffinate from UIX was reacted with sodium carbonate to produce an impure intermediate rare earth carbonate precipitate of around 30% TREE. Key impurities in this stream included thorium (400–500 g/t), silicon (~6%), aluminum (6.8%), and iron (0.35%) as well as around 3% zinc. The downstream circuit was designed to separate REE from these impurities.

The intermediate rare earth precipitate was re-leached in pH 1 sulphuric acid to form a ~25 g/L TREE leach solution which was subsequently treated using phosphoric acid and magnesium carbonate to remove the remaining levels of thorium.

Due to the low allowable limit on thorium in the final product, this circuit was required to operate at a relatively high pH of ~4.8, and REE co-precipitation was significant (40–50%). Co-precipitated REE were subsequently recovered in a standalone scavenger leach process, which was capable of re-dissolving (99.8%) the contained REE and Th which in a commercial operation would be returned to the upstream impurity removal circuit. While the overall REE losses were negligible in these circuits, the operation was not efficient and the required scavenger operation led to additional reagent consumption. A post pilot plant series of bench tests aimed to improve this part of the flowsheet.

Thorium free filtrate was subsequently treated in a zinc removal circuit with gaseous hydrogen sulphide to selectively precipitate zinc sulphide. Operation of this circuit was found to be relatively straightforward with high zinc precipitation efficiency of 98% and low REE losses of <0.02%. Zinc removal filtrate was stored during the pilot plant for batch off-line rare earth oxalate precipitation testing.

Batch Production of Final Rare Earth Oxide

Final product was produced by oxalate precipitation at a stoichiometric oxalic acid dosage of 125%. At this addition, the precipitation efficiency was nearly quantitative (99.9% Nd and 99.8% Dy). This led to the production of a 46% TREE precipitate containing 11.5% C. The oxalates were subsequently calcined at 1200 °C to produce a final product that was 99% pure based on the impurity content. Table 3 presents a complete analysis of the final rare earth oxide product.

Batch Leaching of DEEP Fox Sample

Testing of a sample from Deep Fox with a higher grade (1.58% TREE) resulted in 88% TREE (91% Nd, 81% Dy) extraction using a 175 kg/t acid addition and a longer bake time (eight hours instead of four).

Post Pilot Process Improvements

Offline bench scale testwork using slurried RP cakes from the pilot plant was conducted to investigate an alternative secondary thorium removal process route. The objectives of this route were to (1) maintain complete thorium removal (i.e. <0.1 mg/L Th advancing) with minimum REE losses and (2) improve the S/L separation properties. While leaching of RP pulp at pH 1 was chemically successful, the high silicon dissolution led to slow filtration rates. A subsequent pH adjustment using magnesium carbonate to pH 3 resulted in significantly less silicon content in solution (~ 100 mg/L Si instead of 3.1 g/L Si at pH 1, or 97% precipitation) with less than 0.5% REE losses. The reduced silicon decreased RL filtration time, and reduces the chance of silicon fouling in downstream processes.

Solvent extraction (SX) using Primene JMT as the extractant (tridecanol as a modifier and Orfom SX80 used as diluent) was selected as an alternative thorium removal technology and this resulted in near complete removal of thorium with minimal REE co-extraction. Thorium loading was high even at a cumulative phase ratio of 25/1 aqueous/organic (the maximum loading capacity of the organic was

Table 5 Performance summary of the overall pilot plant campaigns

		PP1 & PP2						PP3		Batch	Batch	Overall
Circuit performance		Pug mill (%)	Holo-Flite (%)	Walk in Oven (%)	WL-IR (%)	UIX (%)	RP (%)	RL-ThR (%)	ZnR (%)	REP (%)	Scav (%)	(AB-REP) Calc[a] (%)
Circuit	Nd	100	100	100	83	0.02	100	62	0	100	99.9	82.7
Extraction/ precipitation	Dy	100	100	100	66	0.02	100	67	0	100	99.9	65.7

[a]Assuming 100% of recycled material is recovered and no other effects on performance

Table 6 Reagent consumption summary

Reagent consumption (kg/t ore)	PP1 & PP2						PP3		Batch	Batch	Batch
	Pug mill	Holo-Flite	Walk in Oven	WL-IR	UIX	RP	RL-ThR	ZnR	REP	Scav[a]	Final neut[b]
Reagent ore dosage (kg/t)	H_2SO_4 144	–	–	H_2O 9748	–	Na_2CO_3 19.6	H_2SO_4 21.0	H_2S 1.58	$H_2C_2O_4$ 4.41	H_2SO_4 37	$Ca(OH)_2$ 23.3
Reagent ore dosage	–	–	–	$MgCO_3$ 57.4	–	–	H_2O_2 0.078	–	–	–	–
Reagent ore dosage	–	–	–	H_2O_2 9.4	–	–	H_3PO_4 0.13	–	–	–	–
Reagent ore dosage	–	–	–	–	–	–	$MgCO_3$ 12.0	–	–	–	–

Overall consumption		
H_2SO_4	kg/t	202
H_2O	kg/t	9748
$MgCO_3$	kg/t	69.4
H_2O_2	kg/t	9.50
Na_2CO_3	kg/t	19.6
H_3PO_4	kg/t	0.13
H_2S	kg/t	1.58
$H_2C_2O_4$	kg/t	4.41
$Ca(OH)_2$	kg/t	23.3

[a]Based on 2000 kg H_2SO_4/t ThR residue and 18.4 kg ThR residue/t ore (PP2 performance)
[b]Based on 10.6 kg/t $Ca(OH)_2$ consumed for IR residue neutralization and 12.7 kg/t $Ca(OH)_2$ consumed for ReP oxalate filtrate neutralization

not achieved). Further testing is required to investigate the stripping and regeneration of the organic, as well as any organic cycling effects.

Based on the post pilot plant improvements, a conceptual optimized flowsheet was developed and this is shown in Fig. 2. The revised flowsheet incorporates the thorium SX circuit; based on the negligible REE losses in that circuit, there is no further need to add acid in a scavenger leach process. Instead it is assumed that a loaded thorium strip liquor (not tested) is combined with the IR feed stream and any contained thorium is removed from the process via the IR residue.

Overall Plant Performance and Reagent Consumptions

Overall plant performance is shown in Table 5. These data are based on actual operations and capture a single pass of processed material. An overall extraction is calculated, assuming that all recycled material (i.e. scavenge leach liquor) is recovered. The conceptual flowsheet shown in Fig. 2 identifies opportunities for improved performance via thorium SX, however this was not piloted in integrated fashion.

An overview of all reagents (all based on 100% reagent strength) used in the conceptual flow sheet (as piloted/tested) is shown in Table 6. It is clear that a significant component (18% of overall addition) of the acid consumption is related to recovering co-precipitated REE in the ThR circuit. Replacing the ThR circuit with a ThSX circuit may reduce this.

Conclusions

Search Minerals has rapidly advanced the exploration and development of the Port Hope—Simpson Rare Earth Element District. The Foxtrot deposit contains a variety of rare earth minerals and is rich in heavy rare earth elements with the key revenue elements being Nd, Pr, Dy and Tb. A simple direct leaching process has been developed for treatment of Foxtrot mineralization that represents a breakthrough in the development of this resource.

A 3 tonne sample of Foxtrot material was crushed, mixed with acid, baked, water leached and purified for Fe/Al precipitation and U ion exchange. The solution was then treated with soda ash to precipitate an impure mixed rare earth carbonate. The mixed rare earth carbonate was then acid leached with sulfuric acid and further purified by pH adjustment, zinc removal by solvent extraction and oxalate precipitation. A high purity mixed rare earth oxide was produced by calcination. This product was low in contaminants and ready for refining.

Subsequent work showed that thorium solvent extraction could be used to control Th in the final product and avoid the high REE recycles implicit in Th precipitation and releaching. Additionally, the application of acid baking and water leaching to a surface sample from the Deep Fox deposit returned an overall REE extraction of 88% including 91% Nd and 81% Dy.

Acknowledgements Support was received from the Research & Development Corporation ("**RDC**") of Newfoundland and Labrador and from the Atlantic Canada Opportunities Agency ("**ACOA**") for this development.

References

1. RPA (2016) Search Minerals Inc. Technical Report on the Foxtrot Project in Labrador, Canada. NI 43-101 Report prepared by Masun KM, Weir IC, Goode JR, April 1, 2016

Selective Oxidation of Cerium in Rare Earth Solutions, a Comparison of Four Oxidants

James McNeice and Ahmad Ghahreman

Abstract Separation of rare earth elements (REE) is often achieved by acid leaching of mineral concentrates followed by subsequent solvent extraction (SX), to purify individual rare earth metals (REM). Cerium, the most abundant REE, is usually the largest component of these ores. Demand for less prevalent REE, such as praseodymium and dysprosium; have led to a market saturation and price collapse of cerium. One method of improving the economics of REE separation is to remove cerium from solution prior to SX. This is accomplished via the oxidation of cerium (III) to cerium (IV), which precipitates as cerium hydroxides. This study probes the efficacy of four oxidants; hydrogen peroxide, sodium hypochlorite, potassium permanganate, and Caro's Acid with the goal of improving cerium removal whilst minimizing REE and reagent losses. The effect of oxidant type and dosage, pH, cerium concentration, and temperature on the kinetics of the process will be discussed.

Keywords Cerium · Oxidation · Kinetics · Rare earths · Rare earth elements Permanganate · Hypochlorite · Peroxide · Caro's acid

Introduction

As modern technology has advanced, the rare earth elements (REE) have become crucial components this technology. Necessary for miniaturization and other high-tech applications, in many cases they cannot be substituted. These metals are classified as light rare earths (LREE) or heavy rare earths (HREE) by governments and markets based on weight, abundance, and therefore value. The minerals bastnaesite, monazite, and xenotime compose the majority of the world's exploitable sources of these metals [1, 2]. Geologists classify the LREE ores as 'cerium group'

J. McNeice (✉) · A. Ghahreman
The Robert M. Buchan Department of Mining, Queen's University,
Goodwin Hall, 25 Union Street, room 354, Kingston, ON K7L 3N6, Canada
e-mail: 0jam11@queensu.ca

H. Kim et al. (eds.), *Rare Metal Technology 2018*, The Minerals, Metals & Materials Series, https://doi.org/10.1007/978-3-319-72350-1_2

minerals, due to their approximate 50% cerium oxide by weight. The cerium group minerals monazite and bastnaesite compose the majority of world production, leading to an overproduction of cerium relative to world demand [3]. This ratio of light to heavy REE in deposits has led to a phenomenon referred to as the balance problem [4]. As an example, to produce 1 tonne of europium oxide from a bastnaesite deposit 450 tonnes of cerium oxide must be processed. This means that any push to increase production and access the value of the HREE metals will drive down the value of the LREE metals leading to diminishing returns [3]. These ores can also contain measurable levels of thorium, subjecting them to complex radioactive waste regulations.

Until the turn of the 21st century, the USA produced the majority of rare earth metals (REMs), mostly sourced from the Mountain Pass mine in California. However, due to a combination of market dynamics and regulations, China has dominated the market for the last two decades. The largest Chinese REE mine, Bayan Obo, produces rare earth (RE) concentrate as a byproduct of iron mining. This decreases the relative production costs [5]. This mine alone represents fifty percent of global bastnaesite production [6]. Accelerating global demand for high-tech products has forced Chinese producers to commission other large mines, accounting for >85% of world production by early years of the new millennium [7]. This presents the west with a looming supply risk of these critical metals in the future.

Since 2010, the REE market has been tumultuous. Chinese export restrictions caused rapid price increases, followed by a dramatic drop in prices due to a lifting of these restrictions and increasing supply [3]. This has left western producers that entered the market in the last decade in a poor financial position, culminating in bankruptcy for Molycorp in 2015 [8]. A Chinese led consortium of companies purchased the Mountain Pass property in June 2017 [9]. The last major REE producer operating outside of China is Lynas Corp., of Australia [7]. The low price of LREE metals like cerium, combined with the balance problem has culminated in a large barrier to entry for potential miners of LREE dominant deposits as well as a large incentive to lower the unit costs of cerium removal.

Cerium removal is usually the first step in REE purification, as the largest mass fraction of the concentrate this lowers the throughput of the subsequent steps. In addition, many of the industrial end users require a relatively pure product for the chemical catalysis applications of cerium [4]. Several methods of cerium separation developed based on the relative ease of oxidation of trivalent cerium to its tetravalent form compared to the other lanthanides [5]. Once this change has occurred the cerium ions will exhibit notably different chemical behavior, and much lower solubility, facilitating separation.

Much of the cerium separation achieved at the Mountain Pass concentrator was via the Molycorp process [2, 5, 10]. This process consists of the calcination of bastnaesite flotation concentrate at 620 °C to drive off carbon dioxide and oxidize cerium, followed by treatment with 30% HCl to leach the other trivalent lanthanides. After filtration, this leaves a solid cerium concentrate and a solution that is further refined to separate the other REE present.

Another prevalent method of cerium removal is to dry the RE hydroxides in air at 120–160 °C followed by treatment with nitric acid to leach the trivalent RE hydroxides [5, 11]. The Chinese method consists of a sulphuric acid bake at 500 °C for conversion to RE sulphates, followed by calcination and conversion to hydroxides with NaOH for solvent extraction [5, 10, 12].

Selective precipitation of cerium (IV) from acidic RE (III) leach liquors has been studied using ozone [13], photochemical oxidation [14], wet air [12], hydrogen peroxide [15, 16], hypochlorite [17, 18], and permanganate [19, 20]. The oxidation of cerium in acidic media is of particular interest as acid leaching is usually the first step in the chemical treatment of the previously mentioned high cerium bastnaesite ores. Ideally, this process would eliminate the calcination step.

This work examines the kinetics of cerium oxidation at acidic pH using strong chemical oxidants. Some reagents selected have been previously used in literature but have never been compared directly. The parameters examined were pH, oxidant stoichiometry, cerium concentration, and temperature. After the ideal conditions for each respective selective oxidation, cerium precipitates were washed and collected for future analysis.

Experimental

Cerium oxidation was tested using cerium solutions with concentrations of 2, 8, and 16 g/L. These were produced by dissolving cerium nitrate hexahydrate (99.9%), sourced from Alfa Aesar, in 0.1 M hydrochloric acid (HCl) solution.

Oxidants used in the tests were sodium hypochlorite (13%), hydrogen peroxide (H_2O_2, 37%), potassium permanganate (0.4 M), and Caro's Acid (20–22%). Caro's Acid was prepared by mixing concentrated H_2O_2 with concentrated sulphuric acid (H_2SO_4) at a ratio of 1:3. pH control was achieved using solutions of 1 M HCl and 1 M NaOH.

Oxidation experiments conducted in a glass beaker on a standard laboratory hot plate in a fume hood. The solutions were agitated using a magnetic stir bar at 500 RPM. Target pH was reached by adding sodium hydroxide solution and allowing time to stabilize. The initial volume of all tests was 250 mL unless specified otherwise. The quantities of oxidant required were calculated using the equations presented in the following section. Oxidant solutions were added at fixed rates of 0.30 and 0.50 mmol/min, for permanganate and the other oxidants respectively until depleted. The reactions progressed for total of 180 min. As the oxidation of cerium (III) ions to the cerium (IV) oxidation state proceeded the pH was maintained at the reaction target using hydrochloric acid and sodium hydroxide solutions as needed. To obtain kinetic information on the reaction 5.0 mL of the solution periodically sampled and filtered for analysis.

Optimum stoichiometry for each oxidant was determined at a set target pH of 3.0 and cerium concentration of 8 g/L. Once the stoichiometry was selected pH, temperature, and cerium concentration were varied to observe changes in the reaction and determine the optimal conditions.

Excess oxidant in the solution, if present, was neutralized at the end of each test. Hydrogen peroxide will decompose into oxygen and water upon the addition of manganese dioxide (MnO_2). Excess sodium hypochlorite reacts with the HCl to produce chlorine gas. Excess permanganate was measured by titration in the analysis stage. Caro's Acid decomposes very quickly at room temperature.

After 180 min, the remaining solutions were rotated in a centrifuge at 6000 RPM for 12 min to separate the cerium solids. The remaining solutions were sampled and filtered for analysis. The solids were placed in an incubator at 35 °C until dry. A portion of the solids were washed using a solution of 0.1 M magnesium (I) chloride at pH 3.0 and 70 °C for 30 min. Under these conditions, any cerium (III) that had co-precipitated with the cerium (IV) re-enters solution. The remaining cerium (IV) solids were then centrifuged, washed with water, and dried again. Analysis of the wash solutions for cerium (III) and (IV) completes the mass balance.

Analysis was conducted using an Agilent MP-AES (Microwave Plasma—Atomic Emission Spectrometer) 4200 for determination of total cerium concentration, total REE concentration, and aqueous manganese concentration. Titration using a Metrohm Ti-Touch 916 and ferrous ammonium sulphate as a titrant was used for determination of cerium (IV) concentration. This titrant reduces any cerium (IV) particles remaining in solution, when measuring the potential, a steep endpoint is observed once this process concludes. For permanganate tests, excess permanganate was subtracted from the cerium (IV) concentrations produced by titration.

As cerium (IV) solids are generated in solution, some cerium (III) particles co-precipitate as cerium (III) hydroxide on the surface on the cerium (IV) precipitates. This is measured by analyzing the wash solution samples for total cerium with MP-AES. Kinetic mass balances for cerium oxidation were constructed using the concentration information on cerium in its oxidation states over the course of the reaction.

Results and Discussion

Sodium Hypochlorite

The reaction between cerium and sodium hypochlorite has been proposed in previous papers as Eq. (1) [18]. Sodium hypochlorite also decomposes in an acidic environment to form chlorine gas according Eq. (2) in competition with the oxidation.

$$2CeCl_3 + NaOCl + 6NaOH + H_2O \rightarrow 2Ce(OH)_4 + 7NaCl \quad (1)$$

$$HOCl + Cl^- + H^+ \rightarrow Cl_2 + H_2O \quad (2)$$

Hydrogen Peroxide

Hydrogen peroxide can act as an acid or a base depending on the conditions and can act as a powerful oxidant. The oxidation of Ce(III) to Ce(IV) using H_2O_2 has been investigated on a theoretical basis [16] and experimentally in a nitric acid medium [15]. The reaction proceeds in two competing reactions:

$$2Ce^{3+} + H_2O_2 + 6H_2O \rightarrow 2Ce(OH)_4 + 6H^+ \quad (3)$$

$$2Ce^{3+} + H_2O_2 + 2H_2O \rightarrow 2CeO_2 + 6H^+ \quad (4)$$

Due to the stable nature of hydrogen peroxide in acidic solution manganese dioxide powder was added after sampling to neutralize the remaining H_2O_2 in solution and halt the reaction. This was filtered away with the cerium solids.

Potassium Permanganate

A method for the oxidation of cerium by potassium permanganate in acidic solution has proposed by [19, 20]. The precipitation of cerium is accompanied by the precipitation of manganese dioxide, shown in Eqs. (5) and (6).

$$3Ce^{3+} + MnO_4^- + 10H_2O \rightarrow 3Ce(OH)_4 + MnO_2 + 8H^+ \quad (5)$$

$$2MnO_4^- + 8H^+ + 6Cl^- \rightarrow MnO_2 + 3Cl_2 + 4H_2O \quad (6)$$

This method requires further processing if a pure cerium product is desired.

Caro's Acid

Caro's Acid is a strong oxidizing acid, created by the mixture of hydrogen peroxide and sulfuric acid, based on Eq. (7). The oxidizing power of Caro's Acid is utilized in industrial cyanide destruction processes due to the speed of the reaction [21].

$$H_2SO_4 + H_2O_2 \rightarrow H_2SO_5 + H_2O \quad (7)$$

The oxidation of cerium (III) by Caro's acid has not been reported in literature previously. Based on the mechanism of cerium oxidation using the other oxidants, the proposed reaction, Eq. (8), was created as a basis for the experiment.

$$2Ce^{3+} + H_2SO_5 + OH^- + 2H_2O \rightarrow 3CeO_2 + SO_4^{2-} + 7H^+ \quad (8)$$

Stoichiometry Tests

The optimal stoichiometry for each oxidant was determined using an 8 g/L synthetic cerium chloride solution. Temperature was ambient and pH was set at 3.0. Figure 1 illustrates cerium oxidation over time. As the pH of cerium chloride solution is raised to 3.0 some cerium (III) ions precipitated as $Ce(OH)_3$. Oxidation was measured as cerium was leaving the solution after pH adjustment, defined as time 0. All of the oxidation reactions occur quickly and show improved recovery when dosage approached 150% theoretical stoichiometry. No cerium (IV) was detected by titration in the remaining solution after solid-liquid separation; therefore all cerium remaining in solution was of the cerium (III) form. Complete oxidation of cerium from solution was achieved using potassium permanganate at oxidant dosages in excess of 125%. Caro's acid oxidized 92–96% of the cerium present in solution at all dosages, with the reaction proceeding faster at higher dosages. In contrast, the other two reagents only oxidized 40% or less of the cerium in solution. Produced cerium solids were white-yellow, orange, and brown when using hypochlorite, peroxide or Caro's acid, and permanganate, respectively.

pH Tests

After selecting oxidant stoichiometry of 1.5 times theoretical, pH was varied to observe changes in yield. Figure 2 represents cerium precipitation over time at pH 1.0–4.0. Initial cerium concentration and stoichiometry were held constant at 8 g/L and 150%. A similar relationship between pH and cerium oxidation arises with all reagents, namely that an increase in pH improves yield. This should not be a surprise; however, it is clear this reaction is strongly inhibited by the decomposition of reagents at pH 1 and under. This is most pronounced in (Fig. 2c) in which there is little difference in the reactions above pH 3. Caro's acid was the most effective oxidant below pH 3.0. The speed of the reactions in the hypochlorite pH tests was relatively constant, indicating that the loss in yield is due primarily to the greater speed of Eq. (2) relative to Eq. (1).

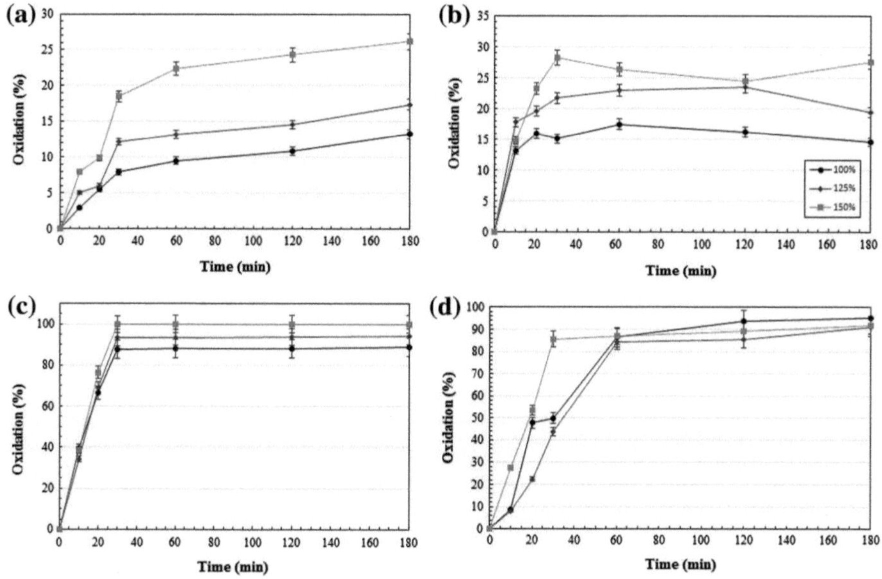

Fig. 1 Cerium Precipitation versus time using **a** sodium hypochlorite, **b** hydrogen peroxide, **c** potassium permanganate and **d** Caro's acid. Each line represents the percentage of cerium that has precipitated from solution by oxidation, at a given time during the respective test

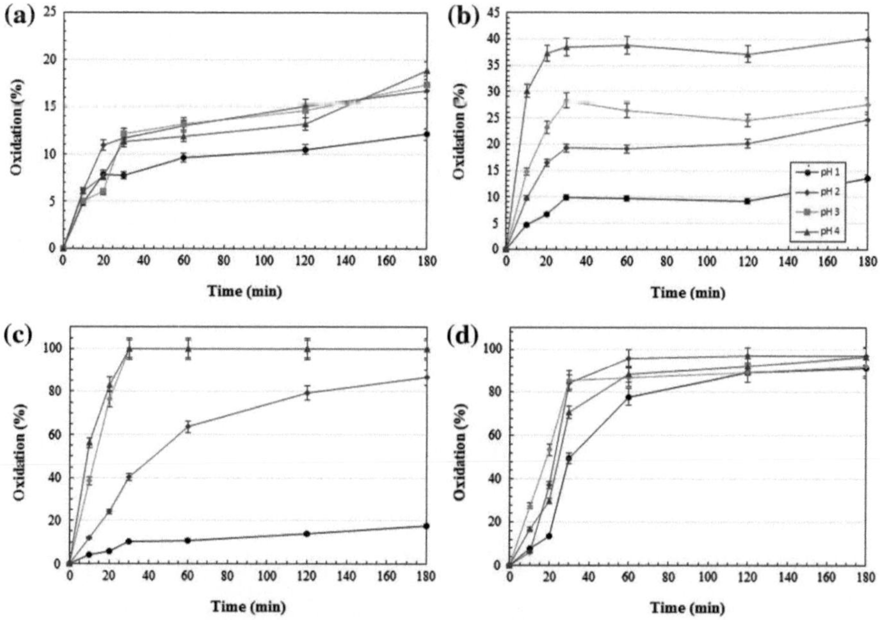

Fig. 2 Cerium precipitation versus time over various pH using **a** sodium hypochlorite, **b** hydrogen peroxide, **c** potassium permanganate and **d** Caro's acid

Caro's acid shows the least sensitivity to pH, oxidizing over ninety percent of cerium present across the test range. The hypochlorite will evolve into chlorine gas at greater rates as pH is lowered, leaving a smaller amount to react with cerium ions. This can be seen in the gradual decrease in oxidation as pH decreases. Hydrogen peroxide will decompose in solution in the presence of other ions, it is ineffective for oxidizing cerium below pH 2.0. Potassium permanganate is relatively stable in solution above pH 2.5 in this system. The ions are reduced to Mn2 + at higher concentrations of HCl, reducing effective dosage at low pH.

Cerium Concentration Tests

Following the selection of 1.5 and 3.0 as ideal stoichiometry and pH respectively, the start concentration of cerium was varied to observe any change in yield. The following Fig. 3 represents these tests.

An increase in cerium concentration was associated with a slight increase in cerium oxidation in tests with both hypochlorite and permanganate. This is most likely attributable to the co-precipitation of cerium as cerium (III) and (IV) hydroxides in the presence of CeO_2 particles. Hydrogen peroxide, on the other hand, was able to oxidize almost twice as much cerium from solution in the dilute

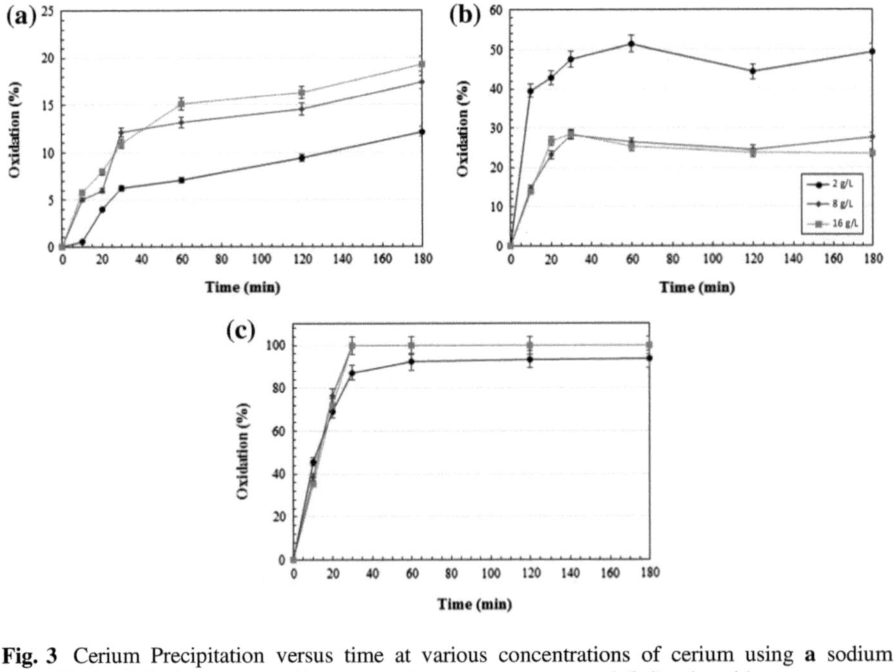

Fig. 3 Cerium Precipitation versus time at various concentrations of cerium using **a** sodium hypochlorite, **b** hydrogen peroxide, **c** potassium permanganate and **d** Caro's acid

solution tests. This result suggests that H_2O_2 may be able to achieve more efficient cerium oxidation in dilute solutions where there is a lower charge density in solution.

Temperature Tests

Once the oxidation of cerium by the three oxidants was established at room temperature, the effect of heat was studied. pH 3.0 and 1.5 times calculated stoichiometry were used for each reagent in these tests. Temperature held constant at 60 °C during the entire reaction and the initial concentration of cerium was set at 8 g/L. Figure 4 shows the result of these tests.

The heating produced faster reaction kinetics for both hypochlorite and permanganate tests. Yield nearly doubled in the case of hypochlorite, indicating that heat is an important factor when oxidizing cerium using this reagent. Hydrogen peroxide did not show a response in terms of oxidation yield with the addition of heat, however, the precipitate collected was white-grey instead of the orange produced at room temperature. The difference in structure of these precipitates will be discussed in a future study.

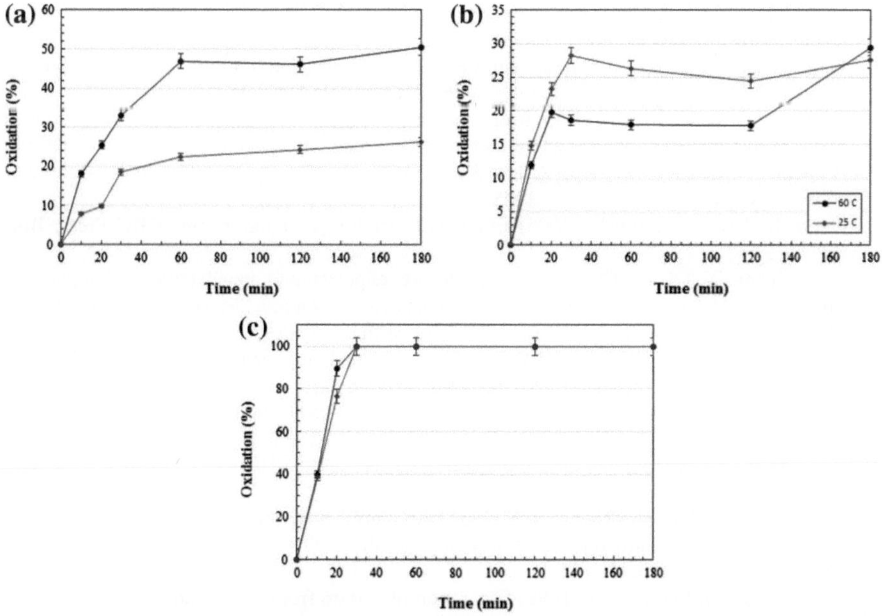

Fig. 4 Cerium oxidation versus time at different temperatures using **a** sodium hypochlorite, **b** hydrogen peroxide and **c** potassium permanganate

Conclusion

The kinetics of cerium oxidation using sodium hypochlorite, hydrogen peroxide, potassium permanganate, and Caro's acid were investigated at a range of acidic conditions. Oxidation of over ninety percent of aqueous cerium was achieved using Caro's acid under all tested conditions. Complete oxidation of cerium can be achieved using in excess of 25% the stoichiometric requirement of potassium permanganate at any temperature in solutions with pH above 2.0. Sodium Hypochlorite followed similar trends but was inferior to permanganate in its ability to oxidize cerium, with a maximum of 51% oxidation achieved at 150% stoichiometry and 60 °C. Hydrogen peroxide produced the lowest oxidation yield of all reagents under all conditions excluding initial cerium concentration, showing an increase to 52% oxidation in dilute systems, indicating the need for further study on this trend. The next steps in the authors' research on cerium oxidation will concern the selectivity for cerium of each oxidant in the presence of other REE.

Acknowledgements The authors would like to acknowledge the financial support for this study through Kingston Process Metallurgy (KPM) and Ontario Centre of Excellence (OCE).

References

1. Haxel G, Hedrick J, Orris G, Stauffer P, Hendley II J (2002) Rare earth elements: critical resources for high technology. USGS
2. Parker J, Baroch C (1971) The rare earth elements, yttrium and thorium—a materials survey. Bureau of Mines Information Circular 8476, Washington, D.C., 1971
3. Zepf V (2016) An overview of the usefulness and strategic value of rare earth metals. In: Rare earths industry—technological and environmental applications, Waltham, Maryland, Elsevier
4. Binnemans K, Jones P, Van Acker K, Blanpain BMB, Apelian D (2013) Rare-earth economics: the balance problem. TMS
5. Gupta K, Krishnamurthy N (2005) Extractive metallurgy of rare-earths. CRC Press, Boca Raton
6. Li L, Yang X (2014) China's rare earth ore deposits and benefaction techniques. In: Proceedings of the 1st European rare earth resources conference, Milos Island, Greece
7. Mancheri N (2016) An overview of chines rare earth export restrictions and implications. In: Rare earth industry—technological and environmental applications, Waltham, Maryland, Elsevier
8. Miller J, Zheng A (2015) Molycorp files for bankruptcy protection. The Wall Street Journal, New York
9. Brickley P (2017) Mountain pass mine approved for sale to JHL, QVT, Shenghe. The Wall Street Journal, New York, NY
10. Fuerstenau D, Khan PL, Raghayan S (1982) An alternate reagent scheme for the flotation of mountain pass rare earth ore. In: Proceedings of the 14th international mineral processing congress, Toronto, Canada
11. Luan H, Hao D, Mioduski T (1988) Separation of cerium from other lanthanides by leaching with nitric acid rare earth (III) hydroxide-cerium (IV) oxide mixtures. J Radioanal Nucl Chem 105–113

12. Zou D, Chen J, Yu L, Deqian L (2014) Wet air oxidation kinetics of cerium (III) of rare earth hydroxides. Ind Eng Chem Res 2014(53):13790–13796
13. Shiloy V, Gogolev A, Fedoseev A, Perminoy V (2014) Mechanism of cerium (III) oxidation with ozone in sulfuric acid solutions. Radiochemistry 56(4):339–441
14. Donohue T (1978) Photochemical oxidation of cerium from rare earth mixtures in aqueous solution. Chem Phys Lett 601–604
15. Nechaev A, Sibiley A, Smirnov A (2016) A rational approach to processing cerium—containing raw materials. Theor Found Chem Eng 50:863–866
16. Yu P, Hayes S, O'Keefe M, Stoffer J (2006) The phase stability of cerium species in aqueous systems: II. The formula systems. Equilibrium considerations and pourbaix diagram calculations. Electrochem Soc 153(1):C74–C79
17. Ali M, El-Alfy M, Zayed M, El-Hazek N, Rabie, KA, Aly H (1996) Separation of cerium (III) from Egyptan monazite by oxidation to cerium (IV). In: Proceedings of the third Arab conference on the peaceful uses of atomic energy, Damascus
18. Ho E, Wilkins D, Soldenhoff K (2014) Recovery of cerium from chlorine solution by oxidation with sodium hypochlorite. In: Proceedings of the 7th international symposium on hydrometallurgy 2014
19. Morais C, Abreu R (2010) Purificaation of rare earth elements from monazite sulphuric acid leach liquor and the production of high purity ceric oxide. Miner Eng 23:536–540
20. Morais C, Benedetto J, Ciminelli V (2003) Recovery of cerium by oxidation/hydrolysis with $KMnO_4$–Na_2CO_3. In Proceedings of the fifth conference in Honor of Professor Ian Ritchie—volume 2: electrometallurgy and environmental hydrometallurgy
21. Castrantas HM, Manganaro JL, Rautiola CW, Carmichael J (1995) Treatment of cyanides in effluents with Caro's acid. U.S. Patent 5, 397, 482

A Study on the Effect of Crystal Habit Modifiers on the Co-precipitation of REE with Gypsum

Farzaneh Sadri, Zhi Yang and Ahmad Ghahreman

Abstract Impurity removal is a very important stage of the hydrometallurgical treatment of rare earth elements (REE) processing circuit, which is performed after water leaching of the acid baked REE concentrate. Lime neutralization has a high efficiency in removal of impurities and is the preferred process due to its economic advantage; however, the process has its own problems, such as the production of large volumes of gypsum and co-precipitation of REE with gypsum, i.e. REE loss. This work deals with the effect of several additives (cetryltrimethylammonium bromide (CTAB), polyvinylsulfonic acid (PVS) and polyacrylic acid (PAA)) on the uptake of REE by gypsum from synthesized acidic solutions. The main purpose of this study is to investigate the changes in the crystalline morphology of gypsum precipitated by the addition of crystal habit modifiers and to determine the crystal changes' effects on the uptake of REE by gypsum. In this regard, different concentrations (0.2, 2 and 5 g L^{-1}) of above-mentioned additives were studied at room temperature.

Keywords Rare earth elements (REE) · Gypsum · Precipitation CTAB · PVS · PAA

Introduction

Acid baking or digestion is one of the favorite industrial processes to digest REE-containing concentrates and ores, which the sands are digested by treatment with an excess amount of concentrated H_2SO_4 at elevated temperatures [1–3]. The

F. Sadri (✉) · Z. Yang · A. Ghahreman
Department of Mining Engineering, Queen's University, 25 Union Street, K7L 3N6 Kingston, ON, Canada
e-mail: Farzaneh.sadri@queensu.ca

Z. Yang
e-mail: carrieyang@yahoo.com

A. Ghahreman
e-mail: ahmad.g@queensu.ca

H. Kim et al. (eds.), *Rare Metal Technology 2018*, The Minerals, Metals & Materials Series, https://doi.org/10.1007/978-3-319-72350-1_3

resulted pasty material is then treated with cold water to dissolve the REE into solution, leaving the silica, rutile, zircon and undigested concentrate as a residue. After filtering the solution from the gangue minerals, a clear REE sulfate solution is obtained. This solution contains some impurities, such as iron (Fe), aluminum (Al) and thorium (Th) that can create some difficulties for further processing such as solvent extraction as well as additional costs. Therefore, the solution is then treated with a neutralizing agent to remove the impurities from the solution [4].

Lime neutralization has a high efficiency in removal of dissolved heavy metals and the principle of lime neutralization process lies in the insolubility of heavy metals in alkaline conditions. Due to the low lime costs in comparison to other chemicals, it is the preferred process, but has its own problems, such as requiring high dosages of lime to increase the pH and production of gypsum [5]. When the solubility of sulfate and the calcium added from lime exceeds the saturation limit, gypsum starts to precipitate out of the solution [5]. Lime neutralization is the cheapest process for impurity removal stage in hydrometallurgical treatment of REE ores, but is not widely in use due to the fact that a proportion of REE is lost to gypsum, while removing other impurities, such as Fe, Al, Th [4].

Many industrially relevant crystallization processes are influenced by the presence of impurity elements. The negative effects of impurities range from crystal product contamination and alteration of morphology to the formation of other than the intended phase [6].

It is well-known that impurities present in solution may transfer to the crystallizing solid. The transfer may take place via interstitial uptake between regular lattice positions, co-precipitation, or isomorphous substitution of one of the ions in the host lattice by the impurity ion or by adsorption [7].

The morphology of dihydrate crystals can be controlled by many methods, one of which is the use of the specific additives. Inorganic and organic crystal habit modifiers have been shown to have an influence on the crystal habit and growth kinetics, but their effectiveness strongly depends on the system, which they are introduced in. They selectively adsorb on the surface of the crystal or nuclei and depending on their concentration, retard or slow down the growth process. They might be effective if there is a good extent of lattice matching between them and the crystal, which means that there should be a good match between the ion spacing of the crystallizing phase and the distances of the functional groups of the additive [8, 9]. The hydroxyl groups of organic additives adsorb on special planes and form hydrogen bond layer, which hinders the touch of growth units with that plane and as a result, the growth would be promoted along other directions [10].

The scope of this study is to develop a novel process to diminish REE co-precipitation with gypsum, while removing the impurities from the solution. Accordingly, a synthesized solution that simulates the leaching solution after acid baking of REE concentrate is used to investigate the effect of morphology changes in the precipitated gypsum crystals by adding $Ca(OH)_2$ on the percentage of REE that co-precipitate with it at low pH values. For this purpose, three crystal habit modifiers, which have been shown [11] to be effective in changing the morphology of calcium sulfate dihydrate precipitates in highly acidic solutions, are used to inspect the changes in the REE uptake by precipitated gypsum.

Experimental

Materials and Reagents

The materials and reagents used in these experiments are presented in Table 1. All chemicals were of analytical grade and de-ionized water was used in all experiments. All the experiments were performed in duplicate as a check on the experimental technique and precision.

Experiments

Gypsum saturated solution pre-adjusted in pH = 1 by sulfuric acid was first prepared and cerium, neodymium, erbium and ytterbium sulfate salts were added to the solution to make a 4 g/l REE-containing gypsum saturated solution. This synthesized solution was used as the basic solution for all experiments.

All titration experiments were performed in a Pyrex beaker on a magnetic stirrer (Arex digital, VELP Scientifica) and the pH of the solutions were continuously measured by a pH meter (Fisher Scientific XL600 accumet). The titrant used in these experiments was 5% calcium hydroxide ($Ca(OH)_2$) pulp with the pH value of 13. To investigate the effect of crystal modifiers, three different concentrations of

Table 1 Materials and reagents used in this study

Reagent	Grade	Company
Ce(III) sulfate octahydrate	99.9%	Alfa Aesar
Nd(III) sulfate octahydrate	99.9%	Alfa Aesar
Er(III) sulfate octahydrate	99.9%	Alfa Aesar
Yb(III) sulfate octahydrate	99.9%	Alfa Aesar
Hexadecyltrimethylammonium bromide (CTAB)		Sigma
Polyacrylic acid (PAA)	35 wt% in H_2O	Aldrich
Polyvinylsulfonic acid (PVS)	25 wt% in H_2O	Aldrich
Calcium sulfate dihydrate	98%	Acros
Calcium hydroxide		Fisher Scientific
Sulfuric acid	95–98%	Fisher Scientific
Nitric acid	68–70%	BDH
Ce, Nd, Er and Yb standard solutions for MP-AES and ICP	1000 ppm	Alfa Aesar

CTAB, PVS and PAA (0.2, 2 and 5 g/L) were added to different solutions at the beginning of the experiments. By adding $Ca(OH)_2$ to the solution, pH was gradually changed from 1 to 9 and samples were taken from the solution at the beginning and ending of each experiment (pH 1 and 5). Solutions were filtered in pH 5 and precipitates were collected for characterization. One experiment was conducted with no additive and after filtering the solution at pH 5, the titration was continued to pH 8.5 to determine the complete precipitation of REE in the solution.

The crystal morphology of the collected precipitates were analyzed with a FEI FEG-Nova NanoSem scanning electron microscope (SEM) after sputter coating the samples with a thin layer of Cr. X-ray powder diffraction analysis was carried with a Bruker AXS powder diffractometer. Co Kα radiation with a wavelength of 1.79 Å was employed. Chemical analysis was conducted with an Agilent Technologies 4200 MP-AES apparatus. X-ray photoelectron spectroscopy (XPS) measurements were conducted on a Thermo Scientific K-Alpha instrument, using an Al Kα X-ray source at 1486.6 eV. Spectra were generated at a perpendicular takeoff angle, using a pass energy of 20 eV and steps of 0.1 eV. During analysis, the pressure was in the order of $\approx 1.33 \times 10^{-9}$ Pa. As an internal reference for the absolute binding energies, the Au (4f7/2) peak was used. The experimental spectra were deconvoluted after subtraction of the Shirley background using the XPSPEAK41 program.

Results and Discussion

The most common way to separate impurities from the sulfate solution is by selective precipitation through controlled acidity. Figure 1 illustrates the titration curve for all four REE used in this study in a regular titration test with no additive. It is clear that around pH 7, REE precipitate out of the solution by increasing pH. But then, in lower pH values, a decrease in the amount of REE ions in the solution is met. This refers to the REE that co-precipitated with gypsum during titration. As it can be seen, the heavy REE are less prone to co-precipitate with gypsum and it can be due to the fact that they have a higher ionic radius difference with Ca^{2+} ions. Therefore, their substitution with Ca ions in the structure of gypsum is more energy consuming than the light ones. Substitution of REE^{3+} ions with Ca^{2+} ions in each plane forms a doped layer in the structure of calcium sulfate dihydrate and the REE cations in the solution are more inclined to substitute the Ca atom on the planes that need less substitution energy.

Table 2 presents the percentage of the four REE investigated in this study in the solution at pH 5. This pH was selected as the target pH at which most of the impurities in the solution, such as Fe, Al and Th have completely precipitated before reaching this pH, meanwhile the REE in the solution have not yet started to precipitate. However, it is obvious that a proportion of REE have precipitated in low pH values, where the impurities are being removed from the solution.

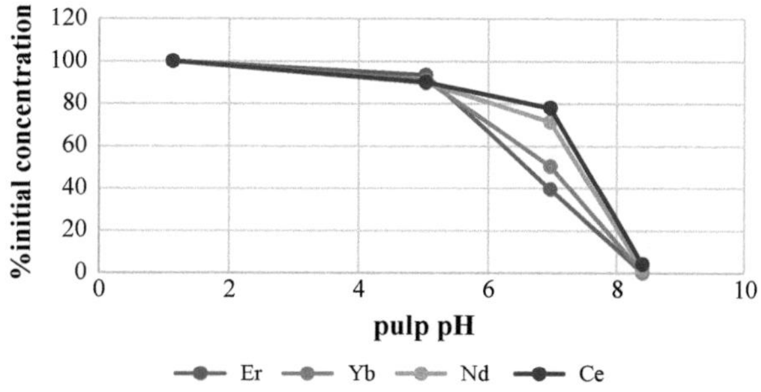

Fig. 1 Titration curve for Er, Yb, Nd and Ce in this study

Table 2 The percentage of Er, Yb, Nd and Ce in the solution at pH 5

Additive	g/L	Er (%)	Yb (%)	Nd (%)	Ce (%)
CTAB	0.2	98.8	97.6	94.8	95.1
	2.0	99.9	99.0	94.7	95.6
	5.0	98.4	97.9	95.1	96.5
PVS	0.2	95.8	95.7	91.5	91.5
	2.0	95.5	94.6	94.8	94.3
	5.0	92.9	91.7	85.4	87.0
PAA	0.2	94.5	93.3	89.2	91.7
	2.0	71.2	64.5	59.7	68.8
	5.0	26.1	20.2	17.4	26.7
No additive	0.0	94.5	93.2	90.7	91.1

Table 2 demonstrates that in the absence of any additive, the solution at pH 5 contains about 94.5, 93.16, 90.7 and 91.13% Er, Yb, Nd and Ce, respectively. This means that about 6% Er, 7% Yb, 10% Nd and 9% Ce have precipitated from pH 1 to 5 and a total of 8% of the REE in the solution are lost to gypsum. This proportion of REE can be either adsorbed on the surface of gypsum or entered the structure of gypsum by substitution with Ca^{2+} ions. The filtrate of this experiment was characterized with different characterization methods, which are presented in the following.

Crystal Structure

Figure 2 shows the XRD pattern of the precipitates prepared by $Ca(OH)_2$ 5% titration from pH 1 to 5. It can be seen that the precipitates were composed of calcium sulfate dihydrate (gypsum) phase by the strong peak at $2\theta = 13.5148°$. These were confirmed to be dihydrate and no hemihydrate was detected by XRD, which means that no complete or partial phase transformation occurred in these experiments and gypsum did not go under phase transformation in the presence of REE ions and modifiers (results for the other samples are not provided to avoid repetition).

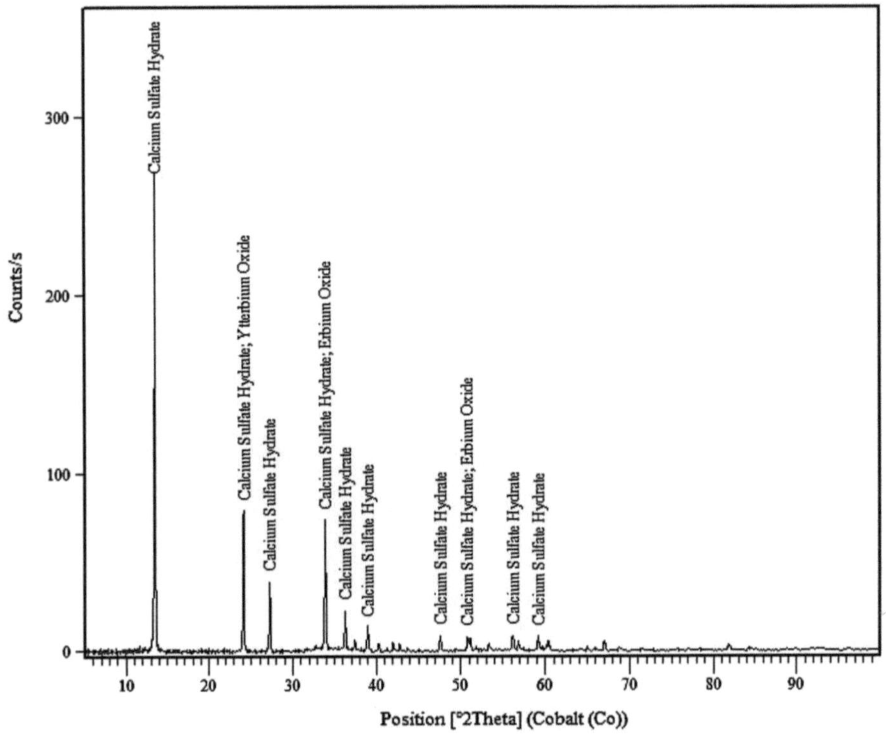

Fig. 2 XRD pattern of the precipitates of titration experiment with 5% $Ca(OH)_2$ at pH value of 5

The XRD results gave no information about the REE species as a result of low concentration of these ions in the structure or amorphous phase formation, except for the heavy ones, erbium oxide and ytterbium oxide that have been detected as minor phases in the structure of gypsum which defines the co-precipitation of heavy REE by means of surface adsorption on specific planes in gypsum crystal structure. This might have happend because of lower tendency of Er and Yb ions to substitute with Ca ions in gypsum crystal structure than the light REE such as Nd and Ce.

X-ray Photoelectron Spectroscopy (XPS)

XPS was carried out to measure the elemental composition of the precipitates. The Shirley method [12] of background removal was used in the least squares fitting. XPS measurements of the binding energy for precipitates are provided in Fig. 3. It shows that Ce, Nd, Er and Yb elements do exist on the surface of the gypsum precipitates. A comparison of the average S2p level binding energy measured by fitting Lorentzian-Gaussian functions to the data gave two peaks with 170.17 and 169.14 eV values. The latter confirmed the presence of $CaSO_4 \cdot 2H_2O$ (binding

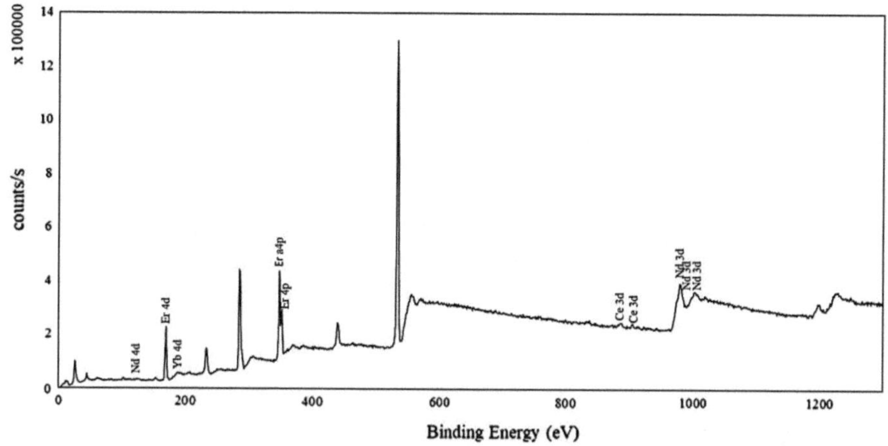

Fig. 3 XPS spectra of the precipitates of titration experiment with 5% $Ca(OH)_2$ at pH value of 5

energy of 169.5 eV) as the calcium sulfate phase which is generated in the course of these experiments as well as the probable presence of $Ce(SO_4)_2$ (binding energy of 169 eV [13]). The former binding energy value (170.17 eV) confirms the presence of $Er_2(SO_4)_3$, $Nd_2(SO_4)_3$ and $Yb_2(SO_4)_3$ (binding energy of 170.10 eV) which validates the isomorphous substitution of Ca^{2+} in the host lattice by the REE ions in the solution. Analyzing the peak values for Yb4d, Nd4d, Nd3d, Er4p, Er4d and Ce3d confirmed the presence of $Yb_2(SO_4)_3$ (E_b = 187.30 eV), $Nd_2(SO_4)_3$ (E_b = 122.5 eV), Nd_2O_3 (E_b = 982 eV), Er_2O_3 (E_b = 168.80 eV), CeO_2 (E_b = 907.30 eV) and Ce_2O_3 (E_b = 885.80 eV).

Effect of Modifiers

Table 2 shows the effect of three additives, CTAB, PVS and PAA in three different concentrations on the amount of REE that co-precipitated with gypsum. It is clear from the table that the presence of CTAB in the solution has improved the results; in other words, at pH 5, more REE are present in the solution compared with the regular experiment with no additive. It means that less proportion of REE precipitated with gypsum and REE loss has decreased about 4% for all the REE available in the system. Based on the results, 2 g/L CTAB can be chosen as the optimum condition that is capable of keeping the most percentage of REE in the solution in comparison with the other concentrations of CTAB and other additives. PVS is shown to have a very slight positive effect on the process when it is introduced to the system in 0.2 and 2 g/L, but 5 g/L PVS has shown a negative effect on the process and increases co-precipitation of REE with gypsum. Finally, PAA is shown to have a tremendous negative effect on the process and has increased REE uptake by gypsum.

The different behavior of REE in the presence of the said additives can most likely be due to the different gypsum morphologies, which are obtained when additives are introduced to the system. To investigate the morphology differences in more details, gypsum was precipitated from a REE-free gypsum saturated solution in the presence and absence of the additives. Figure 4 illustrates the morphology of gypsum precipitated from gypsum-saturated solutions in sulfuric acid by changing its pH from 1 to 9 using $Ca(OH)_2$ in the presence and absence of additives. As it is shown, gypsum precipitates in thick elongated plate-shape from the acid solution in the absence of additives and impurities (Fig. 4a). Such plate-like (slab) morphology was also observed in other studies involving the crystallization of calcium sulfate dihydrate in the absence of impurities. Adding CTAB, PVS and PAA to REE-free gypsum-saturated solution caused different gypsum morphologies after precipitation (Fig. 4b–d).

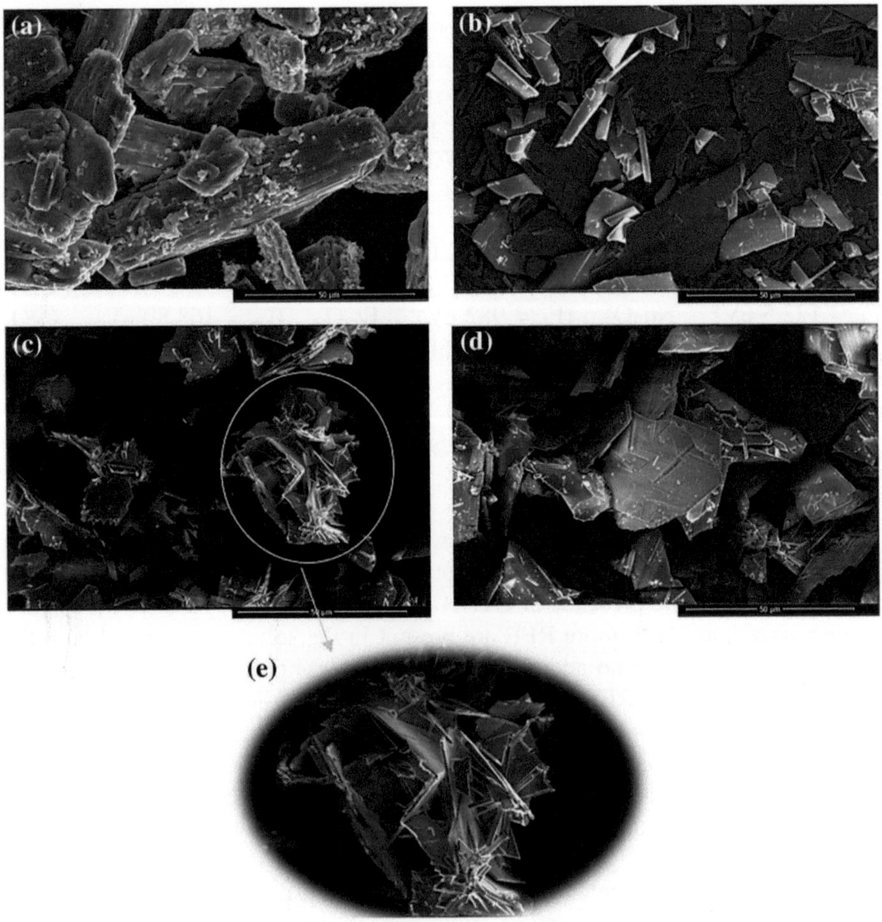

Fig. 4 Morphology of **a** non-modified, **b** CTAB, **c** PVS and **d** PAA-modified gypsum

An evaluation of SEM images showed that CTAB favored formation of slab (plate-like) crystals but thinner and much smaller than non-modified gypsum (Fig. 4b). CTAB is assumed to act through forming electrostatic ion-pairs between CTAB and surface sulfate groups, namely, preferred adsorption on certain crystal faces that interferes with the crystal growth process [11]. This is the reason of formation of smaller and thinner gypsum crystals in the presence of CTAB in the solution. This suggests that CTAB is probably adsorbed on most of the crystal faces, which are the active sites for REE adsorption on gypsum and results in less REE loss to gypsum while its nucleation and growth process.

Figure 4c represents the morphology of modified gypsum with PVS. As it can be seen, PVS has totally changed the crystal morphology of precipitated gypsum. By comparing Fig. 4c, a we see that very small, elongated crystals have been formed in the presence of PVS in the acidic solution and it is obvious that PVS has strongly retarded growth of dihydrate crystals in almost all directions and some crystals with unknown shapes have been formed. Crystal modification mechanism of PVS is very complex and strongly depends on the crystallization environment. Generally, PVS affects the nucleation and growth process and changes the morphology of the precipitated crystals. PVS can act via two mechanisms; (i) interrupting the periodic incorporation of Ca^{2+} and SO_4^{2-} ions in the structure of gypsum by forming calcium sulfonate complexes instead of calcium sulfate groups which can interfere with the nucleation process or adsorption of the sulfonate groups on the surface of the formed crystals and retarding growth [14]; (ii) the polymeric chain in the structure of PVS may cause steric effects and hence modify the crystal morphology. However, it's been reported by Feldmann and Demopoulos [11] that the steric effect is the main mechanism that PVS modifies the crystal structure of dihydrate crystals in strong acid solutions (in that case HCl) and the detrimental effect seems to originate from the polymeric chain of this additive rather than the sulfonate functional group. The dual behavior of PVS in different concentrations is probably due to the complex mechanism of this additive and it might block the preferred adsorption sites for REE on gypsum, but at the same time generate some active adsorption sites by making sulfonate functional groups on the surface of gypsum or the steric effect of the polymeric chain; therefore, the outcome of these two effects can determine the total REE which is adsorbed on gypsum to be more or less than the regular precipitates.

Figure 4d shows the morphology of modified gypsum in the presence of PAA as a surfactant. Compared with dihydrate morphology in the absence of additives, PAA resulted in the formation of thinner and elongated crystals. PAA is postulated to affect the nucleation and growth process by slowing down growth in all faces of the crystal except {011} and $\{\bar{1}11\}$. Carboxylic groups ($-COO^-$) of PAA tend to form polymeric networks in the presence of Ca^{2+} ions; therefore, each Ca^{2+} binds to two PAA ions. This can change the availability of Ca^{2+} ions and subsequently affect the nucleation process as well as the growth process by its preferential adsorption on crystal faces. PAA has a very high affinity towards divalent ions such as Ca^{2+} in any pH value [15]. It forms monodentate complexes with Ca^{2+} even in highly acidic environment that it has a very low dissociation [16]. In addition, the difference between the bond types and energies of Ca^{2+}–SO_4^{2-} ions in dihydrate crystal lattice in the absence and presence of any other ions or additives in the solution that leads to formation of new bonds between ions, can modify the surface energies of specific crystal planes and changes the nucleation and growth kinetics depending on the type of bonds and surface energies. The presumption is that the more PAA available in the system (5 g/L compared to 0.2 g/L), the more Ca^{2+} ions are engaged in binding with the carboxylic groups of PAA and making the network; as a consequence, available REE ions in the system have a higher chance to get into the structure of gypsum and substitute with Ca^{2+} ions. Hence, a high proportion of REE co-precipitate with gypsum before reaching the acidity at which their precipitation occurs.

Conclusion

Selective precipitation of some ions such as Fe, Al and Th from REE-containing leaching solution of acid baked concentrate by means of controlled acidity is a common way to separate these impurities from REE ions in the solution. Application of lime or limestone as a reagent is the cheapest way to change the acidity of solution but has its own disadvantages, one of which is generation of gypsum precipitate and consequently, co-precipitation of a proportion of REE with gypsum. This leads to loss of REE to precipitated gypsum and as a result, lime and limestone have been substituted by much more expensive reagents to avoid the said problem. A novel process was investigated in this paper to decrease the uptake of REE by precipitated gypsum via addition of additives. CTAB additive was shown to improve the REE loss by its preferred adsorption on certain crystal faces that interferes with the crystal growth process. Adding PVS to the system was shown to have a very slight improvement in lower concentrations (0.2 and 2 g/L) and a negative effect in its high concentration (5 g/L). PAA showed a detrimental effect on the process, made a high proportion of REE co-precipitate before the solution reaches the acidity at which their precipitation was supposed to occur. It is presumed to be due to a high engagement of Ca^{2+} ions in making polymeric networks with carboxylic acid groups of PAA that gives the present REE in the solution a higher chance to be involved in the structure of gypsum.

Acknowledgements The authors would like to acknowledge the financial support for this study through NSERC.

References

1. Gupta CK, Mukherjee TK (1990) Hydrometallurgy in extraction processes. CRC Press, Florida

2. Sadri F, Nazari AM, Ghahreman A (2017) A review on the cracking, baking and leaching processes of rare earth element concentrates. J Rare Earths 35(8):739–753. https://doi.org/10.1016/S1002-0721(17)60971-2
3. Sadri F, Rashchi F, Amini A (2017) Hydrometallurgical digestion and leaching of Iranian monazite concentrate containing rare earth elements Th, Ce, La and Nd. Int J Miner Process 159:7–15
4. Cuthbert FL (1958) Thorium production technology. Addison-Wesley, Massachusetts
5. Licong D, Yifei Z, Fangfang C, Shaotao C, Shaowei Y, Yan L, Yi Z (2013) Reactive crystallization of calcium sulfate dihydrate from acidic wastewater and lime. Chin J Chem Eng 21(11): 1303–1312. https://doi.org/10.1016/S1004-9541(13)60626-6
6. Feldmann T, Demopoulos G (2013) Ind Eng Chem 52:6540–6549. https://doi.org/10.1021/ie302933v
7. Martynowicz E, Witkamp G, van Rosmalen G (1991) Uptake of metal-ions in calcium-sulfate hemihydrate crystals. Paper presented at the 3rd European conference on crystal growth, Budapest, Hungary, 5–11 May 1991
8. Söhnel O (1992) Precipitation: basic principles and industrial applications. Oxford
9. Feldmann T (2013) Crystallization kinetic investigations of calcium sulfate phases in aqueous $CaCl_2$-HCl solutions. Ph.D. thesis. McGill University
10. Xin Y, Hou S, Xiang L, Yu Y (2015) Adsorption and substitution effects of Mg on the growth of calciumsulfate hemihydrate: an ab initio DFT study. Appl Surf Sci 357:1552–1557
11. Feldmann T, Demopoulos G (2014) Effects of crystal habitmodifiers on the morphology of calcium sulfate dihydrate grown in strong $CaCl_2$-HCl solutions. J Chem Technol Biotechnol 89:1523–1533
12. Shirley D (1972) High-resolution X-ray photoemission spectrum of the valence bands of gold 5:12–15
13. Ferrizz R, Gorte R, Vohs (2002) TPD and XPS investigation of the interaction of SO_2 with model ceria catalysts. Catal Lett 82:123–129
14. Freyer D, Voigt W (2003) Crystallization and Phase Stability of $CaSO_4$ and $CaSO_4$-based salts. J Monatshefte für Chemie 134:693–719
15. Wall F, Drenan J (1951) Gelation of polyacrylic acid by divalent cations. J Polym Sci 7:83–88
16. Miyajima T, Mori T, Ishiguro S (1997) Analysis of complexation equilibria of polyacrylic acid by a Donnan based concept. J Colloid Interface Sci 187:259–266

Beneficiation and Leaching Study of Norra Kärr Eudialyte Mineral

Victoria Vaccarezza and Corby Anderson

Abstract Due to the demand for rare earth elements for everyday technology and applications, there has been much research initiation into the extraction and recovery of rare earth elements. An otherwise unknown mineral, eudialyte is a zircon silicate consisting of rare earth oxides, specifically the heavy REO yttrium (III), with only trace amounts of thorium and uranium. The first part of this project aimed at finding an efficient way to liberate and upgrade eudialyte to concentrate the REO content. A combination of gravity and magnetic separation techniques were employed, resulting in limited success. The second part looked at the hydrometallurgical processing of eudialyte through various leaching experiments with sulfuric acid. In all, this project will discuss why certain processes were utilized over others, as well as the difficulties that arose during processing.

Keywords Eudialyte · Rare earth elements · Beneficiation · Leaching
Gravity separation · Magnetic separation

Introduction

The term "rare earth elements" refers to the 17 metallic elements comprising of the lanthanides, yttrium and scandium [1]. These elements have been referred to as a group because of their chemically similar properties. The light rare earth group consists of the first eight elements of the lanthanide series (atomic numbers 57–64) and sometimes scandium. The heavy rare earth group consists of the rest of the elements in the lanthanide series (atomic numbers 65–71) and yttrium. Yttrium and cerium are the rare earth elements of importance with respect to eudialyte, since they are the most abundant in the mineral. Yttrium is essential for fluorescent light phosphors, computer and television displays, as well as to stabilize zirconia in

V. Vaccarezza (✉) · C. Anderson
Department of Metallurgical and Materials Engineering, Kroll Institute for Extractive Metallurgy, Colorado School of Mines, Golden, CO, USA
e-mail: vmvaccarezza@mymail.mines.edu

H. Kim et al. (eds.), *Rare Metal Technology 2018*, The Minerals, Metals & Materials Series, https://doi.org/10.1007/978-3-319-72350-1_4

thermal plasma sprays used on the surfaces of aerospace components [1, 2]. Cerium can be used in a variety of applications, including: as a polishing agent in precision optical polishing of glass, mirrors, optical glass and disk drives; as a sensitizer in ceramics; in catalytic converter, and many other areas. With the consumption of rare earths expected to continue to grow, especially in the energy, electronics and optoelectronics sectors, demand for these elements is also expected to rise in accordance [3]. These elements and their compounds are necessary for the development of many modern technological devices that consumers have become heavily dependent on in a daily basis.

Eudialyte is a potential source for yttrium and other rare earth elements with the added advantage of low concentrations of the radioactive elements thorium and uranium. The eudialyte mineral used in this project originates from the Norra Kärr deposit in southern Sweden. The Norra Kärr deposit is a zirconium and rare earth element enriched peralkaline nepheline syenite intrusion which hosts the eudialyte group minerals. The first step in many extractive processes is to try to create an enriched preconcentrate that has been removed of unwanted materials or materials that are valuable for another process. This beneficiation step is not only important for the future hydrometallurgical processing of the ore, but economically as well, since these impurities may hinder the final recovery and grade of the desired element.

There is little known about the flotation characteristics of eudialyte, since previous experiments conducted on eudialyte are limited. Russian literature reports eudialyte recovery via flotation using sodium oleates and oleic acid as collectors [4, 5]. Flotation experiments were also done on the eudialyte ore from the Lovozero deposit at the Kola Peninsula in the former USSR. On average, the eudialyte ore contains 13.5 wt% ZrO_2 and 2.5 wt% rare earths. A reversible flotation flowsheet was developed, where fatty acid collectors were used to first float aegirine [6].

Magnetic and flotation beneficiation work was also conducted on the Norra Kärr eudialyte mineral by RWTH Aachen University. Dry and wet high intensity magnetic separators were used with the goal to produce a separation between the major gangue and eudialyte in one step. However, the magnetic susceptibilities between aegirine and eudialyte overlap enough to hinder the ability to create a clean separation. Focus shifted to flotation concentration, with the goal being, again, to avoid a two-step flotation process, as the one suggested for the Lovozero eudialyte mineral. Three different eudialyte samples were tested as raw ore feeding into the circuit, while three other eudialyte samples were pre-concentrates from a magnetic separation step. Overall, the pre-concentrate eudialyte samples used in the flotation step yielded the highest upgrade ratios and recoveries in the 80 s [7].

The Norra Kärr project in Sweden was undertaken by Tasman Metals Ltd., in consultation with ANZAPLAN, with the intention on determining the most suitable beneficiation route for the Norra Kärr mineralized material. Different techniques were investigated, such as spiral concentration, electrostatic separation, high-G separation, magnetic separation and froth flotation [8]. Results showed that aegirine could be selectively floated, but co-flotation of non-liberated particles concluded that a direct flotation of eudialyte would be unsuccessful [9]. High recovery values were recorded for eudialyte via WHIMS, but with no significant upgrade in the rare earth concentration [10, 11]. The literature survey regarding eudialyte beneficiation experiments indicate that at least a multiple step process is necessary for separation of the eudialyte mineral from its gangue components.

The name eudialyte derives from the Greek word for "well-decomposable" in acid [12]. However, the issues with the leaching of eudialyte lie with the co-dissolved silica. This silica forms a gelatinous phase hindering the filtering processing for rare earth element extraction [13]. The current goal of processing eudialyte is to achieve a reasonable recovery of leached rare earth elements while minimizing or eliminating the formation of the silica gel. There is a discussion regarding the efficiency of leaching the rare earth elements and zirconium as sulfate or chloride ions in terms of the solubility. Also, which acid minimizes the silica gel formation when used in a concentrated manner. It has been suggested by Voßenkaul et al., that the recovery of rare earth elements is more favorable in chloride systems. In using hydrochloric acid, rare earth chloride salts are developed and are typically more soluble in water than the sulfate salts from employing the sulfuric acid. The solubility of rare earth element sulfate salts in water decreases proportional to the decrease in atomic number of the rare earth element, except for cerium and praseodymium. Thus, the heavy rare earth elements stay in solution, while the light rare earth elements are precipitated [1]. Since yttrium and the heavy rare earth elements are soluble, double-sulfate precipitation is not possible. Double-sulfate precipitation is used for separating rare earth elements by their light or heavy respective groups. Equation 1 shows the double-sulfate precipitation chemical reaction:

$$2\mathrm{REE}^{3+} + 4\mathrm{SO}_4^{2-} + 2\mathrm{Na}^+ \leftrightarrow \mathrm{REE}_2(\mathrm{SO}_4)_3{\cdot}\mathrm{Na}_2\mathrm{SO}_4{\cdot}\mathrm{H}_2\mathrm{O}\,[1] \tag{1}$$

However, in terms of minimizing the formation of the silica gel, the use of sulfuric acid may have a greater advantage than the hydrochloric acid. Apart from its low cost, volatility and corrosive activity, sulfuric acid has better solubility in water at room temperatures than hydrochloric acid. Concentrated sulfuric acid can be found at 98 w/w%, while hydrochloric acid strength is 37 w/w%. To minimize the silica gel formation, the solution must not have access to large amounts of water. The reasoning lies in the thermodynamics and kinetics of the silica-water system. With the addition of acid to a silicate, shown by the chemical Eq. 2, the silicate will acidify to form the weak monosilicic acid (Fig. 1):

$$(\mathrm{SiO}_2)_x(\mathrm{s}) + 2\mathrm{H}_2\mathrm{O}(\mathrm{l}) \leftrightarrow (\mathrm{SiO}_2)_{x-1}(\mathrm{s}) + \mathrm{Si(OH)}_4(\mathrm{aq})\,[14] \tag{2}$$

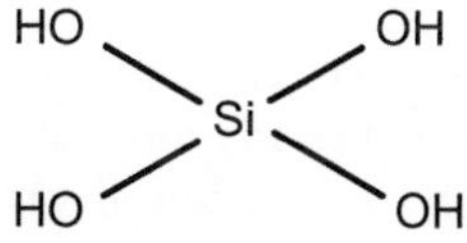

Fig. 1 Molecule of monosilicic acid

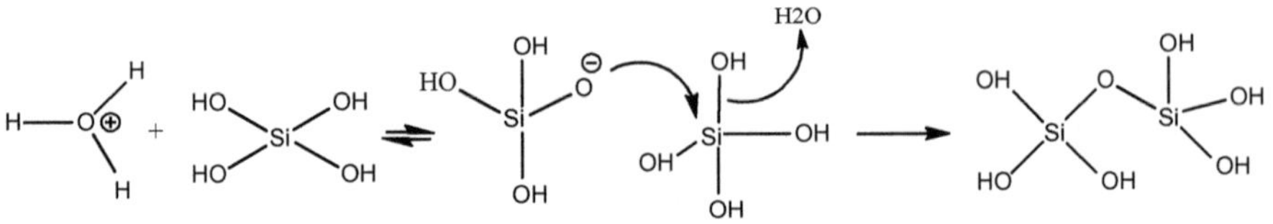

Fig. 2 Polymerization mechanism for the development of silica gel [16]

Once the silicic acid is formed, a polymerization reaction occurs analogous to a condensation polymerization reaction. The presence of water aids in the polymerization process. The polymerization process is shown below [14] (Fig. 2).

The polymerization proceeds forward to maximize the formation of siloxane linkages (Si–O–Si), essentially forming a gel with internal siloxane linkages and external SiOH groups [15]. To minimize or eliminate the silica gel formation, it is concluded that the system needs a to be deprived of water during the acidic leach since the exposure to water is driving the polymerization following the acidification of the silicate. A recent approach in seeking to prevent the formation of the silica gel involves a "dry digestion" of the eudialyte mineral with hydrochloric acid. The process provides just enough acid to wet the mineral sample allowing the silica to precipitate. The amount of acid to "wet" the mineral should be around the stoichiometric or slightly below that amount. However, due to the small volume available, the precipitates should grow to larger particles that can be separated from the valuable metals [16]. The varying parameters in these experiments include varying acid concentration, retention time in acid and amount of water used to leach the elements.

Methods and Materials

Norra Kärr Eudialyte Mineral

Eudialyte is a zirconium silicate mineral, notable for its high concentration of the heavy rare earth elements, specifically yttrium. The crystal structure comprises of a nine-membered silica ring and a six-membered ring of calcium octahedra that is held together by zirconium octahedra and three-membered silica rings (Table 1).

Description of Beneficiation Method

The beneficiation consisted of a two-stage process via gravity and magnetic separation. The goal of the gravity separation/HLST was to disassociate the major iron-bearing mineral aegirine from the eudialyte and remaining gangue in the sink and float, respectively (Fig. 3).

Table 1 Eudialyte and gangue mineral characteristics [12, 18]

Mineral	Specific gravity	Chemical composition	Magnetic properties
Eudialyte	2.70–3.10	$Na_4(Ca, Ce)_2(Fe, Y, Mn)ZrSiO_{22}(OH, Cl)_2$	Paramagnetic
Aegirine	3.50–3.60	$NaFeSi_2\ O_6$	Paramagnetic
Potassium feldspar	2.50–2.60	$KAlSi_3\ O_8$	Non-magnetic
Nepheline syenite	2.55–2.60	$(Na, K)AlSiO_4$	Non-magnetic

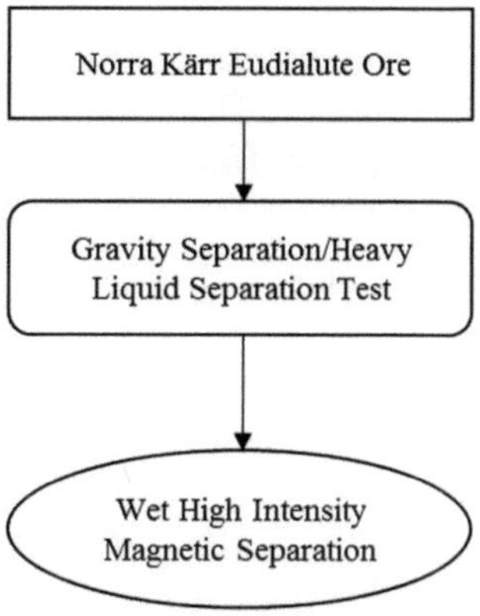

Fig. 3 Preliminary beneficiation flowsheet for Norra Kärr eudialyte sample

Initially, the non-toxic sodium polytungstate was used for HLST at three different specific gravities (2.7, 2.95, 3.08) and at the mesh size fractions. After conducting the HLST with the sodium polytungstate, it was found that a higher specific gravity value would need to be obtained to effectively separate the aegirine from the eudialyte and other lighter gangue. Another HLST was administered at Hazen Research, Inc., at a specific gravity of 3.2 with methylene iodide. All the HLST conducted utilizing a centrifuge in order to expedite the settling process. The float material was then processed through the WHIMS for separation between the paramagnetic eudialyte and non-magnetic gangue. After initial test work was done, an advanced flowsheet was developed for investigating different parameters through beneficiation. HLST were done at four different size fractions[1]: as-received ore sample, pulverized sample and screen material at ±400 mesh. The sample was pulverized to achieve a better degree of liberation in comparison to the as-received sample. The ore sample was also screened to assess a difference between the screened and pulverized material, since pulverizing the sample may have caused

[1]All HLST conducted from this point forward were done at an SG of 3.2 using methylene iodide for best chance of gravity separation.

differences in the surface morphology of the sample. The flowsheets are shown below.

Methylene Iodide Heavy Liquid Separation Test

The HLST done using methylene iodide as the media were conducted by Hazen Research, Inc. Approximately twenty-five grams of as-received ore, pulverized ore and ±400 mesh ore were sent to for the HLST. Each sample was placed in a centrifuge. The dried masses of the float and sink were recorded and sent back for analysis via XRF (Figs. 4 and 5).

Wet High Intensity Magnetic Separation

A laboratory scale WHIMS was used to process the float material obtained from the HLST using methylene iodide. The WHIMS was magnetized at 1 T and the slurry was slowly added to the bowl, where the non-magnetic and magnetic fractions were discharged into respective containers. Both magnetic and non-magnetic fractions were pressure filtered and dried before XRF analysis.

Description of Leaching Study

Two different leaching processes were investigated. In the first process, leaching of the eudialyte sample was done in an excess of acid available to the mineral. In the second process, the eudialyte sample was processed under a starving condition where only enough concentrated acid to wet the sample was added and later leached with DI water. A flowsheet of the process is given in Fig. 6. In each leaching process, a non-magnetic concentrate eudialyte sample was used. By creating a non-magnetic concentrate, the iron content in the sample is reduced. Sulfuric acid was used in leaching process 1 to investigate the recovery of rare earth sulfates. Concentrated sulfuric acid was used in leaching process 2 to minimize water exposure and minimize silica gel formation.

Leaching Process 1

All experiments consisted of a one-liter solution of 0.1 M sulfuric acid, approximately 100 g of non-magnetic concentrate sample and an agitator to keep the sample suspended. Figure 6 shows the experimental set-up for leaching process 1.

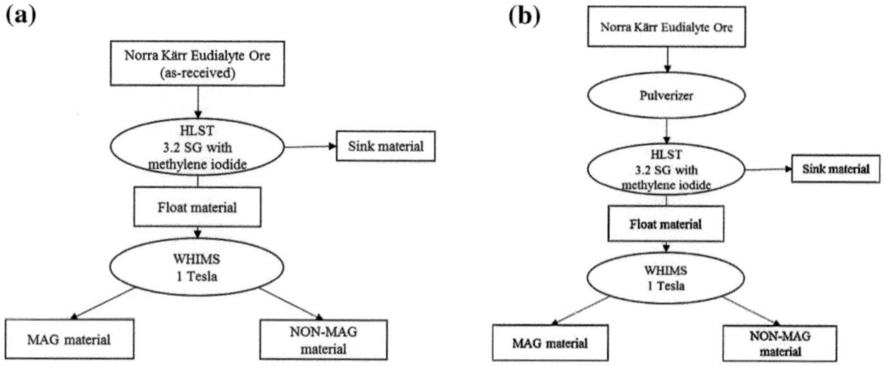

Fig. 4 **a, b** Flowsheet for as-received and pulverized Norra Kärr eudialyte sample

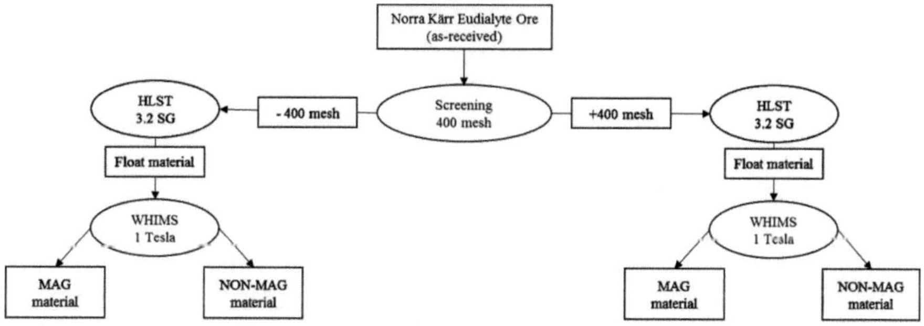

Fig. 5 Flowseet for screened Norra Kärr eudialyte sample

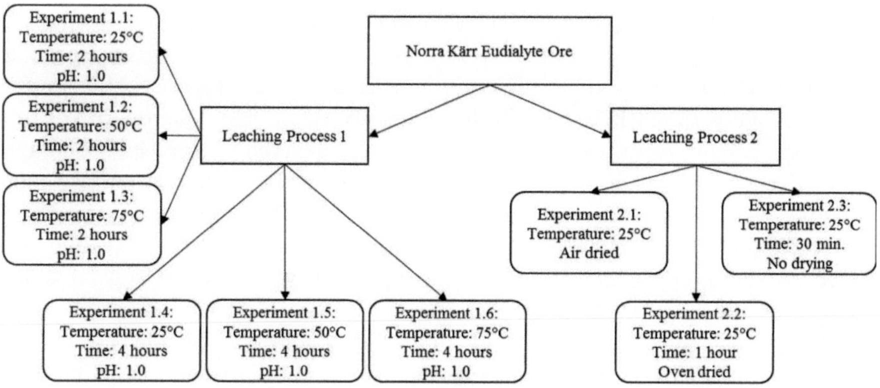

Fig. 6 Flowsheet for leaching processes for Norra Kärr eudialyte sample

Free acid titrations were also conducted after sample addition and every half an hour after to assess free sulfuric acid concentration. Experiments 1.1–3 were all conducted for 2 h at 25, 50 and 75 °C. Experiments 1.4–6 were conducted for 4 h at 25, 50 and 75 °C. Once all the experiments were completed, the leach liquor was vacuum filtered. The filtrate was prepared for ICP-MS analysis and the filter cake was dried and weighed before XRF analysis.

Leaching Process 2

All experiments consisted of just enough 98% pure sulfuric acid addition to wet the sample, approximately 25 g of non-magnetic concentrate sample and 150 mL of DI water were used to obtain the sulfates in solution. In experiment 2.1, the DI water was added once the sample dried. In experiment 2.2, the sample was left for one hour and then placed in a furnace at 60 °C to dry. Once dried, DI water was added. Finally, in experiment 2.3, the sample was left to sit for 30 min before immediate DI water addition. The leach liquors were then vacuum filtered. The filtrate was prepared for ICP-MS analysis and the filter cake was dried and weighed before XRF analysis.

Results and Discussion

Methylene Iodide Heavy Liquid Separation Test Results

Methylene iodide was used in this experiment to reach a density of 3.2 g/cm^3. Figure 7a, b, shows the recovery of TREEs and zirconium for four size fractions at the specific gravity of 3.2. As the average particle size increases from 29 μm to +400-mesh, the recovery of the TREEs and Zr decreases. Figure 7, illustrate a greater recovery of TREEs and Zr for the pulverized size fraction of 29 μm, indicating a certain degree of liberation is achieved when the as-received ore sample is pulverized. This degree of liberation may be due to the change in surface morphology by pulverizing the ore, where a single particle is crushed/ground into smaller particles, thus revealing a new surface. This differs from screening the as-received ore sample because in screening, the smaller particles are separated, not produced by crushing or grinding.

Table 2 shows the elemental upgrade ratios for the four size fractions at a density of 3.2 g/cm^3. As can be seen by the upgrade ratios and variance values for each elemental group, there is not a lot of difference in upgrade ratios for each size fraction. However, there is a greater upgrade ratio for iron in all size fractions in this heavy liquid separation test.

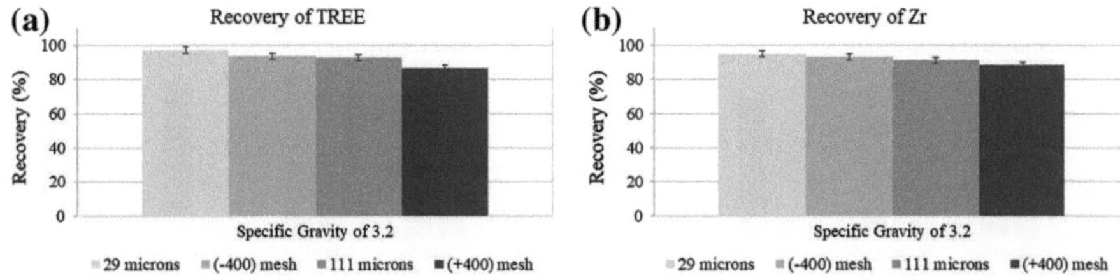

Fig. 7 **a**, **b** Recovery of total rare earth elements and zirconium for the four size fractions

Table 2 Upgrade ratios for four size fractions at a specific gravity of 3.2

	Upgrade ratio	
	TREE	Zr
29 μm	1.7	0.9
−400 mesh	1.3	1.0
111 μm	1.2	1.1
+400 mesh	1.1	1.0

This result provides several conclusions. Firstly, there is no dependence on the size fraction of the ore to upgrade the eudialyte ore due to small variance values. Secondly, the sole step of gravity separation may not be an effective method to upgrade the ore in terms of its rare earth element and zirconium concentrations, but may be an effective method for the removal of iron. Finally, the ore may not be well liberated at 29 μm for adequate separation.

Results of WHIMS on Float Products

WHIMS was done on the float products of the heavy liquid separation tests done using methylene iodide. Figure 8 display the recovery of the TREEs and Zr for each size fraction, respectively. It was expected for the recovery of the TREEs and Zr to be high in the magnetic fraction since the float product consisted of the paramagnetic eudialyte and non-magnetic gangue. However, the results show the TREEs and Zr reporting to the non-magnetic fraction. Table 3 shows the upgrade ratio values for TREEs and Zr in magnetic fraction for each size fraction. Again, there is no significant upgrade of the eudialyte ore, as indicated by these values. These values also do not show dependence on particle size for better separation.

Due to the complex and variable chemical composition of eudialyte, it may be suggested that this Norra Kärr eudialyte ore does not exhibit paramagnetic behavior compared to other eudialyte minerals studied in literature. However, before such a conclusion can be made regarding this specific eudialyte sample, more magnetic separation tests should be conducted at higher magnetic strengths. Magnetic

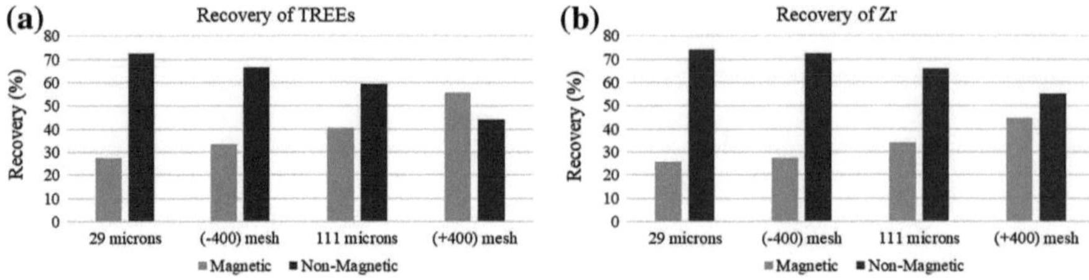

Fig. 8 **a**, **b** Recovery of TREE and Zr in magnetic fraction from float products at 1 T

Table 3 Upgrade ratios for magnetic fraction on float product at 1 T

	Upgrade ratio	
	TREE	Zr
29 μm	1.1	1.0
−400 mesh	1.8	1.5
111 μm	1.4	1.2
+400 mesh	1.6	1.3

separation of eudialyte may also be highly variable with respect to which fraction the mineral will report to (magnetic or non-magnetic) due to the iron present.

Results of Leaching Process 1

Figures 9a, b, show the recovery of TREEs and Zr at each temperature for both time intervals. Taking into consideration the unreasonable TREEs and Zr recovery values due to gelation, there is an increase in recovery as leaching time increases. Increased leaching time allows the acid to permeate more particle surfaces, since the ore is in solution for a longer period of time. However, at an increased temperature, such as in experiment 1.6, this increased exposure to the acid and water will lead to gel formation. In all, Fig. 9, shows a linear relationship between temperature and time for the recovery of total rare earth elements and zirconium.

The acid consumption values in Table 4, show the experiments conducted at 25 °C have the highest amount of acid consumption. With the exception of experiment 1.4, the leaching experiments conducted at four hours have a higher acid consumption than those done at two hours.

Results of Leaching Process 2

Upon the addition of the DI water to the sample in experiment 2.3, the solution became cloudy and bubbled. Since the sample in experiment 2.2 was oven dried

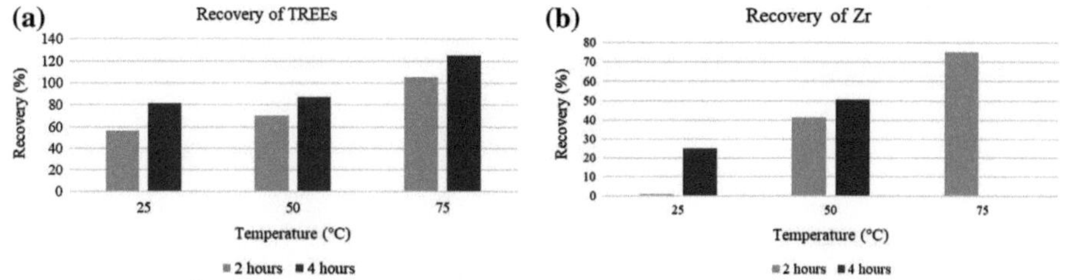

Fig. 9 **a**, **b** Recovery of total rare earth elements and zirconium as a function of time and temperature for all experiments

Table 4 Consumption of sulfuric acid for leaching experiments at two (experiments 1.1–3) and four (experiments 1.4–6) hours

Experiment	Acid consumed (g)
Experiment 1.1	7.7
Experiment 1.2	3.5
Experiment 1.3	3.5
Experiment 1.4	7.6
Experiment 1.5	5.6
Experiment 1.6	5.6

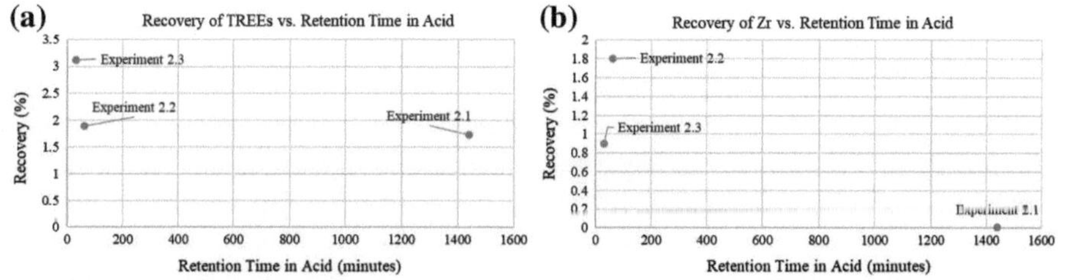

Fig. 10 **a**, **b** Recovery of total rare earth elements and zirconium versus retention time in acid for experiments 2.1–2.3, respectively

after an acid retention time of one hour, the addition of DI water to the sample did not immediately bubble like in experiment 2.3, but some bubbles were observed as the solution was left sitting for about ten minutes. In experiment 2.1, the acid-wet sample was left to air dry for 24 h before DI water was added. Similar to the oven-dried sample, this solution did not immediately bubble until after about ten minutes. While neither of these experiments showed gelation during the acidification or after the DI water was added, nor did the filtrate solutions gel, the addition of water did cause a reaction to occur, evident by the bubbling (Fig. 10 and Table 5).

Table 5 Solution pH of each experiment after the addition of DI water ore sample

Experiment	Solution pH
Experiment 2.1	0.85
Experiment 2.2	1.11
Experiment 2.3	1.20

Conclusion

Based on previous work and characterization of the ore, the initial goal of this research was to upgrade the eudialyte ore through a combination of gravity and magnetic separation. These physical beneficiation techniques were applied to four different size fractions in hopes of achieving better separation via liberation of the mineral. However, it was concluded an efficiently clean separation of the eudialyte mineral from its gangue was not possible through these methods. Further investigation into the beneficiation of eudialyte should focus on the mineralogy and developing a process specific to the eudialyte mineral.

The two leaching processes explored two extremes with regard to hydrometallurgical treatment of eudialyte. As discussed before, the gelation experienced while leaching eudialyte is an important process and economical hindrance, so it was essential to discover a method to minimize this gelation. Recovery of the TREEs and Zr was achieved in the first leaching process, obtaining an optimum at 50 °C and four hours. Since the gelation of the filtrate leach solution scales with the high temperature, there is a window between filtering and gelation (upwards of a few days) in which the solution can be processed without risk of any gelation. While the experiments in the second leaching process succeeded in eliminating the formation of the silica gel, there was no significant recovery of the TREEs or Zr.

This work was supported by the Critical Materials Institute, an Energy Innovation Hub funded by the U.S. Department of Energy, Office of Energy Efficiency and Renewable Energy, Advanced Manufacturing Office.

References

1. Gupta C, Krishnamurthy N (2016) Extractive metallurgy of rare earths. CRC Press, Boca Raton, FL
2. Fischetti M Global demand stresses limited supply. Sci Am 305(4):64–65, Academic Search Premier, EBSCO*host*
3. Rare Earth Consumption is Growing (2003) Power Electronics Technology, General Reference Center GOLD. http://www.powerelectronics.com/industry/rare-earth-consumption-growing
4. Zakharova I, Nayfonov T (1974) A study on floatability of eudialyte and associated minerals. Russ J Nonferrous Met 1:12–16 (in Russian)
5. Ferron CJ, Bulatovic SM, Salter RS (1991) Beneficiation of rare earth oxide minerals. Mater Sci Forum 70–72:251–270

6. Naiphonov TB, Beloborodov VI, Zaharova IB, Kulakov AN, Zorina TA (1991) Flotation technology for beneficiation of eudialyte ore. In: XVII international mineral processing congress, Dresden, vol 4, pp. 131–138
7. Stark T, Silin I, Wortuba H (2017) Mineral processing of eudialyte ore from Norra Kärr. J Sustain Metall 3:32–38
8. Forrester K, Reveley S, Short M, Saxon M, Holmström H, Leijd M (2015) Norra Kärr Project Development. In: Proceeding of Alta Uranium-REE Sessions, Perth, Australia
9. Forrester et al. K Beneficiation of rare earth element enriched eudialyte from the Norra Kärr peralkaline intrusion with wet high intensity magnetic separation. In: Conference of metallurgists, Canadian Institute of Mining, Metallurgy and Petroleum, Vancouver
10. Tasman Metals, Ltd. (2015) Prefeasibility Study-NI 43-101-Technical report for the Norra Kärr Rare Earth Element Deposit
11. ENP Newswire (2014) Tasman precipitates heavy rare earth element concentrate and provides flowsheet development update for Norra Kärr project, Sweden. COMTEX News Network, Inc.
12. mindat.org
13. Davris P, Stopic S, Balomenos E, Panias D, Paspaliaris I, Friedrich B (2016) Leaching of rare earth elements from eudialyte concentrate by suppressing silica gel formation. Mineral Eng
14. Iler RK (1979) The chemistry of silica: solubility, polymerization, colloid and surface properties, and biochemistry. Wiley
15. Encyclopedia of Chemical Technology, 4th edn, vol 21. Wiley, New York, pp 983–984; 993–995 (1991)
16. Voßenkaul D, Birich A, Müller N, Stoltz N, Friedrich B (2016) Hydrometallurgical processing of eudialyte bearing concentrates to recover rare earth elements via low-temperature dry digestion to prevent the silica gel formation. J Sustain Metall
17. Sjöqvist ASL, Cornell DH, Andersen T, Erambert M, Ek M, Leijd M (2013) Three compositional varieties of rare-earth element ore: eudialyte-group minerals from the Norra Kärr Alkaline Complex, Southern Sweden. Minerals 3:94–120
18. Wyss G (2016) Characterization of eudialyte process samples. Center for advanced mineral and metallurgical processing (CAMP). Montana Tech of the University of Montana

Review on the Processes for the Recovery of Rare Earth Metals (REMs) from Secondary Resources

Archana Kumari, Manis Kumar Jha and D. D. Pathak

Abstract Treatment of secondary/waste to recover rare earth metals (REMs) is gaining importance due to its increasing global demand, lack of availability of high grade natural resources and huge generation of secondaries. Present paper reports the critical review on available processes for recovery of REMs from secondaries viz. manufacturing scraps, e-waste, industrial residues such as red mud, fly ash, waste water, etc. Hydrometallurgical processes with combination of Beneficiation/ Pyro-/Electro techniques are discussed to recover REMs effectively. The recommended processes require less energy to deliver high purity yield which is one step towards green environment. Salient findings on various methods are reported with recommendations which will be helpful to researchers working in the area of REMs extraction.

Keywords Rare earth metals · Secondary resources · E-waste
Hydrometallurgy · Red mud · Coal fly ash

Introduction

Rare earth elements (REEs) possess distinctive physicochemical properties which make them a significant constituent of many technologically modernized products viz. magnets, catalysts, batteries, etc. [1, 2]. Consequently, the global demand for REEs is accelerating gradually in the international market [3]. More than 50% of the REEs reserves are located in China which dominates the world by reducing their export and completely controlling ~90% of the total production, making global

A. Kumari · M. K. Jha
Metal Extraction and Recycling Division, CSIR-National Metallurgical Laboratory, Jamshedpur 831007, India

A. Kumari (✉) · D. D. Pathak
Department of Applied Chemistry, Indian Institute of Technology (ISM), Dhanbad 826004, India
e-mail: vats.archana@gmail.com

H. Kim et al. (eds.), *Rare Metal Technology 2018*, The Minerals, Metals & Materials Series, https://doi.org/10.1007/978-3-319-72350-1_5

supply chain susceptible [4]. According to the U.S. Mineral Commodity Summaries of 2017, the global mine production of REMs is 1,26,000 metric tons while the total reserves is found to be 120,000,000 metric tons [5]. Commercial processing of high grade natural ores (bastnaesite, monazite, etc.) to extract REMs involves multi-step processes with huge investment and also evolves large quantity of toxic waste leading to ecological destruction. Each ton of REEs production is reported to release ~9,000–12,000 cubic meters of toxic gases [1]. Thus, to cope up with the supply and demand imbalance of REMs and the environmental damage; economical exploitation of secondary resources viz. e-waste, red mud, waste waters, etc. can serve as promising alternatives as REEs content in these wastes are much higher than natural sources [3].

In view of environmental concern and economic benefits, R&D sector has adopted three different areas including greater efficiencies in materials use; REEs substitutes and their recycling from various resources to accumulate REMs for future use [6]. But successful recycling is the only solution adopted for valuable recovery and is also helpful towards green environment. Significant investigations and technologies to extract REMs from waste materials have been developed but still some considerable optimization is required before commercialization.

In this regard, we have published a review paper in TMS-2014 on recovery of REMs from primary and secondary resources [7]. But keeping need of more emphasis on secondaries and in order to have enhanced vision on the state-of-the-art to explore improved possibilities for REMs extraction from these waste resources, a review of the prevailing technologies is desirable. Thus, present review is based on the processes adopted/reported to recover REMs from various alternative resources i.e. secondaries/waste. For each waste, comparative investigations on reported processes have been made and concluding remarks for the best one is presented. This review will help to enhance knowledge about REEs contained in different secondary resources and their separation technique that will be feasible from financial, environmental and social point of view.

Recycling of REMs from Secondary Resources

Non-existence of a viable and operational high grade natural REEs deposits is compelling many countries to emphasis on REMs recovery from secondary sources via recycling techniques. A recent evaluation has differentiated three major fields for recycling secondaries: (i) *direct recycling* of manufacturing scrap/residues; (ii) *urban mining* of end-of-life products and (iii) *industrial waste* (solid/liquid) [8, 9]. Different challenges and opportunities are faced during their processing. A detailed study is described below.

Manufacturing Scrap/Residues

During the manufacturing processes, many unused and rejected samples (production scraps) are left which contain substantial amount of metals, and thus are treated as a valuable resource. These scraps gather attention due to relatively high metal concentration as well as definite and continuous source of waste formation in large volumes. Metal recovery from these manufacturing scraps is comparatively easier than end-of-life products.

REMs containing magnetic materials are brittle and thus, ~20–30% of magnets get easily scrapped in form of swarf and fines during their manufacturing process [10]. REMs scraps can be recovered in both its oxidised and unoxidised state. The re-melting processes with low yield bring the scrap in its unoxidised state. REMs recovery from these unoxidised scraps is reported to be economical compared to oxidised scraps [6, 11]. The composition of these scraps and residues are similar to the composition of electronic devices. In general, recycling of magnetic scraps is performed by magnet manufacturing companies using either pyro- or hydrometallurgical methods as presented in Fig. 1 [12]. However, few details regarding the actual recycling processes have been disclosed yet [10]. Thus, selective extraction seems to be the most accepted method from economic and environmental viewpoint.

The only problem causing restriction to the recycling of pre-consumer magnetic scrap is the requirement of both regional and centralized collection and processing facilities due to their high magnetic field [6]. Some other issues such as low yields, contamination, expensive re-processing, etc. are also in queue to be solved [13].

End-of-Life Products

Equipments which are not fit for their original intended use and have reached their end-of-life are generally termed as E-waste. Due to lack of suitable recycling techniques and infrastructure, processing of these scraps for REMs recovery are quite complicated [6]. But absence of efficient and equipped primary deposits has enhanced the dependence of REMs production on these scraps for future supply. Emphasis has been made on magnets, catalysts, fluorescent lamps and hydride batteries for REMs recovery and their investigations are discussed below.

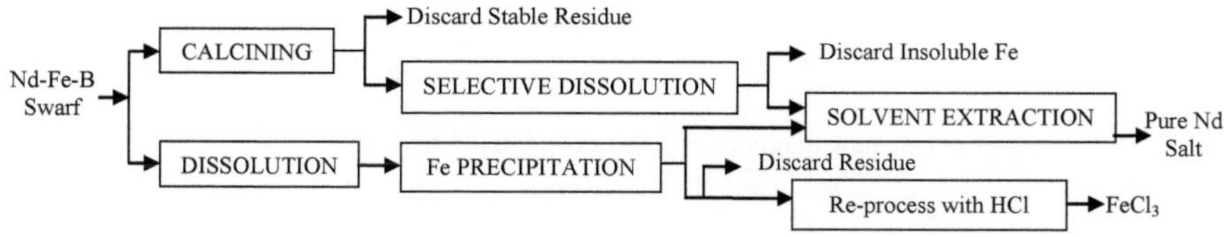

Fig. 1 Process developed to recycle REE magnets and their component materials [12]

Magnets

Neodymium-iron-boron (Nd–Fe–B) permanent magnets are the best available magnets due to their superior energy product and are widely used in wind turbines, household electrical appliances, computer hard disk drives, etc. These magnets have different life cycle depending on their applications in different field viz. from 2–3 years in consumer electronics to 20–30 years in wind turbines. Nd–Fe–B magnets contain ~31–32 wt% REEs, mainly Nd and Pr with minor amount of Dy, Tb and Gd [14]. Some scientific issues like contamination, varying magnet composition as well as magnet corrosion are associated with recycling processes [6, 15]. Processing cost and difficulty in collecting and dismantling products make REMs recovery complicated but now automatic dismantling processes are available. Hitachi has developed a dismantling machine for 100 Nd–Fe–B magnets per hour from hard disks and compressors [16]. Industrial shredding process for end-of-life products and Nd–Fe–B magnets is illustrated in Fig. 2 where ferrous scrap contains ~130–290 gm Nd/ton [14, 17]. Use of molten salts [18, 19], hydrometallurgical processes [20–24], extraction with liquid metals [25, 26], melt spinning, formation of slags, and re-sintering are few methods used for magnet recycling [27–30]. Comparative studies of the metallurgical methods for magnet recycling were also made [9, 14]. Thus, a combination of metallurgical processes will be successful for proficient recovery of REEs.

Apart from Nd–Fe–B magnets, samarium cobalt (SmCo) alloys are also used in manufacturing magnets which contain admixtures of different metals [9]. Studies for the recovery of Sm from waste SmCo magnet, their sludge, swarf, etc. via roasting, leaching, solvent extraction, precipitation, etc. is also made [31, 32]. But its lower energy product makes their share in the market comparatively smaller than Nd–Fe–B magnets.

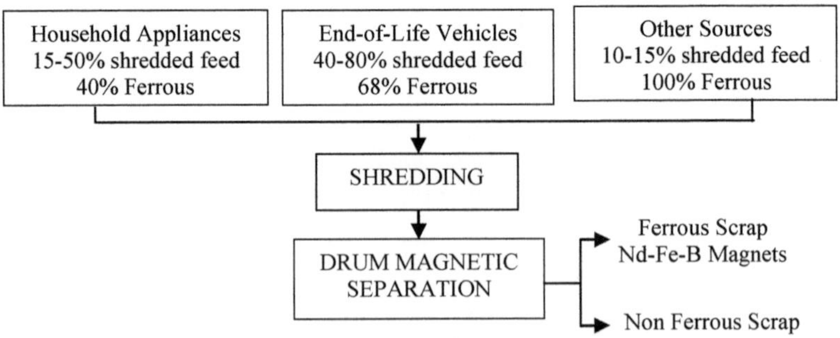

Fig. 2 Shredding of permanent magnets and the associated REEs [14]

Fluorescent Lamps

Longer life expectancy and low energy utilization has significantly escalated the use and consequently the generation of fluorescent lamps. These lamps are gas discharge tubes coated with a thin layer of phosphor powder rich in REMs which constitutes ~3% of the total weight of these lamps [2]. Recycling of REMs from these phosphor powders is carried out in three different manners: (i) *Direct re-use*: It is a simple method where no chemical processing is required but is applicable only to one type of fluorescent lamp. (ii) *Individual separation of phosphor components*: It is a relatively simple process which consumes negligible amount of chemicals but it is very difficult to obtain pure phosphor fractions by this method as separation process changes the particle size of phosphor (iii) *Chemical attack for REMs recovery*: This method is generally applicable to all types of phosphor mixtures and is able to provide pure oxides of REMs. But the involvement of multi step process consumes a lot of chemicals and generate large amount of effluents [33].

Initially, REMs recycling from lamp phosphors was limited only to large and compact fluorescent lamps with no investigation from phosphors used in LCD backlights or in white LEDs [9, 34]. But now research has touched each and every possibility to obtain REMs, some of the recent studies and the results achieved are presented in Table 1.

Table 1 Various processes reported to recover REMs from scrap fluorescent lamps

Targeted REMs	Processes used	Results achieved	References
Ce and Tb	Alkali mechanical activation	~85% Ce and 89.8% Tb was leached due to the changes caused by the destruction of spinel structure	Ling et al. [35]
Y, Eu, Tb, La, Gd and Ce	Thermal pre-treatment	Thermal pretreatment used improves the efficiency of hydrometallurgical processes and recovered ~99% Y as well as Eu and 80% Tb, 65% La, 63% Gd and 60% Ce	Ippolite et al. [36]
Y and Eu	Leaching + Solvent extraction	Selective leaching followed by extraction and separation results in 99.96% pure oxides of rare earths	Tunsu et al. [37]
Tb	Leaching	Mathematical model is deduced which verified Tb dissolution at 90 °C using 2 M HCl and 5% pulp density	Innocenzi et al. [38]
Eu and Y	Leaching	Functionalized ionic liquid was used and ~100% Eu and Y was selectively recovered from waste phosphor	Dupont and Binnemans [39]
Tb, Eu, Y	Mechanical activation + leaching	Mechanical activation followed by leaching was carried out recovering 89.4% Tb, 93.1% Eu and 94.6%Y	Tan et al. [40]

Catalysts

Fluid catalytic cracking catalysts, broadly used in the petrochemical industries is one more alternative source containing oxides of rare earths (3.5 wt%), mainly La and small amount of Ce, Pr as well as Nd [9, 41]. The global mass flow of these catalysts is estimated to be 600,000 tons annually and thus, is considered to be a probable source of REMs [42]. Although, recovery of platinum group metals from automotive catalysts is being actively recycled using a method developed by Umicore, no efficient work have been done for REMs extraction. This is due to the loss of REMs in form of slags from smelting processes [43, 44]. Limited studies for recuperating REMs via acidic leaching is reported but development of a technically feasible and economically suitable recycling method for processing fluid cracking catalysts is still required [45, 46]. Lots of research is going on but a question still persists whether catalyst recycling is viable from economic point of view.

Nickel-Metal Hydride Batteries

Entrenched methods for Ni, Cr and Fe recovery from different types of batteries are available in the market but recuperation of REMs is still under progress. Nickel metal hydride batteries are rich source of REMs containing 8–10% misch metal (La, Ce, Pr and Nd) which ends up in slag during processing [47]. Researchers investigated the dissolution of REMs from hydride batteries using different mineral acids and 4 M HCl was reported to be the best [48]. However, increase in temperature up to 95 °C decreases the requirement of HCl from 4 to 3 M [49]. Use of H_2SO_4 also recovered sufficient amount of REMs from battery scraps. Thus, a lot of studies to dissolve REMs from hydride batteries using HCl and H_2SO_4 is reported [50–52]. Further, salts of REMs are also obtained using precipitation and calcinations methods [53–55]. Umicore and Rhodia have developed a process to recycle nickel hydride batteries using ultra high temperature smelting technology [9].

However, wastes such as the slurry obtained from glass polishing powder containing Ce, optical glasses used in camera lenses, microscopes, binoculars, etc. containing La along with Y and Gd; phosphors present in cathode ray tubes, etc. is also gathering attention [7, 9]. Hence, viable steps must be taken so that these wastes can also be a part in fulfilling the future requirements of REMs.

Industrial Waste (Solid and Liquid)

Despite high REEs content, availability of end-of-life resources is relatively low for direct recycling. Thus, massive generation of industrial residues viz. red mud, coal fly ash, waste water, etc. has gathered attention. Although, these waste streams are found to contain comparatively less amount of REMs, possibilities are continuously being explored to recover them from imperative industrial waste streams [56]. Some of them are discussed below.

Red Mud

Highly alkaline residual mixture (pH > 10) is generated during bauxite treatment for alumina production using NaOH (Bayers process) is generally termed as bauxite residue or red mud. About 1.5–2.5 ton of red mud is generated for per ton of alumina produced with the global generation reported to be ~120 million tons annually. Enormous production makes their storage complicated and is also potentially harmful for the environment [57, 58]. Thus, feasible management strategy is required for proper utilization of red mud which contains several minor elements including REMs. The concentration of REEs is found to be comparatively richer in bauxites depositing on carbonate rocks [59]. Bulk application of bauxite residues are now scarcely used in concrete production and much attention is being paid towards REMs extraction but lack of detail studies still persist [56, 57].

Direct acid leaching was mainly used to recover REMs from the bauxite residue. Studies were carried out to digest red mud using dilute H_2SO_4 which selectively dissolve REEs, leaving Fe and Ti in the residue [60, 61]. Dilute HNO_3 was used at lab scale as well as at pilot scale to leach out Sc from red mud [62, 63]. HCl in different conditions was also used and ~21.2 mol HCl is reported to be consumed for treating 1 kg of red mud to leach out more than 85% of Sc [64, 65]. Later, use of different acids viz, HCl, HNO_3, H_2SO_4, CH_3COOH, etc. was studied. But until yet, HCl leaching is reported to be best with ~70–80% REEs recovery at 25 °C [58]. Among REEs, Sc recovery is mainly focused and is found to be almost doubly enriched in red mud compared to the original ore [66, 67]. Neither much work is done for leaching out major elements nor REEs, except few. It was observed that low extraction of REMs can be improved by increasing acid concentration but the same is suitable for Fe dissolution also, which is not beneficial from separation point of view. Thus, smelting was selected for Fe removal to obtain concentrated REEs. But requirement of high temperature and large amount of fluxes increases the consumption of energy during smelting which is not economically feasible [59]. Thus, in view of research made till date, pure hydrometallurgical route seems to be the best for REMs recovery from red mud.

Coal Fly Ash

In developing countries, coal is a reliable and secure source of energy thus; its global consumption is increasing day by day. Therefore, huge amount of fly ash is generated. Due to the stringent pollution regulations globally, the recycling/proper disposal of fly ash is necessary. Coal fly ash has been investigated to be a possible source of many elements but only few researchers examined it to contain extractable amount of REEs. On an average global basis, coal fly ash is estimated to contain 445 ppm of REEs [68, 69]. Laboratory scale experiments have been carried out to see the possibilities to recover REMs but pilot scale studies and commercial processes for REMs recovery using this waste has not been found. It may be due to the availability of REMs in trace level and feasibility concerns [70]. Various investigations have been carried out to evaluate exact content of REEs present in different coal fly ash but very limited work has been done for its extraction [69–71].

Waste water

Nowadays, the waste stream generated during REMs extraction and separation has the possibility to contain significant amount of these metals in it. Moreover, the outflow of heavily contaminated acidic water (commonly known as acid mine drainage) from metal or coal mines are also rich in REMs content [72]. Investigation shows that during these acidic water treatment, REEs present are either adsorbed onto the Fe–Al–Mn oxide/hydroxide colloids forming co-precipitates or they directly precipitates as $RE(OH)_3$ [73]. The sludge generated during this treatment is an imperative feedstock for REMs extraction. Figure 3 presents the treatment of acid mines drainage and its sludge separation containing major ions and REEs from discharged water [74].

Large volume of sludge is accumulated in many acid mine drainage sites. Out of which 46% contain REMs more than 300 ppm while 31% exceeds 1000 ppm. Some richest acid mine drainage sludge sources is presented in Table 2 [74]. Although, some investigations to recover different metals from acid mine drainage, coal mining waste water and other waste streams have been done but limited information is available on the possibility to recover REMs [72, 75, 76]. Researchers observed that anthropogenic gadolinium (Gd) and its chelates used for magnetic resonance imaging are toxic to water bodies and human health [77–79]. Use of advanced water treatment plants; nano-$Mg(OH)_2$, mesoporous silica support, chelating resin, biohydrogel and biosorbents has been also reported for REE removal [56, 80, 81].

Apart from these, investigations to extract REMs from phosphogypsum, mine tailing, metallurgically generating slags, electroplating waste, etc. is also going on and they might prove to be a potential secondary source for REMs.

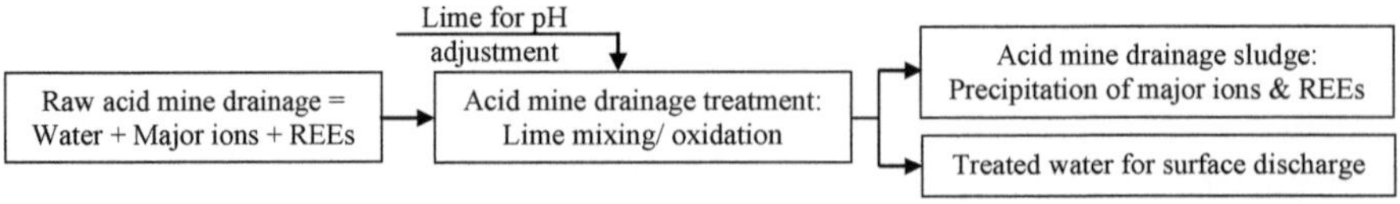

Fig. 3 Schematic diagram for acid mine drainage treatment and its sludge separation from discharged water [74]

Table 2 Rich sludge sources of acid mine drainage [74]

Acid mine drainage treatment plant	Sludge solid (mg/L)	REEs content in sludge (mg/kg)
SaxmanRun	67	3073
Racic	93	2495
Thomas	12	2169
CaledoniaPike	42	1997
Nittanny	339	1824
Morris 2	111	1663
Rogers Mill	43	1595
McVille Pile	364	1510
Pot Ridge 2	114	1462
Antrim	72	1371

Conclusion

Based on the above review, it is concluded that recovery of REMs via recycling secondary resources can diminish the critical challenges as well as the costs involved in mining and processing primary ores. The overviews suggest that REEs extraction from end-of-life products is comparatively feasible and commercially accepted. But combinations of metallurgical processes are required for proficient recovery. Mechanical pre-treatment (dismantling, sorting, etc.) is an essential step prior to extraction process. Studies were carried out globally for separation of REMs from secondary resources using pyro- and hydrometallurgical routes. Due to large amount of energy consumption, pyrometallurgical steps are recommended to be avoided if hydrometallurgy results in better output. Thus, depending upon the REMs content, a variety of acid leaching technologies has been developed. Leach liquor generated is further put to solvent extraction, precipitation, etc. methods for selective recovery of REMs. But recycling processes for these secondaries should be developed with a view of obtaining zero-waste solution. Therefore, modern and lucrative technologies in field of secondary processing are required to be developed.

Acknowledgements Authors are thankful to the Director, CSIR-National Metallurgical Laboratory, Jamshedpur for the permission to publish this paper. One of the authors, Ms. Archana Kumari would like to extend her sincere gratitude to CSIR, New Delhi (Grant: 31/10(60)/2015-EMR-I) for providing Senior Research Fellowship to carry out this research work.

References

1. Kumari A, Panda R, Jha MK, Kumar JR, Lee JY (2015) Process development to recover rare earth metals from monazite mineral: a review. Miner Eng 79:102–115
2. Jha MK, Kumari A, Panda R, Kumar JR, Yoo KK, Lee JY (2016) Review on hydrometallurgical review of rare earth metals. Hydrometallurgy 165:2–26

3. Kumar JR, Lee JY (2017) Recovery of critical rare earth elements for green energy technologies. The Minerals, Metals & Materials Society, Rare Metal Technol 19–29
4. Papangelakis VG, Moldoveanu G (2014) Recovery of rare earth elements from clay minerals. In: 1st European rare earth resources conference, Milos, pp 191–202, 04–07/09/2014
5. U.S. Geological Survey (2017) Mineral commodity summaries 2017: U.S. Geological Survey, 202 p. doi:http://doi.org/10.3133/70180197
6. Panayotova M, Panayotov V (2012) Review of methods for the rare earth metals recycling. In: Annual of the University of Mining and Geology "St. Ivan Rilski", vol 55, pp 142–147
7. Kumar V, Jha MK, Kumari A, Panda R, Kumar JR, Lee JY (2014) Recovery of rare earth metals (REMs) from primary and secondary resources: a review. Rare Metal Technology, TMS (The Minerals, Metals & Materials Society), pp 81–88
8. Zhuang WQ, Fitts JP, Ajo-Franklin CM, Maes S, Alvarez-Cohen L, Hennebel T (2015) Recovery of critical metals using biometallurgy. Curr Opin Biotechnol 33:327–335
9. Binnemans K, Jones PT, Blanpain B, Van Gerven T, Yang Y, Walton A, Buchert M (2013) Recycling of rare earths: a critical review. J Clean Prod 51:1–22
10. Akai T (2008) Recycling rare earth elements. AIST Today 29:8–9
11. Kara H, Chapman A, Crichton T, Willis P, Morley N (2010) Lanthanide resources and alternatives, Department for Transport and Department for Business, Oakdene Hollins Research & Consulting
12. Peterson ES, Jones E (2014) Improving rare earth reuse and recycling. In: Proceedings of 248th American Chemical Society Meeting, San Francisco
13. Schüler D, Buchert M, Liu R, Dittrich S, Merz C (2011) Study on rare earths and their recycling. The Greens/EFA Group in the European Parliament, Darmstadt, Öko-Institut
14. Yang Y, Walton A, Sheridan R, Guth K, Guth R, Gutfleisch O, Buchert M, Steenari B, Gerven TV, Jones PT, Binnemans K (2017) REE recovery from end-of-life Nd-Fe-B permanent magnet scrap: a critical review. J Sustain Metall 3:122–149
15. Goodier E (2005) The recycling and future selection of permanent magnets and power cores. Presentation at Sweift Levick Magnets
16. Clenfield J, Shiraki M, (2010) Hitachi develops machine to recycle rare earths to cut reliance on China
17. Bandara HMD, Darcy JW, Apelian D, Emmert MH (2014) Value analysis of neodymium content in shredder feed: towards enabling the feasibility of rare earth magnet recycling. Environ Sci Technol 48:6553–6560
18. Yamamura T, Mehmood M, Maekawa H, Sato Y (2004) Electrochemical processing of rare earth and rare metals by using molten salts. Chem Sustain Develop 12:105–111
19. Shirayama S, Okabe T (2009) Selective extraction of Nd and Dy from rare earth magnet scrap into molten salt. In: Processing materials for properties, The Minerals, Metals & Materials Society
20. Ellis TW, Schmidt FA, Jones LL (1994) Methods and opportunities in recycling of rare earth based materials. Metals and materials waste reduction, recovery and remediation (TMS), Warrendale, pp 199–208
21. Itakura T, Sasai R, Itoh H (2006) Resource recovery from Nd-Fe-B sintered magnet by hydrothermal treatment. J Alloys Compd 408:1382–1385
22. Xu Y, Chumbley LS, Laabs FC (2000) Liquid metal extraction of Nd from Nd-Fe-B magnet scraps. J Mater Res 15:2296–2304
23. Lyman JW, Palmer GR (1992) Scrap treatment method for rare earth transition metal alloys. US Patent 5 129:945
24. Zhang X, Yu D, Guo L (2010) Test study new process on recovering rare earth by electrical reduction-P507 extraction separation method. J Copper Eng 1:1009–3842
25. Okabe TH, Takeda O, Fukuda K, Umetsu Y (2003) Direct extraction and recovery of neodymium from magnet scrap. Mater Transac 44:798–801
26. Takeda O, Okabe TH, Umetsu Y (2004) Phase equilibrium of the system Ag-Fe-Nd and Nd extraction from magnet scraps using molten silver. J Alloys Compd 379:305–313

27. Itoh M, Masuda M, Suzuki S, Machida K (2004) Recycling of rare earth sintered magnets as isotropic bonded magnets by melt-spinning. J Alloys Compd 374:393–396
28. Kawasaki T, Itoh M, Machida K (2003) Reproduction of Nd-Fe-B sintered magnet scraps using a binary alloy blending technique, Japan. Mater Trans 44(9):1682–1685
29. Zakotnik M, Harris IR, Wiliams AJ (2009) Multiple recycling of Nd-Fe-B type sintered magnets. J Alloys Compd 469:314–321
30. Walton A, Han Y, Speight JD, Harris IR, Williams AJ (2012) The use of hydrogen to extract and re-process Nd-Fe-B magnets from electronic waste. In: Proceeding of the 22nd international workshop on rare earth permanent magnets and their applications, Nagasaki, pp 11–13
31. Sinha MK, Pramanik S, Kumari A, Sahu SK, Prasad LB, Jha MK, Yoo KK, Pandey BD (2017) Recovery of value added products of Sm and Co from waste SmCo magnet by hydrometallurgical route. Sep Pur Tech 179:1–12
32. Stanton C (2016) Sulfation roasting and leaching of samarium-cobalt magnet swarf for samarium recovery. Masters Thesis, Colorado School of Mines
33. Binnemans K, Jones PT (2014) Perspectives for the recovery of rare earths from end-of-life fluorescent lamps. J Rare Earths 32(3):195–200
34. Buchert M, Manhart A, Bleher D, Pingel D (2012) Recycling critical raw materials from waste electronic equipment. Oeko-Institut e.V, Darmstadt, Germany
35. Ling H, Wen J, Yanwei Y, Weimin S (2017) Study on alkali mechanical activation for recovering rare earth from waste fluorescent lamps. J Rare Earths, In Press
36. Ippolite NM, Innocenzi V, De Michelis I, Medici F, Veglio F (2017) Rare earth elements recovery from fluorescent lamps: a new thermal pretreatment to improve the efficiency of the hydrometallurgical process. J Clean Prod 153:287–298
37. Tunsu C, Petranikova M, Ekberg C, Retegan T (2016) A hydrometallurgical process for the recovery of rare earth elements from fluorescent lamp waste fractions. Sep Purif Technol 161 (17):172–186
38. Innocenzi V, Ippolito NM, De Michelis I, Medici F, Veglio F (2016) A hydrometallurgical process for the recovery of terbium from fluorescent lamps: experimental design, optimization of acid leaching process and process analysis. J Environ Manage 184(3):552–559
39. Dupont D, Binnemans K (2015) Rare-earth recycling using a functionalized ionic liquid for the selective dissolution and revalorization of Y_2O_3:Eu^{3+} from lamp phosphor waste. Green Chem 17:856–868
40. Tan Q, Deng C, Li J (2016) Innovative applications of mechanical activation for rare earth elements recovering: process optimization and mechanism exploration. Sci Rep 6:19961
41. Yu ZS, Chen MB (1995) Rare earth elements and their applications. Metallurgical Industry Press, Beijing (P.R. China)
42. Hykawy J (2010) Report on the 6th International rare earths conference, Hong Kong
43. Hagelüken C, Buchert M, Stahl H (2005) Materials flow of platinum group metals-systems analysis and measures for sustainable optimization of the materials flow of platinum group metals. Umicore Precious Metals Refining and Öko-Institut e. V. ISBN 0-9543293-7-6, GFMS Ltd, London, 2005
44. Felix N, Vanriet C (1994) Recycling of electronic scrap at UMS Hoboken Smelter. In: Proceedings of the 18th International precious metals conference, Vancouver, Canada, pp 159–169
45. He H, Meng J (2011) Recycling rare earth from spent FCC catalyst using P507 (HEH/EHP) as extractant. Zhongnan Daxue Xuebao, Ziran Kexueban 42:2651–2657
46. Vierheilig AA (2012) Methods of recovering rare earth elements. US Patent No. 8,263,028
47. Muller T, Friedrich B (2006) Development of a recycling process for nickel-metal hydride batteries. J Power Sources 158:1498–1509
48. Lyman JW, Palmer GR (1995) Hydrometallurgical treatment of nickel-metal hydride battery electrodes. In: Third international symposium on recycling of metals and engineered materials, Alabama (USA), pp 131–144

49. Zhang PW, Yokoyama T, Itabashi O, Wakui Y, Suzuki TM, Inoue K (1998) Hydrometallurgical process for recovery of metal values from spent nickel metal hydride secondary batteries. Hydrometallurgy 77:116–122
50. Poscher A, Luidold S, Antrekowitsch H (2011) Aufbereitung von nickel-metalllhydridakkus zur Wiedergewinnung Seltener Erden, pp 419–446
51. Kaindl M, Luidold S, Poscher A (2012) Recycling of rare earths from nickel-metal hydride batteries with special reference to the acid recovery. Berg-Huetten-maenn. Monatsh 157:20–26
52. Luidold S, Antrekowitsch H (2012) Recovery of rare earth metals from waste material by leaching in non-oxidizing acid and by precipitating using sulphates. EP 2444507
53. Kanamori T, Matsuda M, Miyake M (2009) Recovery of rare metal compounds from nickel-metal hydride battery waste and their application to CH_4 dry reforming catalyst. J Hazard Mater 169:240–245
54. Provazi K, Campos BA, Espinosa DCR, Tenorio JAS (2011) Metal separation from mixed types of batteries using selective precipitation and liquid-liquid extraction techniques. Waste Manage 31:59–64
55. Bertuol DA, Bernardes AM, Tenorio JAS (2009) Spent Ni-MH batteries—the role of selective precipitation in the recovery of valuable metals. J Power Sources 193(2):914–923
56. Binnemans K, Pontikes Y, Jones PT, Gerven TV, Blanpain B (2013) Recovery of rare earths from industrial waste residues: a concise review. In: Proceedings of the 3rd International slag valorisation symposium, Leuven, Belgium, pp 191–205
57. Klauber C, Gräfe M, Power G (2011) Bauxite residue issues: II. Options for residue utilization. Hydrometallurgy 108(1):11–32
58. Borra CR, Pontikes Y, Binnemans K, Gervena TV (2015) Leaching of rare earths from bauxite residue (red mud). Miner Eng 76:20–27
59. Borra CR, Blanpain B, Pontikes Y, Binnemans K, Gerven TV (2016) Smelting of bauxite residue (red mud) in view of iron and selective rare earths recovery. J Sustain Metall 2:28–37
60. Fulford GD, Lever G, Sato T (1991) Recovery of rare earth elements from Bayer process red mud. US Patent 5,030,424
61. Xue A, Chen X, Tang X (2010) The technological study and leaching kinetics of scandium from red mud. Nonferrous Metals Extr Metall 2:51–53
62. Ochsenkühn-Petropulu M, Lyberopulu T, Ochsenkühn K, Parissakis G (1996) Recovery of lanthanides and yttrium from red mud by selective leaching. Anal Chim Acta 319(1):249–254
63. Ochsenkühn-Petropoulou M, Hatzilyberis K, Mendrinos L, Salmas C (2002) Pilot-plant investigation of the leaching process for the recovery of scandium from red mud. Ind Eng Chem Res 41(23):5794–5801
64. Zhang J, Deng Z, Xu T (2005) Experimental investigation on leaching metals from red mud. Light Metals 2:13–15
65. Wang KQ, Yu YB, Wang H, Chen J (2010) Experimental investigation on leaching scandium from red mud by hydrochloric acid. Chin Rare Earths 1:27
66. Bray EL (2010) Bauxite and Aluminium. U.S. Geological Survey Minerals Yearbook—2008, U.S. Department of the Interior
67. Abhilash, Sinha S, Meshram P, Pandey BD, Behera PK, Satpathy BK (2014) Red Mud: a secondary resource for rare earth elements. In: International bauxite, alumina and aluminium symposium, The IBAAS Binder Vol III, Vishakhapatnam, pp 148–162
68. Kertis M, Yudovich Y (2009) Estimations of clarkes for carbonaceous biolithes: world averages for trace element contents in black shales and coal. Int J Coal Geol 78(2):135–148
69. Lin R, Howard BH, Roth EA, Bank TL, Granite EJ, Soong Y (2017) Enrichment of rare earth elements from coal and coal by-products by physical separation. Fuel 200:506–520
70. Dai S, Graham IT, Ward CR (2016) A review of anomalous rare earth elements and yttrium in coal. Int J Coal Geol 159:82–95
71. Phuoc TX, Wang P, McIntyre D (2016) Detection of rare earth elements in powder river basin sub-bituminous coal ash using laser-induced breakdown spectroscopy (LIBS). Fuel 163:129–132

72. Merten D, Bechul G (2004) Determination of rare earth elements in acid mine drainage by inductively coupled plasma mass spectrometry. Microchim Acta 148:163–170
73. Zhao F, Cong Z, Sun H, Ren D (2007) The geochemistry of rare earth elements (REE) in acid mine drainage from the Sitai coal mine, Shanxi Province, North China. Coal Geol 70:184–192
74. Ziemkiewicz P, He T, Noble A, Liu X (2016) Recovery of rare earth elements (REEs) from coal mine drainage. U.S. Department of Energy, National Energy Technology Laboratory
75. Roig MG, Manzano T, Diaz M (1997) Biochemical process for the removal of uranium from acid mine drainage. Water Res 31(8):2073–2083
76. Fu FL, Wang Q (2011) Removal of heavy metal ions from wastewaters: a review. J Environ Manage 92(3):407–418
77. Bau M, Dulski P (1996) Anthropogenic origin of positive gadolinium anomalies in river waters. Earth Planet Sci Lett 143(1–4):245–255
78. Knappe A, Pekdeger A (2005) Positive gadolinium anomaly in surface and ground water of the urban area Berlin, Germany. Chemie der Erde-Geochemistry 65(2):167–189
79. Rogosnitzky M, Branch S (2016) Gadolinium-based contrast agent toxicity: a review of known and proposed mechanisms. Biometals 29:365–376
80. Lawrence MG, Keller J, Poussade Y (2010) Removal of magnetic resonance imaging contrast agents through advanced water treatment plants. Water Sci Technol 61(3):685–692
81. Li C, Zhuang Z, Huang F, Lin Z (2013) Recycling rare earth elements from industrial wastewater with flowerlike nano-$Mg(OH)_2$. App Mater Interfaces 5(19):9719–9725

Selective Reduction and Separation of Europium from Mixed Rare-Earth Oxides Recovered from Waste Fluorescent Lamp Phosphors

Mark L. Strauss, Brajendra Mishra and Gerard P. Martins

Abstract Nearly the entire worldwide production of europium is recovered, as a minor constituent, from mining bastnäsite, monazite ore, or ion absorbing clays. However, this research indicates a process to recycle europium from waste lamp phosphors as a strategy meet the demand for europium. In this process, waste fluorescent lamp powder is retorted, sieved, leached, and precipitated to produce a mixed europium/yttrium rare earth oxide (REO). Europium is separated from yttrium by selectively reducing Eu(III) to Eu(II) using zinc powder and precipitating europium (II) sulfate from solution using sulfuric acid. Screening experiments were conducted observe the effect of pulp density, precipitation time, entrance pH, and stoichiometric ratio sulfate upon the grade and recovery of europium (II) sulfate. The best conditions to maximize grade and recovery of europium (II) sulfate were a 1 h precipitation time, 10× the stoichiometric ratio of sulfate, 100 g/L mixed REO, and the entrance pH equal to 3. The maximum grade of europium sulfate was 95.93%, and the maximum recovery was 73.32%.

Keywords Waste fluorescent lamp · Phosphor dust · Rare earths Recycling · Europium

Introduction

According to USGS [1] 16,000 tons of rare earth products were consumed in the United States in 2016. Of this volume, only a negligible amount of rare earth oxides, including europium, were produced by recycling batteries, magnets and fluorescent lamps. Waste phosphor dust could supply some of the US demand for

M. L. Strauss (✉) · B. Mishra
Worcester Polytechnic Institute, 100 Institute Road, Worcester, MA 01609, USA
e-mail: mlstrauss@wpi.edu

G. P. Martins
Colorado School of Mines, (Kroll Institute of Extractive Metallurgy), 1500 Illinois Street, Golden, CO 80401, USA

H. Kim et al. (eds.), *Rare Metal Technology 2018*, The Minerals, Metals & Materials Series, https://doi.org/10.1007/978-3-319-72350-1_6

europium. According to the DOE 2011 Critical Materials Strategy Report [2], the demand will increase from 200 tons per year to 220 tons per year in 2020. In 2007, 8000 tons of waste phosphor powder was discarded into landfills after removing mercury. Assuming the concentration of 0.63% europium oxide in the waste powder, there is a resource of 50 tons per year in the United States. However, the concentration of europium is the dust should have increased since 2007, due to fewer halophosphate based T12 lamps being disposed and a greater ratio of newer T8 lamps being recycled.

Europium was first isolated from samarium using a series of crystallization by French chemist, Eugène-Anatole Demarçay [3]. Had he known unique chemistry of europium (II) sulfate ($EuSO_4$), he could have isolated europium by precipitation. One of the unique chemical properties of europium were discovered in 1906 when Georges Urbain discovered that yttrium oxide doped with europium created at red color. From this discovery, red phosphor was born. Red phosphors are used in LCDs, LED, flat screens, fluorescent lamps, and cell phones. Eduafo et al. [4] demonstrated how europium and yttrium follow each other into solution from the waste. The purpose of this research is to develop a method to separate europium from yttrium such that the final product is saleable europium oxide, minimum purity 99.9%. In the meantime, the final product of this work is europium (II) sulfate which can be converted to europium oxide by several steps.

Molycorp [5] developed a process to recovery 99.9% pure europium oxide form Eu–Sm-Gd concentrates separated from monazite. More recently, Preston et al. [6], Morais et al. [7], and Rabie et al. [8] demonstrated that zinc metal and sulfuric acid can separate and purity europium from samarium and gadolinium concentrates. The paper proposes using a similar method for europium and yttrium concentrates, as a product of waste lamp phosphor leaching. The equation below demonstrates the equation for converting Eu(III) to Eu(II) via selective reduction.

$$2Eu^{3+} + Zn = 2Eu^{2+} + Zn^{2+} \tag{1}$$

The equilibrium constant, K_c, can be used to describe the thermodynamics of a system. In the equation below the activities of species are replaced with their actual concentrations because there is no simple method to measure the activities of concentrated species in high ionic strength solutions.

$$aA + bB \Leftrightarrow cC + dD$$

$$K_c = \frac{[C]^c [D]^d}{[A]^a [B]^b}$$

In addition, the formation of europium (II) sulfate precipitate, and intermediate in the europium separation experiments is demonstrated in Eq. 2 below.

HSC 5.11 was used to explore the possible reactions for the oxalic precipitation work and selective reduction and precipitation of europium (II) sulfate. The using

Gibbs free energy minimization (“Equilibrium Compositions”) function, HSC identifies the most plausible chloride complexes after the solvation of europium in hydrochloric acid as $EuCl_3(aq), EuCl_2^+, EuCl^{2+}$ as generalized by Eq. (2) below.

$$Eu_2O_3 + 6HCl(aq) = 2EuCl_3(aq) + 3H_2O \quad K_c = 7.36 \times 10^{55}(T = 25\,°C) \quad (2)$$

Similarly, HSC indicates the most probable reduced europium species is $EuCl_4{}^{2-}(aq)$.

$$2EuCl_3(aq) + 4HCl(aq) + Zn = EuCl_4{}^{2-}(aq) + ZnCl_2(aq) + 4H^+ \quad K_c = 1.62 \times 10^{16}(T = 25\,°C) \quad (3)$$

Finally, the typical precipitation reaction is demonstrated below in Eq. (4).

$$EuCl_4{}^{2-}(aq) + H_2SO_4 = EuSO_4 + 2HCl(aq) + 2Cl^- \quad K_c = 1.39 \times 10^{13}(T = 25\,°C) \quad (4)$$

This process uses zinc metal to reduce Eu^{III} to Eu^{II} via Zn^0. Next, the reduced solution is precipitated by the addition of sulfuric acid. As a result, europium (II) sulfate is precipitated and is separated via vacuum filtration. The product was quantified by X-Ray Fluorescence, SEM-EDS, and ICP-OES.

Experimental

I. Materials

Veolia ES Solutions provided the waste phosphor powder which was treated to create concentrate of yttrium and europium oxide ($\sim$91% pure) based on previous research. Zinc metal (99.8% 20–30 mesh) was provided by Alfa Aesar. Sodium hydroxide (Sigma Aldrich, USA) is dissolved in deionized water. The mixed yttrium and europium product was dissolved in hydrochloric acid and deionized water. The pH was adjusted with NaOH (Sigma Aldrich, USA) dissolved in deionized water. ACS grade 18 M sulfuric acid (Sigma Aldrich, USA) was diluted with deionized water.

II. Analysis

X-Ray Fluorescence (XRF) (Thermo Fisher Scientific), scanning electron microscopy electron dispersive spectroscopy (SEM-EDS) were used for quantification and identification, and inductively coupled plasma optical emission spectrometry (ICP-OES) (Fischer Scientific). The data was analyzed using Stat-Ease 9.0.5 Design Expert to create contour plots and conduct statistical analysis.

III. Procedure

A yttrium oxide/europium oxide concentrate was dissolved in concentrated 12 M hydrochloric acid, heated, and diluted with deionized water with a pulp density of either 50 g/L or 100 g/L. The alkalinity of the solution was adjusted with dilute sodium hydroxide to pH 2.5 or 3. Next, the solution was added to magnetically-stirred vessel. Hydrogen gas was bubbled through the system until ORP stabilized. Next, 5 grams of zinc was added to the mixing vessel. After 1 h, the reduced europium solution and dilute solution of 3 M sulfuric acid (either 10 times or 20 times the stoichiometric ratio of sulfuric acid), degassed with hydrogen, were mixed together in a cylindrical vessel, bubbled with hydrogen gas. After 1 or 3 h, the solution was filtered by vacuum filtration with Millipore 47 mm filter in a Pall vacuum filtration setup. The precipitate was washed with a 0.01 M sulfuric acid. The precipitate was analyzed by ICP-OES after lithium borate fusion was conducted.

Results and Discussion

A. Analysis Results

The starting mixed REO concentrate as well as the precipitate were analyzed by ICP-OES. The 10 experiments indicate the various grade and recoveries of europium in the form of europium (II) sulfate (Fig. 1).

The grades of the final product were between 89.54–95.36% europium (II) sulfate. The calculated recovery was lower than desired in the precipitate. The recovery of europium in the filtrate was, in all cases, less than 10%. The lower than expected europium (II) sulfate may have been due to competitive reactions or Eu(II) re-oxidizing due to incomplete isolation from oxygen during transferring or precipitation. However, this logic does not explain the incomplete mass balance for

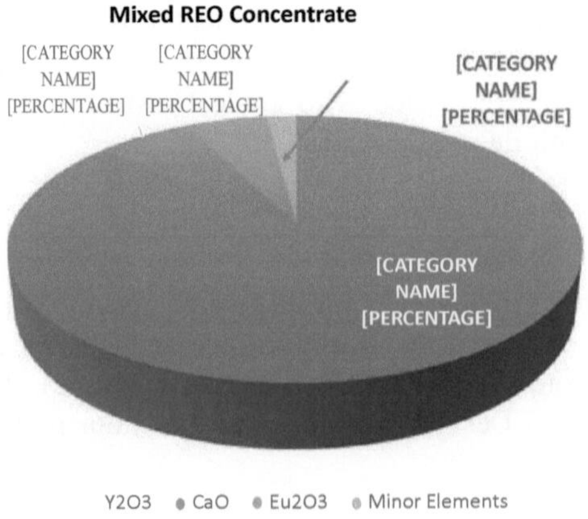

Fig. 1 This chart shows the composition of the starting mixed rare earth oxide concentrate

europium. The purity was determined from ICP-OES by assuming the metals (except for europium (II) sulfate) appear as oxides (Fig. 2).

As demonstrated by Fig. 3, the highest precipitation time (3 h) and lowest amount of sulfate (10×) have the largest effect up the recovery of europium (II) sulfate. However, the effect is not statistically significant—as indicated by the faint color change from blue-green to yellow-green. The F value is equal to 4.45 which corresponds to a p value of 0.1250 which means there is 12.5% chance that the effects are due to noise. According to Le Chatelier's Principle, a larger concentration of reactants (sulfuric acid), should increase the quantity of the products (europium (II) sulfate) in Eq. (4). Preston et al. [6], found that is a correlation between precipitation time and recovery–which is what is aligned the results. Morais et al. [9] suggests that the effect of sulfate concentration faintly negative upon the grade and recovery. The results demonstrate a similar effect (Fig. 4).

The recovery is maximized with the lowest entrance pH (pH = 2.5) and pulp density (100 g/L mixed REO). However, the effect is not considered statistically significant. The F value is equal to 4.45 which corresponds to a p value of 0.1250 which means there is 12.5% chance that the effects are due to noise. The lower the pH, the greater amount of hydronium ions in the products of Eq. (3). However, as the pH increases, the reaction should shift to the left. This intermediate reaction will lead to more reduced europium and ultimately, more recovery europium (II) sulfate. However, this plot shows the opposite behavior. Rabie et al. [10] showed that the higher the pH the higher the recovery of europium (II) sulfate. Regarding pulp density, Le Chatelier's Principle applies. A greater concentration of europium in Eq. (3) should lead to more recovered europium (II) sulfate. Preston et al. [6] found there was a positive relationship between pulp density recovery—which are within the range of europium concentrations that were tested (Fig. 5).

As shown in Fig. 6, the lowest precipitation time has a strong effect upon the grade of europium (II) sulfate. Kinetically, the precipitation of by-products may be favored by longer precipitation times—thereby decreasing the grade. Also, more

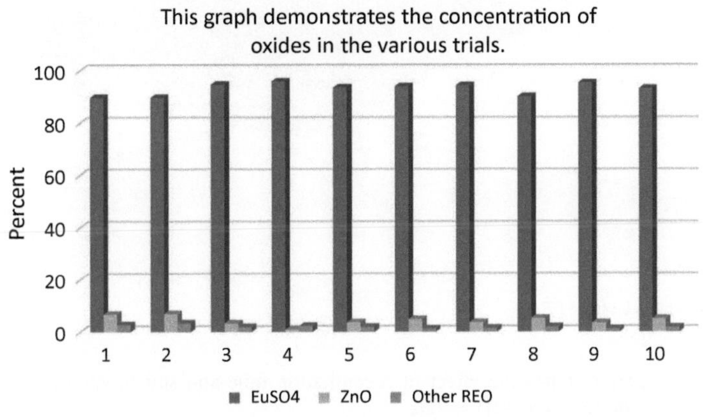

Fig. 2 This chart shows the composition of various solids in the precipitate

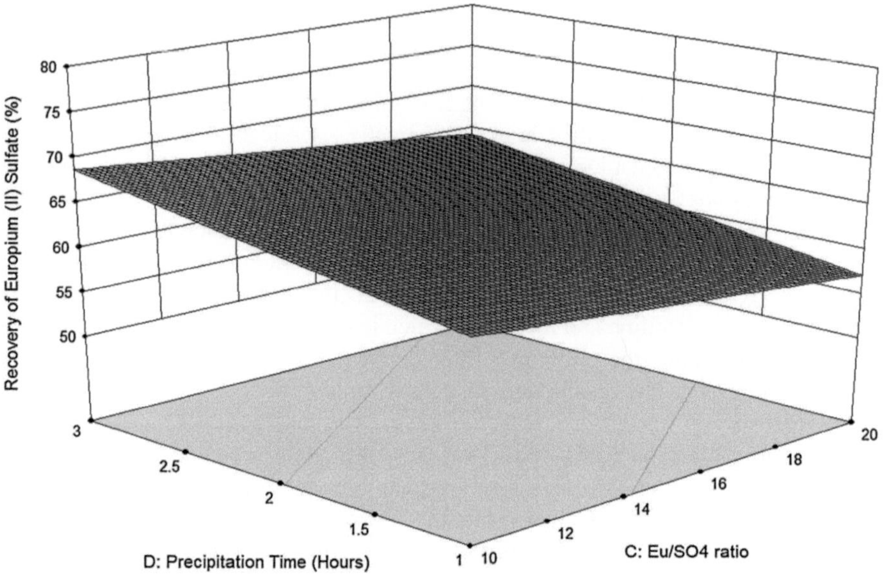

Fig. 3 The 3D plots show the effect of pulp density and entrance pH upon the grade of europium (II) sulfate

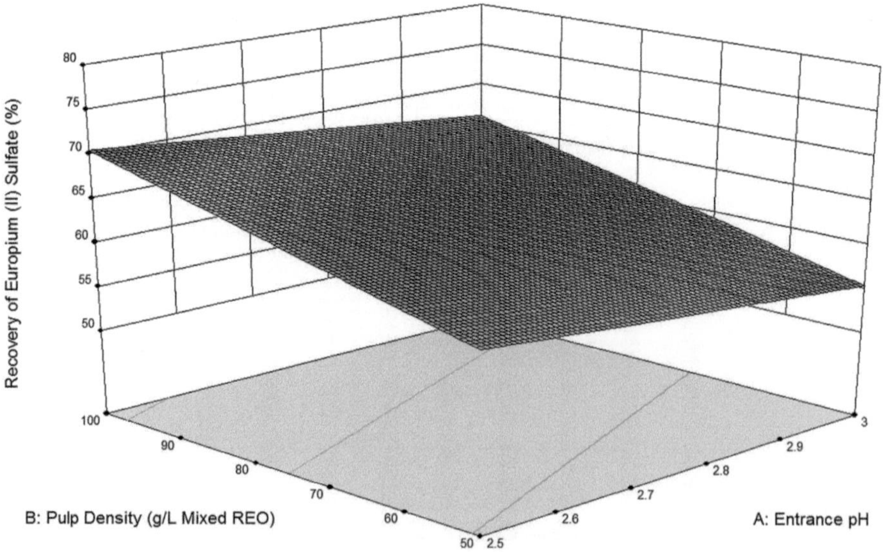

Fig. 4 The plot demonstrates the effect of precipitation time and stoichiometric ratio of sulfate upon the recovery of europium (II) sulfate

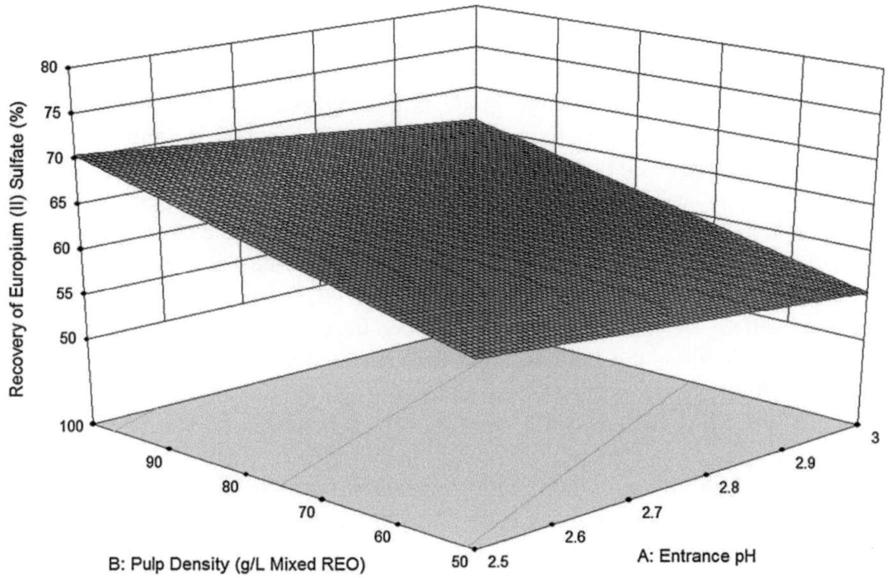

Fig. 5 The plot shows the effect of pulp density and entrance pH upon the recovery of europium (II) sulfate

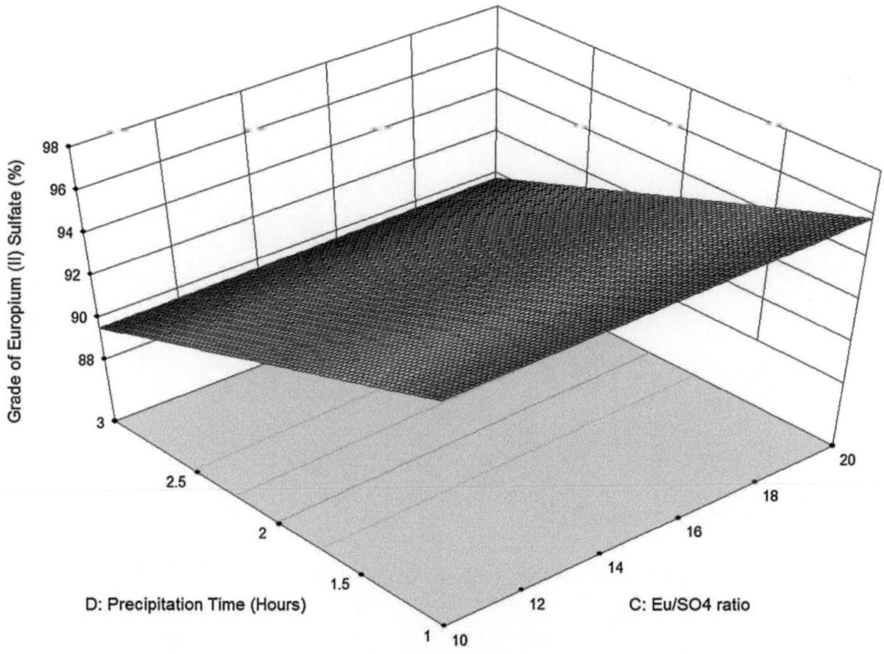

Fig. 6 This plot indicates the effect of precipitation time and sulfate ratio upon the grade of europium sulfate

sulfuric acid slightly improves the grade. Since europium (II) sulfate is the only insoluble precipitate, it is possible that more sulfuric acid inhibits the co-precipitation of insoluble hydroxides—which would increase the grade (Fig. 7).

When the entrance pH = 3, the grade is the highest. This effect is strong as indicated by the blue to red contour transition. In addition, it is statistically significant as the p value is 0.0098—indicating that there is a less than 1% chance that effect was due to noise. The reason for this effect is not self-evident. Based on speciation, a higher pH would favor the precipitation of insoluble hydroxides which would decrease the grade. This mechanism is the opposite from what the results indicate. In addition, the highest pulp density, 100 g/L, has impact on the grade of europium sulfate. A higher pulp density may improve decrease the amount of impurities and increase the grade because the undesired precipitates crowded out by formation of a higher concentration of europium (II) sulfate precipitates.

In addition, SEM-EDS was run on one of the precipitates as shown in Table 1 (Fig. 8).

SEM-EDS can identify the major impurities such as chlorine, but cannot quantify yttrium. The nickel, manganese and aluminum values listed in the table are either due to misidentification or a mistaken introduction of impurities in the in the SEM.

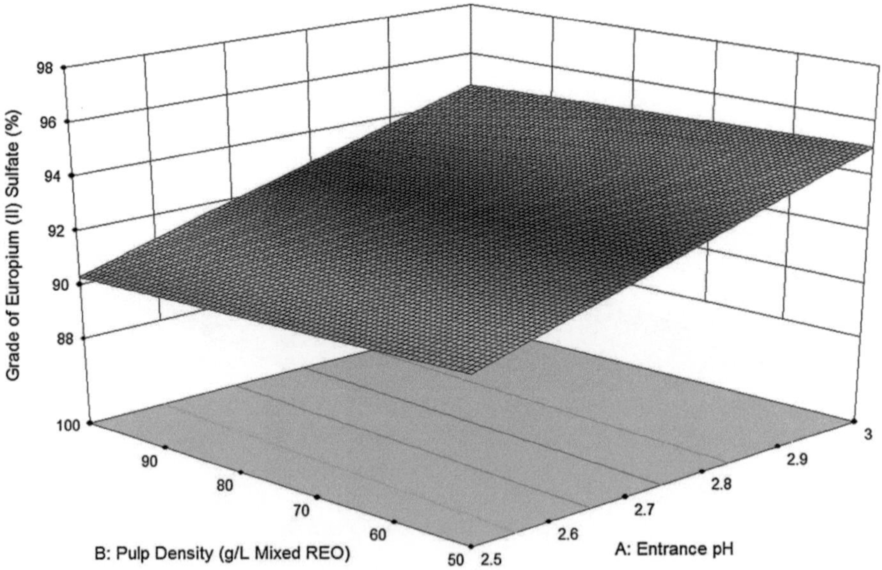

Fig. 7 The plot shows the effect of pulp density and entrance pH upon the grade of europium (II) sulfate

Table 1 This table lists the 4 factors, 2 levels, as well as the response of grade and recovery of europium (II) sulfate

Test #	Entrance (pH)	Pulp Density (g/ L REO)	SO_4/ Eu ratio (SR)	Precipitation Time (h)	Grade of Europium (II) Sulfate (%)	Recovery of Europium (II) Sulfate (%)
1	2.5	100	20	3	89.54	70.79
2	2.5	100	20	3	89.6	68.94
3	3	100	10	3	94.58	73.32
4	2.5	100	20	1	95.93	63.82
5	3	100	10	1	93.54	69.92
6	3	50	20	1	93.9	51.82
7	3	50	20	1	94.32	57.32
8	2.5	50	20	3	90.06	59.45
9	2.5	50	10	1	95.36	54.58
10	3	50	10	3	93.15	55.19

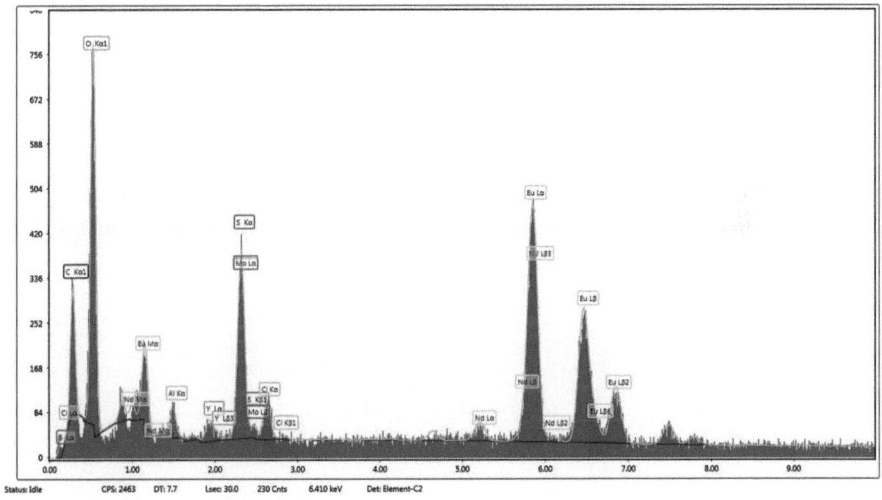

Fig. 8 This SEM-EDS spectrum was analyzed europium (II) sulfate

Conclusions

The selective reduction and precipitation is an effective technique for separating and purifying yttrium and europium concentrates recovered from fluorescent lamps. Europium (II) sulfate may have a purity greater than 95% and recovery greater than 70%. Since the effect of grade was statistically significant but not recovery in this model, grade is the driver of optimization. Based on screening experiments with 4 factors and two levels, the optimum conditions are entrance pH 3, precipitation time is 1 h, pulp density is 100 g/L, and a sulfate to europium ratio is 10 times stoichiometric.

Future work includes discovering conditions to improve the recovery and improve the grade based on the results of this research.

References

1. USGS (2017) Rare earths. U.S. Geological Survey, Mineral Commodity Summaries, January 2016, 2017 [Online]. Available: http://minerals.usgs.gov/minerals/pubs/commodity/rare_earths/mcs-2016-raree.pdf
2. U.S. Department of Energy Critical Materials Strategy, 2011
3. J. Emsley (2001) Nature's building blocks: an A-Z guide to the elements
4. Eduafo PM, Strauss M (2015) Experimental investigation of recycling rare earth metals from waste fluorescent lamp phosphors. Rare Met Technol 253–259
5. Gupta CK (2005) Extractive metallurgy of rare Earths
6. Preston JS, Preez AC (1996) The separation of europium from a middle rare Earth concentrate by combined chemical reduction, precipitation and solvent-extraction methods, pp 93–101
7. Morais CA, Ciminelli VST (2001) Recovery of europium by chemical reduction of a commercial solution of europium and gadolinium chlorides. Hydrometallurgy 60(3):247–253
8. Rabie KA, Sayed SA, Lasheen TA, Salama IE (2007) Europium separation from a middle rare earths concentrate derived from Egyptian black sand monazite. Hydrometallurgy 86(3–4):121–130
9. de Morais CA, Ciminelli VST (2002) Europium recovery by photochemical reduction from Eu And Eu–Gd chloride solutions. Sep Sci Technol 37(14):3305–3321
10. Rabie KA, Sayed SA, Lasheen TA, Salama IEE (2007) Europium separation from a middle rare earths concentrate derived from Egyptian black sand monazite. Hydrometallurgy 86(3–4):121–130

Study of the Mechanochemical Calcification for Mixed Rare Earth Concentrate

Jiang Liu, Ting-an Zhang, Zhihe Dou and Yukun Huang

Abstract The paper aimed to propose that mixed rare earth concentrate was decomposed by pressurized calcification where mechanical milling was applied simultaneously (also known as mechanochemical calcification). The bastnaesite and monazite were both decomposed by OH^-, where released rare earth formed RE $(OH)_3$, released fluorine and phosphorus formed CaF_2 and $Ca_2P_2O_7$. Then, the calcified slag was leached by diluted hydrochloric acid to extract rare earth and both recover fluorine and phosphorus. The phase transformation regularity, the behavioral trends of rare earth, fluorine and phosphorus were investigated. Results showed under the optimal conditions of ball to material weight 80:1, the calcification temperature 250 °C, the liquid-solid ratio 5:1, the leaching yield of rare earth reached to 92.16%, Ce was 86.47%, La was 97.58%, Nd was 99.99% and the mass fraction of F, P entering into calcified liquor were 8.27, 0.21% respectively. The micromorphology of the mineral has a transition from dense to fluffy, and then became dense again.

Keywords Mixed concentrate · Calcification transformation · Acid leaching
Mechanical milling

J. Liu (✉) · T. Zhang · Z. Dou · Y. Huang
Key Laboratory of Ecological Utilization of Multi-metal Intergrown Ores of Ministry of Education, School of Metallurgy, Northeastern University, Shenyang 110819, Liaoning, China
e-mail: 1053831329@qq.com

T. Zhang
e-mail: zta2000@163.net

Z. Dou
e-mail: douzh@smm.neu.edu.cn

Y. Huang
e-mail: 459337531@qq.com

H. Kim et al. (eds.), *Rare Metal Technology 2018*, The Minerals, Metals & Materials Series, https://doi.org/10.1007/978-3-319-72350-1_7

Introduction

The mixed rare earth concentrate (shortened as mixed concentrate) is an important resource of light rare earths, which is composed by bastnaesite and monazite. The mixed concentrate with a dense and refractory surface structure has been decomposed recently by different chemical routes with the aim to strengthen the decomposition reaction [1]. The routes of decomposition described in most of these papers have applied calcination, microwave [2, 3] and autoclave techniques.

The mechanochemical reactions in hydrometallurgy have been reviewed in papers [4–9]. In this type of processing, the chemical reactions and phase transformations take place under a mechanical force. The mechanochemical process can significantly reduce the reaction activation energy, refine the grain, increase the powder activity and strengthen the chemical reaction. Li Honggui et al. studied that tungsten minerals were decomposed by mechanochemical method, which showed the advantages of high decomposition efficiency, short process and low cost. Sun et al. [10] researched mechanochemical decomposition of the monazite, which found mechanochemical process reduced alkali consumption, improved decomposition efficiency and had an obvious economic benefit. As a consequence, it is possible to decompose the mixed concentrate by introducing the mechanochemical route.

At present, the roasting decomposition with sodium hydroxide is an important method to decompose the mixed concentrate in the industry. The method is that the mixed concentrate yield is decomposed to the $RE(OH)_3$ by calcination with sodium hydroxide, and then rare earth chloride is obtained through washing and acid leaching process. However, the roasting process consumes sodium hydroxide 1.8t as 1t mixed concentrate was decomposed. Meanwhile in the consequent washing process, a large amount of waste water containing NaF and Na_3PO_4 as by-products and NaOH as excessive reactants is produced, which increases the burden on the environment. And it is difficult to recover and separate the fluorine and phosphorus, bringing about waste of resources. Therefore, how to reduce the consumption of alkali and recover the fluorine, phosphorus and other valuable components has been the disadvantage and difficulty of the caustic roasting method.

This paper represents a new technology to decompose the mixed concentrate. The NaOH is still used to decompose the mixed concentrate, meantime, a certain amount of CaO is added to promote the chemical reaction and separate the fluorine and phosphorus, the process occurs in the calcification process as follows:

$$REPO_4 + 3OH^- = RE(OH)_3 + PO_4^{3} \quad (1)$$

$$REFCO_3 + 3OH^- = RE(OH)_3 + F^- + CO_3^{2} \quad (2)$$

$$2PO_4^{3-} + 2Ca^{2+} + H_2O = Ca_2P_2O_7 + 2OH^- \quad (3)$$

$$2F^- + CaO + H_2O = CaF_2 + 2OH^- \quad (4)$$

Equations (1)–(2) shows that CaO can promote the decomposition of $REPO_4$ and $REFCO_3$, that is contributing to the transformation of the mixed concentrate. After the reactions, $RE(OH)_3$ and $Ca_2P_2O_7$ were leached into the acid leaching solution and CaF_2 was left in the residue by hydrochloric acid leaching. The paper mainly studied the effect of mechanochemistry on the calcification decomposition of mixed concentrate.

Experimental

Sample Materials

The mixed concentrate tested in this experiment was from the Baotou, China. The particle size of the mixed concentrate was below 74 μm. The chemical compositions of sample are listed in Table 1. The X-ray diffraction pattern of mixed concentrate is given in Fig. 1. The microscopic morphology of mixed concentrate was observed by a scanning electron microscope, and the result is shown in Fig. 2.

Table 1 Chemical composition of mixed rare earth concentrate (wt%)

Chemical composition	RE_xO_y	Ce	La	Nd	F	SiO_2	P
Content/%	48.94	23.2	11.1	6.94	9.18	3.35	3.14

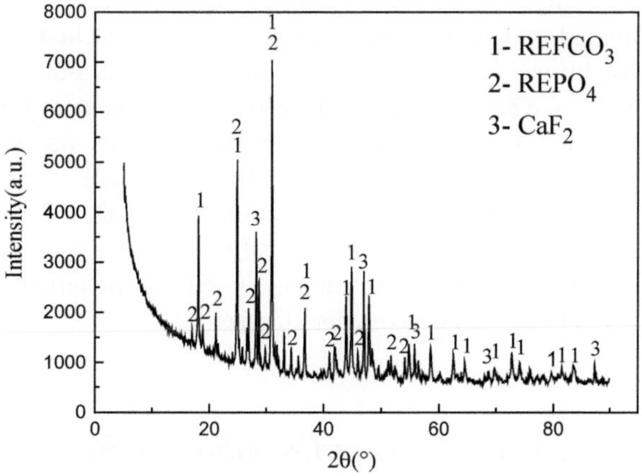

Fig. 1 XRD analysis of mixed concentrate

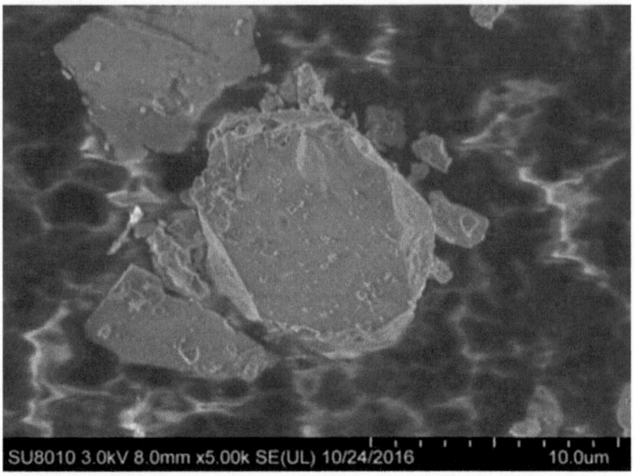

Fig. 2 SEM micrograph of mixed concentrate mixed concentrate

Experimental Process and Analysis

The experiment was carried in a homogeneous reactor, interior of which was equipped with six identical steel spherical reactors. The steel balls (the diameter of 4 mm of 18, 5 mm of 27, 6 mm of 36, 10 mm of 27 and 12 mm of 18) were put in each of the steel spherical reactor, the volume of which was 250 mL. The mixed concentrate was mixed with sodium hydroxide, calcium oxide and a certain amount of deionized water, and the mass ratio of sodium hydroxide and calcium oxide to mixed concentrate were 35 and 20%, respectively. The above sample was put in the steel spherical reactor and followed by the steel spherical reactors which were fitted with the sample were put in the homogeneous reactor. The experiment was carried out at room temperature and the temperature to a set point preserving heat for 3 h. On completion, the steel spherical reactors were cooled in air. Then, the sample was filtered and dried. The calcified slag was leached by hydrochloric acid under the conditions of 3 mol/L hydrochloric acid, 20 mL/g liquid-solid radio, 60 °C and 30 min. The leaching liquid was used to measure the content of rare earths.

The content of rare earths and phosphorus was analyzed by a inductively coupled plasma (ICP) spectrometer; The phase compositions of the sample was determined by an X-ray diffractometer (XRD, PW3040/60) with Cu Kα radiation; the microstructure of the material was observed by a scanning electron microscopy (SEM, SU8010) and spectroscopy (EDS); the amount of fluorine was examined by UV/VIS spectrophotometer (Optizen 2120UV).

Analysis of Working Condition of Steel Balls

The graphs of working condition of steel balls are seen in Fig. 3. The steel balls serve as a core of working during rotating of the reactor, therefore analysis of

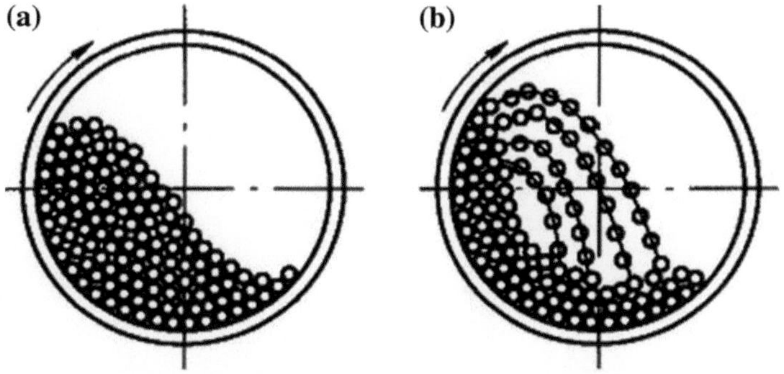

Fig. 3 Working conditions of steel balls (**a**) the first stage; (**b**) the second stage

working condition of steel balls contribute to understanding the reaction process. There are two stages divided for working process of the steel balls: at first, during the rise stage of the balls, that is, steel balls and materials move up along the inner wall with the rotation of the steel spherical reactor. In this stage, steel balls and materials have a relative sliding and rolling with each other; secondly, during the throwing off condition, the steel balls fall by gravity and then rising to the highest point. In this process, the steel balls take advantage of the inertia of falling to make the materials shocked and bumped intensively, which can strengthen the reaction process.

Results and Discussion

Mechanochemical Effect on the Calcification Decomposition of Mixed Concentrate

Ball to Powder Weight Ratio

The mixed concentrate is decomposed by mechanochemical route, so ball to powder weight ratio has an important influence on the leaching yield of the rare earths as seen in Fig. 4. The leaching yield had a trend of increasing at first and decreasing subsequently with the ball to powder weight ratio ranging from 35:1 to 100:1. When it was 80:1, the leaching rares reached the top, Ce 73.36%, La 82.88%, Nd 92.01% and the rare earth (leaching yield of total rare earth elements) 79.16%. The greater ball to powder weight ratio produces the more contact points of reaction, so the efficiency of the reaction is improved. But excessive mill balls increase the ball-ball collision, which decreases the ball-material impact, further weakens the decomposition effect.

Fig. 4 Effect of ball to power weight ratio on the leaching yield of rare earth

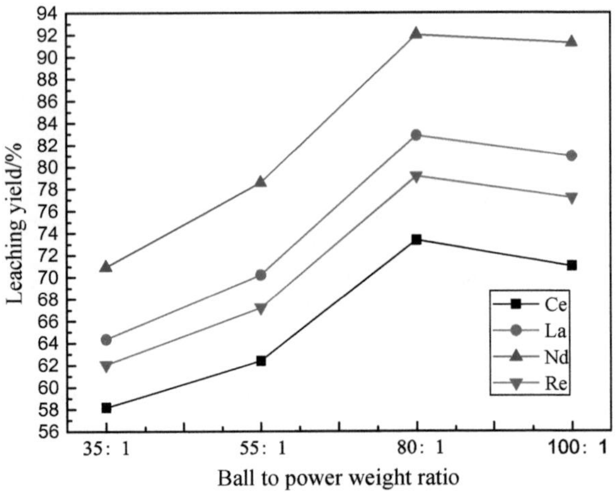

Temperature

The influence of temperature, ranging from 190 to 280 °C, is presented in Fig. 5. As seen in the figure, the leaching yield of total rare earth increased from 70.16 to 87.93% with increasing the temperature from 190 to 250 °C. When the temperature was increased to 280 °C, the leaching yield of rare earth was decreased. In fact, the collision moment of mill balls-materials will lead to local heating, then the water environment can weaken this effect, which makes the heat change caused by mechanical ball mill non-significant. Therefore raising temperature has a great impact on the leaching yield of rare earth, however, further increasing the temperature, some impurities also participate in the reaction, which affects leaching effect of rare earths.

Fig. 5 Effect of temperature on the leaching yield of rare earth

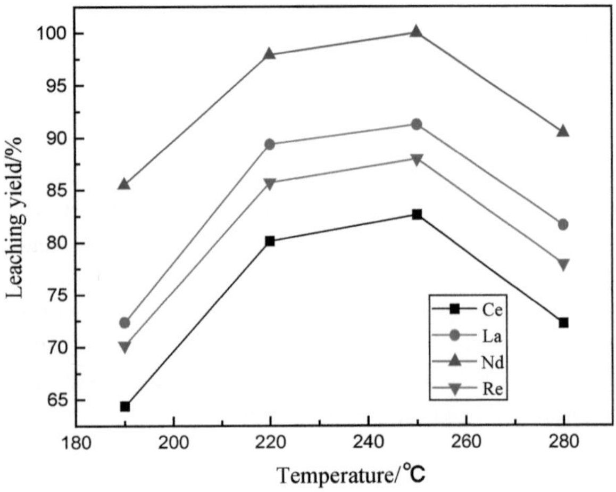

Liquid-Solid Ratio

The influence of liquid-solid ratio ranging from 3:1 to 15:1 on the leaching yields of rare earths is presented in Fig. 6. As seen in Fig. 6, the leaching yield of rare earth had a gradual decreasing trend with the increased liquid-solid ratio and the variations of leaching yields were consistent for three rare earth elements. Compared to the reaction was carried out in the autoclave [11], the leaching yield of rare earth was increased as the liquid-solid ratio was increased. This is because that in the traditional liquid-solid reaction, the flow of the material increased by increasing the liquid-solid, which accelerates the mass transfer process of the reaction, further promoting the decomposition of the mineral. However, for the mechanochemical reaction between liquid and solid phase, the reaction takes place due to shocking of steel balls on the materials, at this point the increased water would hinder the movement of the balls so that the mix effect of the materials was weakened, so the decomposition degree of the mineral was decreased and the leaching yield of rare earth was decreased. When the liquid-solid ratio was 5:1, the leaching yield of rare earth was 92.16%, Ce was 86.47%, La was 97.58%, Nd was 99.99%.

Effect of Liquid-Solid Ratio on the Distribution of Fluorine and Phosphorus

With the change of the liquid-solid ratio, the mass fraction of F, P entering into calcified liquor (the mass of F, P in the calcified liquid divided by the total mass of F, P) measured is shown in Fig. 7. In traditional sodium hydroxide decomposition, the fluorine and phosphorus are converted to sodium fluoride and sodium phosphate respectively into alkali liquor. However in the process added the CaO, the fluorine

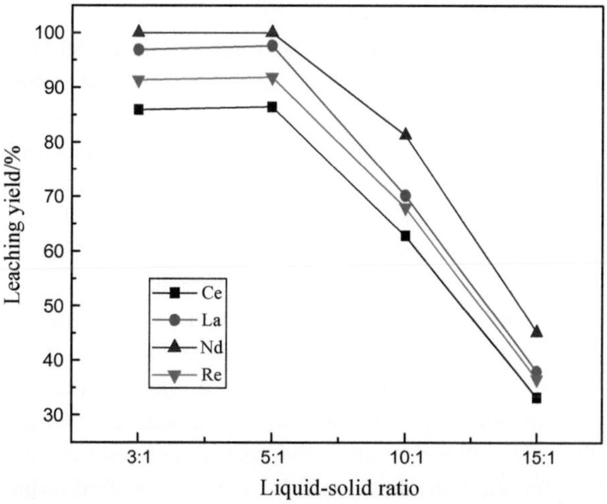

Fig. 6 Effect of liquid-solid ratio on the leaching yield of rare earth

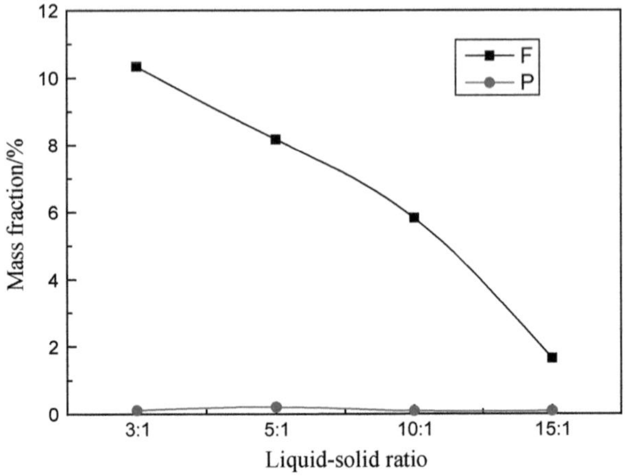

Fig. 7 Effect of liquid-solid ratio on the distribution of fluorine and phosphorus

and phosphorus are transformed to calcium fluoride and calcium pyrophosphate respectively. It is seen from Fig. 2 that about 3–10% of the fluorine entered into the calcified liquor because of the low dissolution of the calcium fluoride in alkaline solution, and with decreasing liquid-solid ratio, the content of fluorine in the calcified liquor was increased gradually since the degree of decomposition of the mixed concentrate was increased as known from Fig. 1, released fluorine from the mineral was increased so more fluorine centered into the liquid. But different liquid-solid ratio had little effect on the running of the phosphorus content. When the liquid-solid ratio was 5:1, the mass fraction of F, P entering into calcified liquor were 8.27, 0.21% respectively.

Analysis of XRD and SEM to the Products

Ball to powder weight ratio 80:1, temperature 250 °C and liquid-solid ratio 5:1 are the best decomposition condition already described in the above, so XRD and SEM of the calcified slag and the leaching residue obtained under the condition were analyzed. The diffraction patterns of the samples are shown in Fig. 8. It can be seen that there were obvious phase transformations before and after calcification for the mixed concentrate. After calcification, the phases of $REFCO_3$ and $REPO_4$ in the mixed concentrate disappeared, and the phases of CaF_2, $Ca_2P_2O_7$ and $RE(OH)_3$ were identified. Further, the calcified slag was leached by hydrochloric acid to extract the rare earths and phosphorus, only CaF_2 was left in the leaching residue as shown in Fig. 8b. Therefore this process achieves the separation of the fluorine and phosphorus.

The SEM micrographs of the samples are presented in Fig. 9. Compared to the smooth and dense surface of the mixed concentrate, it was seen that there was obvious difference after the mixed concentrate was decomposed. As shown in Fig. 9a, the particle size of the sample calcified reduced obviously attributing to the

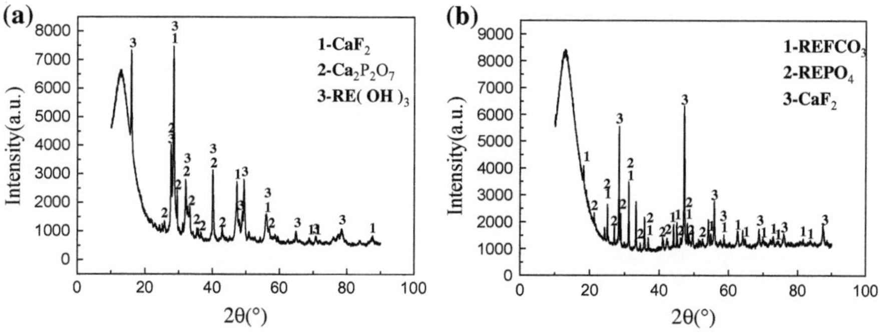

Fig. 8 XRD patterns of (**a**) the calcified slag and (**b**) the leaching residue

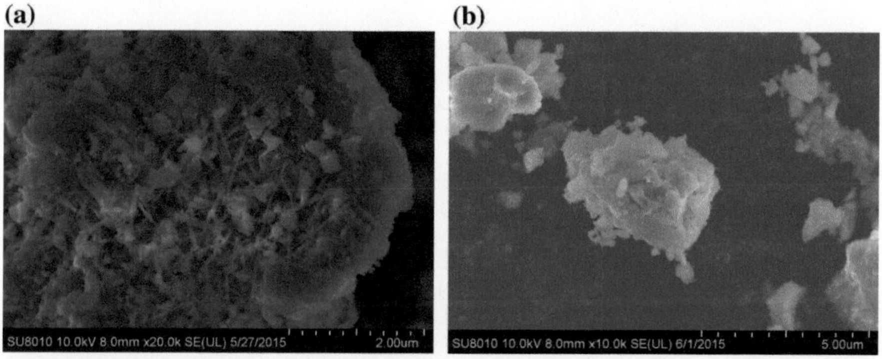

Fig. 9 SEM patterns of (**a**) the calcified slag and (**b**) the leaching residue

mechanochemical process. And the surface of the sample became fluffy with a lot of floccule enriched. After hydrochloric acid leaching the calcified sample, the microstructure was transformed again (Fig. 9b). The surface of the sample after leaching became smooth and dense, that suggests the floccule in the calcified slag was leached.

Conclusion

The results of this study showed the ball to material weight, the temperature and the liquid-solid ratio had the main impact on the mechanochemical decomposition of the mixed concentrate. With increasing ball to material weight, rising to temperature and decreasing liquid-solid ratio, the leaching yield of the rare earth increased rapidly and reached 92.16% under the conditions: ball to material weight 80:1, the calcification temperature 250 °C, the liquid-solid ratio 5:1. At the moment the mass fraction of F, P entering into calcified liquor were 8.27, 0.21% respectively.

According to the XRD analysis, the mixed concentrate was decomposed by mechanochemical calcification as the decomposition products were $RE(OH)_3$, CaF_2 and $Ca_2P_2O_7$. That contributed to extract the rare earths and separate the fluorine and phosphorus. The SEM micrographs showed that the microstructure of the mixed concentrate had obvious difference through mechanochemical calcification and leaching process. That is the surface had a transition from dense to fluffy, and then became dense again.

Acknowledgements The authors gratefully acknowledge the financial support received from the National Basic Research Program of China (973 program, No. 2012CBA01205) and the basic research universities special fund operations (N162505002).

References

1. Huang X, Li H, Xue X, Zhang G et al (2006) Development status and research progress in rare Earth hydrometallurgy in China. J Chin Rare Earth Soc 24(2):129–133
2. Huang Y, Zhang T, Dou Z et al (2016) Influence of microwave heating on the extractions of fluorine and rare Earth elements from mixed rare earth concentrate. Hydrometallurgy 162:104–110
3. Huang Y, Zhang T, Dou Z et al (2016) Decomposition of the mixed rare earth concentrate by microwave-assisted method. J Rare Earths 34(5):529–535
4. Balaz P, Achimovicova M (2006) Mechano-chemical leaching in hydrometallurgy of complex sulphides. Hydrometallurgy 84(1–2):60–68
5. James SL, Adams CJ, Bolm C et al (2012) Mechanochemistry: opportunities for new and cleaner synthesis. Chem Soc Rev 41(1):413–447
6. Bowmaker GA, Chaichit N, Pakawatchai C et al (2008) Solvent-assisted mechanochemical synthesis of metal complexes. Dalton Trans 22:2926–2928
7. Billik P, Plesch G (2007) Mechanochemical synthesis of nanocrystalline TiO_2 from liquid $TiCl_4$. Scripta Mater 56(11):979–982
8. Abdel Rehim AM (2005) A new technique for extracting zirconium form Egyptian zircon concentrate. Int J Miner Process 76(4):234–243
9. Kumar R, Alex TC, Jha MK et al (2004) Mechanochemistry and the Bayer process of alumina production. In: Paper presented at the 133rd TMS annual meeting, Charlotte, NC, 14–18 March 2004
10. Sun P, Li H, Li Y et al (1998) New technology on decomposition of monazite by mechanical activation and caustic decomposition. J Central South Univ 29(1):36–38
11. Tang F (2014) Study on efficient extraction of the rare earth from bastnaesite concentrate by calcification transition process. Master's thesis, Northeastern University

Part II
Rare Earth Elements II and Platinum Group Metals

Electrochemical Formation of Tb Alloys in Molten LiCl–KCl Eutectic Melts and Separation of Tb

Hirokazu Konishi, Hideki Ono, Tetsuo Oishi and Toshiyuki Nohira

Abstract The electrochemical formation of Tb-Ni alloys was investigated in a molten LiCl–KCl–$TbCl_3$ (0.50 mol% added) at 723 K. Open-circuit potentiometry was conducted using a Ni electrode after electrodepositing Tb metal at 0.20 V (vs. Li^+/Li) for 300 s. There were four potential plateaus at (a) 0.66 V, (b) 0.80 V, (c) 0.95 V and (d) 1.56 V, respectively. Alloy samples were prepared by potentiostatic electrolysis at 0.60 and 0.70 V at 723 K. The alloy phase was identified as only $TbNi_2$. Anodic dissolution of Tb from the formed $TbNi_2$ was conducted at 0.90, 1.20 and 1.60 V, respectively. In the sample obtained at 0.90 V for 3 h, the existence of $TbNi_3$ was identified by the XRD. Phase of the sample obtained at 1.20 V for 3 h was $TbNi_5$. The sample obtained at 1.60 V for 3 h was Ni. Alloy samples were prepared by potentiostatic electrolysis at 0.50–0.80 V for 1 h using Ni plate cathodes in a molten LiCl–KCl containing $TbCl_3$ (0.50 mol%) and $NdCl_3$ (0.50 mol%). The highest mass ratio of Tb/Nd in the alloy sample was 56 at 0.70 V.

Keywords Rare earth · Electrolysis · Molten salt · Separation

Introduction

The use of rare earth (RE)-iron group (IG) alloys has increased significantly in a number of industrial fields over the past few decades. In particular, the demand for Dy-added Nd–Fe–B magnets is rapidly increasing because these magnets are

H. Konishi (✉) · H. Ono
Osaka University, 2-1 Yamadaoka, Suita, Osaka 565-0871, Japan
e-mail: konishi@mat.eng.osaka-u.ac.jp

T. Oishi
National Institute of Advanced Industrial Science and Technology, 16-1 Onogawa, Tsukuba, Ibaraki 305-8569, Japan

T. Nohira
Kyoto University, Gokasho, Uji, Kyoto 611-0011, Japan

H. Kim et al. (eds.), *Rare Metal Technology 2018*, The Minerals, Metals & Materials Series, https://doi.org/10.1007/978-3-319-72350-1_8

indispensable for high-performance motors in electric vehicles (EVs) and hybrid electric vehicles (HEVs). These magnets need to possess sufficient thermal stability for use in such motors in high-temperature environments. The addition of Dy is necessary to improve the thermal stability of Nd–Fe–B magnets. However, there is the concern about a shortage of RE metals because of the uneven distribution of RE resources. Against this background, it is necessary to develop an inexpensive and environmentally friendly recovery/separation process for RE metals, especially the recovery of Nd and Dy from magnet scraps.

We proposed a new separation and recovery process for RE metals from scraps using molten salt and an alloy diaphragm as shown in Fig. 1 [1–3]. RE containing scrap is used as the anode. A RE-transition metal (TM) alloy is used as the diaphragm, which functions as a bipolar electrode. During electrolysis, all the RE metals in the anode are dissolved in the molten salt as RE ions. One or several specific RE ions are selectively reduced to form RE-TM alloys on the alloy diaphragm according to their formation potentials and/or alloying rates. Subsequently, the RE atoms chemically diffuse through the alloy diaphragm and are dissolved into the molten salt as RE ions in the cathode room. The permeated RE ions are finally deposited on the Mo or Fe cathode as RE metals. The RE ions remaining in the anode room can be collected by electrolysis using another cathode in the anode room. Almost all impurities remain in the anode room as residue or anode slime.

This new process for RE metals from scraps was first applied to a molten LiCl–KCl–$DyCl_3$–$NdCl_3$ system [4, 5]. The present study focused on Tb as a scarce element in scraps. As a first step, the electrochemical formation of Tb–Ni alloys was investigated in a molten LiCl–KCl–$TbCl_3$ system at 723 K. Moreover, the separation of Tb from Nd was conducted by electrolysis.

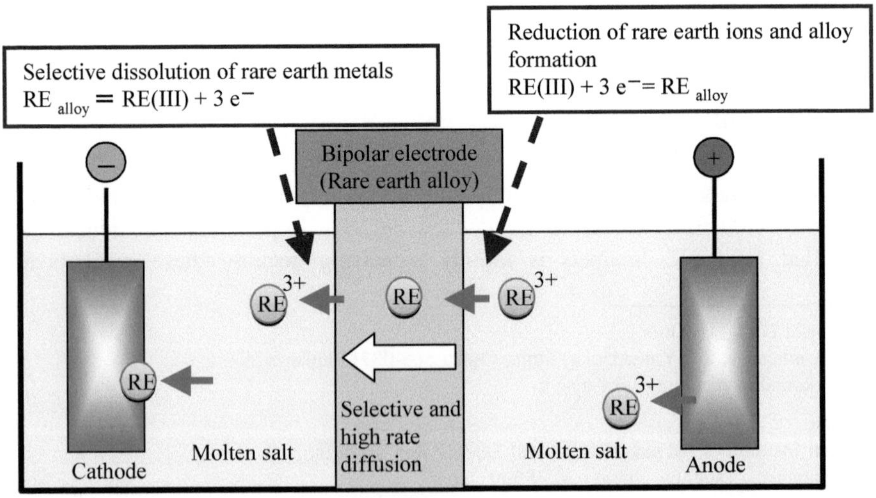

Fig. 1 Schematic drawing of the process for separation and recovery of rare earth metals

Experimental

All chemicals were anhydrous reagent grade. The LiCl–KCl eutectic (LiCl: KCl = 58.5: 41.5 mol%; Wako Pure Chemical Co., Ltd.) was placed in a high purity alumina crucible, and kept under a vacuum for more than 24 h at 473 K to remove water. All experiments were performed in LiCl–KCl eutectic melts under a dry Ar atmosphere at 723 K. $TbCl_3$ (99.9%, Kojundo Chemical Laboratory Co., Ltd.) 0.50 mol% was added directly to the melts. A chromel-alumel thermocouple was used for temperature measurements. The working electrodes for the investigation of electrochemical behavior were Mo (5 mm × ϕ 1 mm, 99.95%, Nilaco Co., Ltd.) and Ni (5 mm × ϕ 1 mm, 99%, Nilaco Co., Ltd.) wires. For the formation of alloy samples, rectangular shaped Ni plates (10 mm × 5 mm × 0.2 mm, 99%, Nilaco Co., Ltd.) were used as the working electrodes. The reference electrode was an Ag wire immersed in LiCl–KCl containing 1 mol% of AgCl, placed in a Pyrex glass tube with a thin bottom to maintain electrical contact with the melt. The potential of this reference electrode was calibrated with reference to that of a Li^+/Li electrode, which was prepared by electrodepositing Li metal on a Mo wire. All the potentials given hereafter are referred to the Li^+/Li electrode potential on a Mo wire. The counter electrode was a glassy carbon rod (50 mm × ϕ 5 mm, Tokai Carbon Co., Ltd.). A potentio/galvanostat was used for cyclic voltammetry and open-circuit potentiometry. The alloy samples were prepared by potentiostatic electrolysis using the same apparatus. After electrolysis, the samples were rinsed with distilled water, and analyzed by X-ray diffractometer (XRD). Cross-sections of these samples were also observed by scanning electron microscopy (SEM).

Results and Discussion

Cyclic Voltammetry

As shown in Fig. 2, the phase diagram of the Tb–Ni system shows the presence of eight intermetallic compounds (Tb_3Ni, Tb_3Ni_2, TbNi, $TbNi_2$, $TbNi_3$, Tb_2Ni_7, $TbNi_5$ and Tb_2Ni_{17}) at 873 K [6]. Therefore the presence of eight intermetallic compounds was suggested at the experimental temperature of 723 K.

Taking into account the possibilities of the formation of these various Tb–Ni intermetallic compounds, cyclic voltammetry was conducted. Before the investigation of Tb–Ni formation, the electrochemical behavior of Tb (III) was studied using a Mo or Ni wire as the working electrode. Figure 3 shows the obtained cyclic voltammograms for Mo and Ni electrodes after addition of 0.50 mol% $TbCl_3$ at 723 K. The scanning rate was set at 0.1 V s^{-1}. The broken curve represents the voltammogram at the Mo electrode. During the scan in the negative direction, a significant increase in the cathodic current was observed at 0.48 V (vs. Li^+/Li). Since Mo does not form alloy with Tb, the cathodic current is considered to

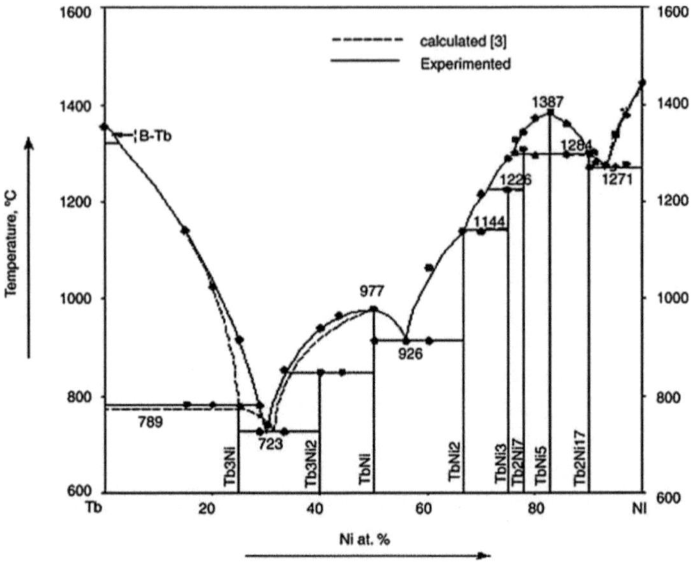

Fig. 2 Phase diagram of the Tb–Ni system [6]

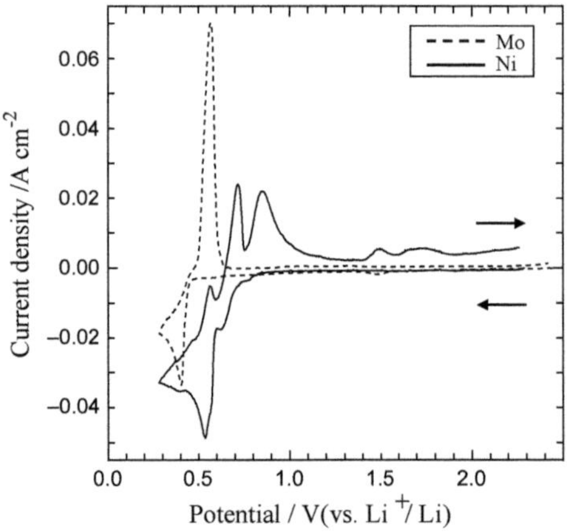

Fig. 3 Cyclic voltammograms for Mo and Ni electrodes in LiCl–KCl–$TbCl_3$ (0.50 mol% added) at 723 K. Scan rate: 0.1 V s^{-1}

correspond to Tb metal deposition. After reversing the scan direction at 0.30 V, an anodic current peak was observed, which should be the result of anodic dissolution of the Tb metal. The solid curve represents the voltammogram at the Ni electrode. A cathodic current was observed from 0.70 V. Since this potential is more positive than that for Tb metal deposition (0.48 V), the cathodic current is regarded to correspond to the formation of Tb–Ni alloys. When the potential scan direction was reversed at 0.30 V, several anodic peaks were observed. These anodic peaks suggest the Tb dissolution from the different Tb–Ni alloy phases.

Open Circuit Potentiometry

Open-circuit potentiometry was carried out to investigate the formation potential of Tb–Ni alloys. Figure 4 shows the open-circuit potential transient curve for a Ni electrode after electrodepositing Tb metal at 0.20 V for 300 s in LiCl–KCl–$TbCl_3$ (0.50 mol% added) at 723 K. There were four potential plateaus observed at about (a) 0.66 V, (b) 0.80 V, (c) 0.95 V and (d) 1.56 V, respectively. These potential plateaus are considered to correspond to different coexisting Tb–Ni phases. Here, it is suggested that the potential plateaus (a), (b), (d) are related to the anodic peaks in Fig. 3. However, concerning the plateau (c), no corresponding anodic peak was observed in the voltammogram. The anodic peak corresponded to the potential plateaus (b) might be divided in two peaks.

Electrochemical Formation of Tb–Ni Alloys

To confirm the formation of Tb–Ni alloys, potentiostatic electrolysis was conducted at 0.60 V for 4 h using a Ni plate cathode at 723 K in LiCl–KCl–$TbCl_3$ (0.50 mol% added). This potential value is more positive than that of the Tb metal deposition (0.48 V). The sample was analyzed by XRD, and their cross-section was observed by SEM. Figure 5a shows the XRD pattern of the sample obtained at 0.60 V for 4 h; the alloy phase was identified as $TbNi_2$. The cross-sectional SEM image of the sample is shown in Fig. 5b. This $TbNi_2$ film was dense and coherent with a thickness of 46 μm. Moreover, potentiostatic electrolysis was conducted at 0.70 V for 2 h. From the XRD pattern and cross-sectional image of the sample, the alloy phase was also identified as $TbNi_2$. The thickness of $TbNi_2$ film was found to be

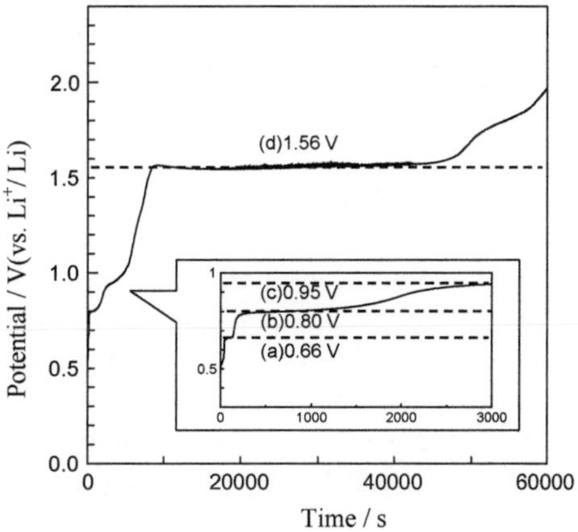

Fig. 4 Open-circuit potential transient curve for a Ni electrode after electrodepositing Tb metal at 0.20 V for 300 s in LiCl–KCl–$TbCl_3$ (0.50 mol% added) at 723 K

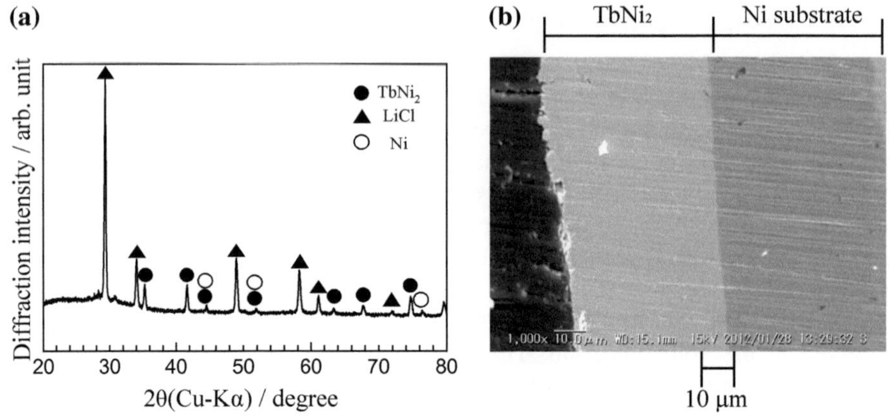

Fig. 5 (**a**) XRD pattern and (**b**) cross-sectional SEM image of the sample prepared by potentiostatic electrolysis using a Ni plate cathode at 0.60 V for 4 h in LiCl–KCl–$TbCl_3$ (0.50 mol % added) at 723 K

20 μm. From this result, the potential plateau (a) 0.66 V could not be corresponded to the coexisting Tb–Ni phase. Since transformation rate of a Tb–Ni alloy initially formed at 0.60 V might be very fast, the formed Tb–Ni alloy is considered to transform to $TbNi_2$.

Anodic Dissolution of Tb from the Formed $TbNi_2$

In order to form various Tb–Ni phases, anodic dissolution of Tb from the formed $TbNi_2$ was conducted at 0.90, 1.20 and 1.60 V, respectively. The $TbNi_2$ was prepared by potentiostatic electrolysis at 0.60 V for 1 h at 723 K. These potential values were chosen as intermediate values between the adjoining plateaus. Figure 6 shows the XRD pattern of the sample obtained at 0.90 V for 3 h; the original $TbNi_2$ was transformed to the Tb-poor phase ($TbNi_3$) at 0.90 V for 3 h from the XRD analysis. From this result, the potential plateau (b) was corresponded to the coexisting ($TbNi_2$ + $TbNi_3$) phase. Figure 7 shows the XRD pattern of the sample obtained at 1.20 V for 3 h; the sample obtained at 1.20 V for 3 h was found to be the $TbNi_5$ film. Therefore, the potential plateau (c) corresponded to the coexisting ($TbNi_3$ + $TbNi_5$) phase. Figure 8 shows the XRD pattern of the sample obtained at 1.60 V for 3 h; the identified phase of sample obtained at 1.60 V for 3 h was Ni. Thus, the potential plateau (d) was based on the coexisting ($TbNi_5$ + Ni) phase.

From these results, it was found that $TbNi_2$ are changed to other phases, i.e., $TbNi_3$, $TbNi_5$ and Ni, depending on applied potential in LiCl–KCl–$TbCl_3$(0.50 mol%

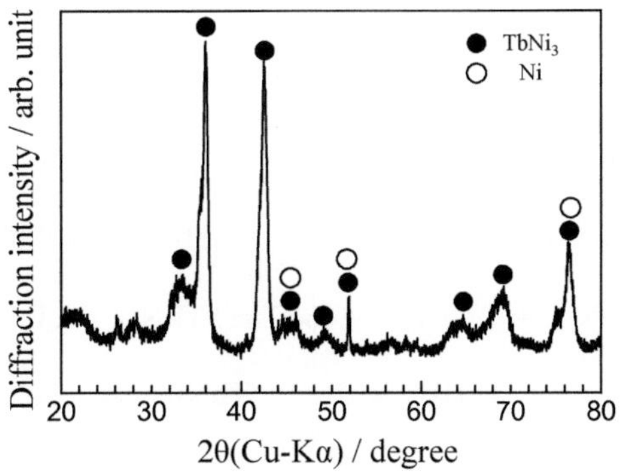

Fig. 6 XRD pattern of the sample prepared by potentiostatic electrolysis using the formed $TbNi_2$ cathode at 0.90 V for 3 h in LiCl–KCl–$TbCl_3$ (0.50 mol% added) at 723 K

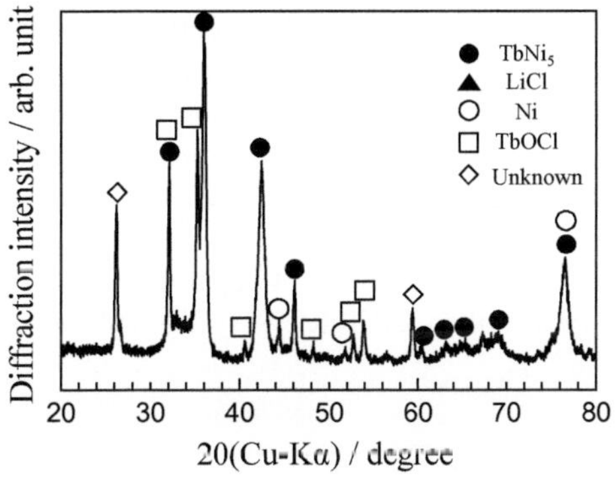

Fig. 7 XRD pattern of the sample prepared by potentiostatic electrolysis using the formed $TbNi_2$ cathode at 1.20 V for 3 h in LiCl–KCl–$TbCl_3$ (0.50 mol% added) at 723 K

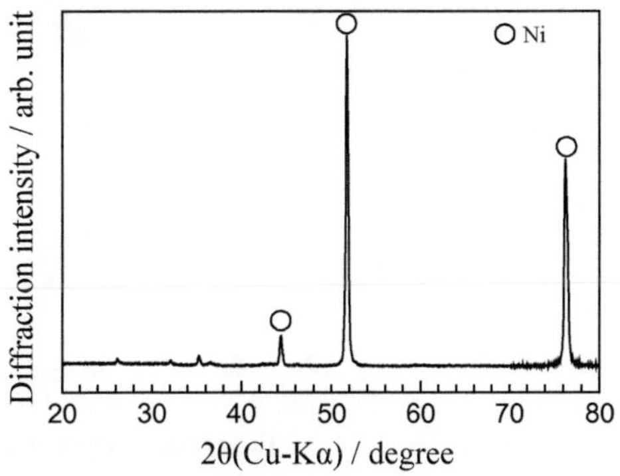

Fig. 8 XRD pattern of the sample prepared by potentiostatic electrolysis using the formed $TbNi_2$ cathode at 1.60 V for 3 h in LiCl–KCl–$TbCl_3$ (0.50 mol% added) at 723 K

added) at 723 K. The potential plateaus obtained by open-circuit potentiometry corresponded to equilibrium potential of different coexisting Tb–Ni phases. The several transformation reactions and the corresponding equilibrium potential are summarized as follows.

$$2TbNi_3 + Tb(III) + 3e^- = 3TbNi_2 \quad \text{at } 0.80V \tag{1}$$

$$3TbNi_5 + 2Tb(III) + 6e^- = 5TbNi_3 \quad \text{at } 0.95V \tag{2}$$

$$5Ni + Tb(III) + 3e^- = TbNi_5 \quad \text{at } 1.56V \tag{3}$$

Separation of Tb from Nd

Based on the obtained results, the separation of Tb from Nd was investigated using Ni electrodes cathodes in a molten LiCl–KCl containing $TbCl_3$ (0.50 mol%) and $NdCl_3$ (0.50 mol%) at 723 K. The alloy samples were prepared by potentiostatic electrolysis at 0.50–0.80 V for 1 h using Ni plates cathodes. Figure 9 shows the mass ratio of Tb/Nd in the alloy samples measured by inductively coupled plasma-atomic emission spectroscopy (ICP-AES). The highest mass ratio of Tb/Nd in the alloy sample, 56, is found at 0.70 V.

These results indicated the possibility of separating Tb from Nd by controlling electrolysis potential in a molten LiCl–KCl–$DyCl_3$–$TbCl_3$–$NdCl_3$ system.

Fig. 9 The amounts of deposited Tb and Nd and the mass ratio of Tb/Nd, in a molten LiCl–KCl–$TbCl_3$ (0.50 mol% added)–$NdCl_3$ (0.50 mol% added) at 723 K

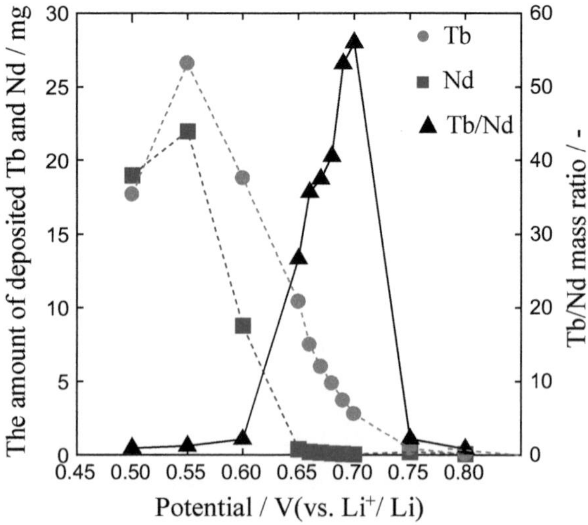

Conclusions

The electrochemical formation of Tb-Ni alloys was studied in LiCl–KCl–$TbCl_3$ (0.50 mol% added) at 723 K. The results can be summarized as:

1. Potentiostatic electrolysis at 0.60 V for 4 h using a Ni electrode resulted in formation of the dense and coherent $TbNi_2$ film having thickness of about 46 μm.
2. The formed $TbNi_2$ were transformed to other phases, i.e., $TbNi_3$, $TbNi_5$ and Ni, by anodic dissolution of Tb, depending on the applied potential.
3. The several transformation reactions and the corresponding equilibrium potential for the Tb–Ni alloys were clarified.
4. Alloy samples were prepared by potentiostatic electrolysis at 0.50–0.80 V for 1 h using Ni plate cathodes in a molten LiCl–KCl containing $TbCl_3$ (0.50 mol%) and $NdCl_3$ (0.50 mol%). The highest mass ratio of Tb/Nd in the alloy sample was 56 at 0.70 V. Tb could be highly separated from Nd by controlling electrolysis potential.

References

1. Oishi T et al (2010) Kagaku Kogaku Ronbunshu, vol 36. Soc Chemical Eng Japan, Bunkyo-ku, Tokyo, pp 299–303
2. Kobayashi S et al (2011) J Electrochem Soc 158:E142–146 (Pennington, NJ: Electrochemical Soc Inc.)
3. Kobayashi S et al (2012) J Electrochem Soc 159:E193–197 (Pennington, NJ: Electrochemical Soc Inc.)
4. Konishi H et al (2011) Molten salts 54:21–28 (Suita, Osaka: Molten Salt Committee of The Electrochemical Society in Japan)
5. Konishi H et al (2012) ECS Trans 50:463–472 (Pennington, NJ: Electrochemical Soc Inc.)
6. Yao Q, Wang Y, Zhou H (2005) J Alloys Comp 395:98–100 (Lausanne, Vaud: Elsevier Science Sa)

Electrochemical and Spectroscopic Study of Eu(III)/Eu(II) Couple in the 1-Ethyl-3-Methylimidazolium Bis(Trifluromethanesulfonyl)Imide Ionic Liquid

David Bengio, Thomas Dumas, Eric Mendes, Pier-Lorenzo Solari, Richard Husar, Michel Schlegel, Philippe Moisy and Stéphane Pellet-Rostaing

Abstract Ionic liquids (ILs) are molten salts composed of an organic cation that are liquid below 100 °C. They demonstrate unique physico-chemical properties such as good conductivity, negligible vapor pressure and non-flammability. As a consequence, they are often regarded as green solvents and could become an alternative to the use of both high-temperature molten salts and volatile organic solvents in rare-earth elements processing and recycling. The diversity of interactions existing in an ionic liquid allows the solubilization of both polar and apolar compounds. Moreover, coordinating functions on their composing ions can lead to

D. Bengio (✉) · T. Dumas · E. Mendes · R. Husar · P. Moisy
CEA Marcoule, BP 17171, Bagnols-sur-Cèze cedex 30207, France
e-mail: david.bengio@cea.fr

T. Dumas
e-mail: thomas.dumas@cea.fr

E. Mendes
e-mail: eric.mendes@cea.fr

R. Husar
e-mail: r.husar@hzdr.de

P. Moisy
e-mail: philippe.moisy@cea.fr

P.-L. Solari
Synchrotron SOLEIL, Saint-Aubin 91190, France
e-mail: pier-lorenzo.solari@synchrotron-soleil.fr

M. Schlegel
CEA Saclay, Gif-sur-Yvette cedex 91191, France
e-mail: michel.schlegel@cea.fr

S. Pellet-Rostaing
ICSM, BP 17171, Bagnols-sur-Cèze cedex 30207, France
e-mail: stephane.pellet-rostaing@cea.fr

H. Kim et al. (eds.), *Rare Metal Technology 2018*, The Minerals, Metals & Materials Series, https://doi.org/10.1007/978-3-319-72350-1_9

stabilization of some species. For instance, in non-aqueous media, europium exists in the (II) oxidation state which is not stable in aqueous solutions. Understanding the mechanism of Eu(III) electrochemical reduction to Eu(II) and the stabilization of Eu(II) species in IL media could be of major interest for the development of innovative recycling processes. Using transient electrochemistry and UV-Vis spectroscopy we could get valuable information on the redox behavior of the Eu(III)/(II) couple in the ionic liquid 1-ethyl-3-methylimidazolium bis(trifluoromethanesulfonyl)imide ([EMIm][NTf_2]). Reduction of Eu(III) to Eu(II) and stability of Eu(II) was also studied thanks to a XAS-spectro-electrochemistry set up used to follow in situ the evolution of the XANES spectrum around the L_3 edge of Eu during electrolysis.

Keywords Lanthanides · Europium · Electrochemistry · Ionic liquids
XANES

Introduction

Ionic liquids are salts composed of an organic cation and an organic or inorganic anion with a melting point below 100 °C. Due to their negligible vapor pressure, low flammability, wide electrochemical window and high conductivity, their use as solvents for chemistry and electrochemistry has been extensively studied for the past decades [1]. The great diversity and complexity of solvent-solvent and solvent-solute interactions allow them to solubilize both polar and apolar compounds [2]. Thus, they are often regarded as green solvents that could be an alternative to the use of volatile organic solvents and high-temperature molten salts for rare-earth elements processing and recycling. Depending on the types of anion and cation used and since both the anion and the cation can be functionalized, an incredible number of different ionic liquids can be synthesized and it is theoretically possible to find a suitable ionic liquid in order to complete a specific task [3–5]. Among the numerous types of ionic liquids, the ones comprising the bis(trifluoromethanesulfonyl)imide anion (NTf_2^-) (c.f. Fig. 1) are often used for they form air and water stable ILs which are generally liquid at room temperature and display relative hydrophobicity, low viscosity, high conductivity and high thermal and electrochemical stability [6].

Most interestingly in our case, ILs allow the use of electrochemical techniques to apply redox-controlled conditions and access specific oxidation states.

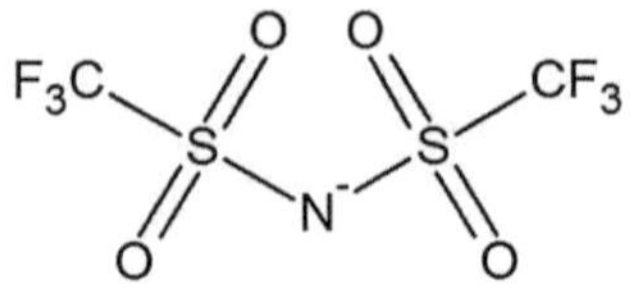

Fig. 1 The bis(trifluoromethanesulfonyl) imide anion (NTf_2^-)

Moreover, they tend to stabilize some species that are otherwise difficult to isolate. For instance, in aqueous solutions, europium is stable at the oxidation state (III). The (II) oxidation state exists but is difficult to stabilize. Yet, in non-aqueous media, stable forms of the (II) oxidation state exist [7]. Thus, as part of research on innovative processes for rare-earths recycling, the study of the redox behaviors of several III/II lanthanide couples in ionic liquids could be of major interest. It could lead the way to an efficient separation and recovery of certain strategic [8] rare-earth elements [9]. We decided to study europium first because its divalent state is already well described in the literature in several media but not in ionic liquids [7, 10, 11]. Accessing the apparent redox potentials of the Eu(III)/Eu(II) couple in ionic liquid media and the structures of the complexes formed with the solvent ions would help us determine the feasibility of selective lanthanide separation in these media since most other lanthanides do not have a stable (II) oxidation state.

To control and describe the different oxidation states of Eu we used transient electrochemistry, i.e. cyclic voltamperometry and chronoamperometry along with X-ray Absorption Spectroscopy (XAS). The latter is a very powerful tool to study the coordination of a metal ion in diverse media [12], including in ionic liquids [13]. It gives access to the coordination geometry thanks to X-ray Absorption Near Edge Structure (XANES) and to the local environment (nature of neighboring atoms, coordination number and bond distances) thanks to Extended X-ray Absorption Fine Structure (EXAFS). These techniques have been commonly coupled to electrochemical techniques in order to study specific oxidation states of diverse radionuclides in aqueous media [14, 15]. Even though simple XAS measurements have been performed in ionic liquids [16], to the best of our knowledge, this coupling of XAS and electrochemistry had never been implemented in ionic liquid media.

We worked with the 1-ehtyl-3-methyl-imidazolium bis(trifluromethanesulfonyl) imide ionic liquid which shows particularly high conductivity and low viscosity [17]. Thanks to XAS, we followed in situ the electrolysis from Eu(III) to Eu(II). XAS spectro-electrochemistry was performed on the MARS beamline of Synchrotron SOLEIL using a spectro-electrochemical cell co-developed by the teams of MARS beamline and CEA (Marcoule and Saclay). The cell has already given access to published information and has been validated for studies on radioactive material [12]. We benefit from the features that allow its use with radioactive material since the highly hygroscopic ionic liquids need to be handled in an oxygen [18] and water-free environment. Previous to XAS measurements, the experiment was conducted in our lab and monitored by UV-Vis spectroscopy (UVS).

The present paper shows the first results obtained in UVS and XANES spectroscopy. A method to obtain the XANES spectrum of Eu(II) after non-exhaustive electrolysis is presented which shows the feasibility of an in situ XAS monitoring of the electrochemical reaction.

Experimental Section

Reagents

The ionic liquid, 1-ethyl-3-methylimidazolium bis(trifluoromethanesulfonyl)imide, [EMIm][NTf_2] (99.9%, Solvionic) was used as received. When necessary ([H_2O] > 50 ppm), the ionic liquid was dried by bubbling pure argon (Alphagaz 2, Air Liquide) in the solution. Wet ionic liquid was obtained by contacting the ionic liquid with ultrapure water. Europium was added as a bis(trifluoromethanesulfonyl) imide salt $Eu(NTf_2)_3$ provided by Solvionic. Unless otherwise stated, all chemicals were handled in a glove-box under argon atmosphere ([O_2] < 10 ppm, [H_2O] < 10 ppm).

Karl-Fischer Titration

Water content of the ionic liquid was determined before and after all experiments by Karl-Fischer titration using a Metrohm 917 KF operated with a one-compartment cell (generator electrode without diaphragm) filled with Hydranal Coulomat AG (Fluka). The volume of test samples was calculated to ensure a significant titration (at least 50 μg of water has to be injected).

Analytical Electrochemistry

All electrochemical measurements were performed using a BioLogic VSP potentiostat/galvanostat run with EC-Lab V11.01 software. The experiments were carried out using a jacketed three-electrode cell. A Pt disc electrode (2 mm diameter), provided by Radiometer Analytical (EDI101 tip) was used as working electrode. Prior to use, it was activated by electrochemical cycling in a H_2SO_4 0.5 M aqueous solution. A platinum coiled wire (0.5 mm diameter, Goodfellow) was used as a counter-electrode. The reference electrode was prepared as suggested in the literature [19] with an Ag wire (0.5 mm diameter, Goodfellow) in contact with a solution of 0.017 M AgOTf (Sigma Aldrich) in [EMIm][NTf_2] separated from the working solution by a ceramic junction (provided by Radiometer Analytical). An apparent potential of −0.42 V against this Ag/Ag^+ reference electrode was measured for the Fc^+/Fc couple by cyclic voltammetry in a solution containing 30 mM Fc in [EMIm][NTf_2]. To facilitate comparison with the literature, all potentials presented in this work are referenced versus Fc^+/Fc. The potential of this prepared reference electrode was found to be stable during several months. All the experiments were performed at 40 °C in a glove-box under argon atmosphere ([O_2] < 10 ppm, [H_2O] < 10 ppm) and the cell was purged with dried argon during the experiment. It is to be noted that the potentials were not corrected for the electrolyte resistance. Given the small current we are dealing with, it seems to be a reasonable approximation.

UV-Vis Spectroscopy

UV-Vis spectra were recorded with a Varian Cary 50 spectrometer. Static acquisitions were performed in standard 10 mm quartz cuvette. In situ experiments were performed in a 3 mL spectro-electrochemical equipped with quartz windows. The geometry of this cell was similar to the XAS one described in the next section. Spectra were recorded on the same apparatus thanks to a pair of 10 m UV-Vis optic fibers (Hellma) and an optical interface.

Spectro-Electrochemical Cell

In order to work in anhydrous conditions with our hygroscopic ionic liquid, an airtight electrochemical cell allowing XAS measurement was needed. We used an electro-XAS setup co-developed by the teams of MARS beamline and CEA. This device was initially designed for the study of actinides. The whole structure is made of PEEK and the inner cell (700 µL–2.5 mL) is enclosed in a secondary containment enabling its use for radioactive material and assuring the airtightness. It allows working in both fluorescence and transmission modes. The three electrodes used in the set up are:

- The working electrode consists in a piece of glassy carbon foam occupying the whole volume of the cell and for which electrical contact was made thanks to a platinum wire. This material was chosen for its extended surface area, necessary to the completion of the exhaustive electrolysis in a reasonable time.
- The reference electrode is an Ag/AgOTf 0.01 M in [EMIm][NTf_2] electrode. It was made out of a PTFE tube ended with a ceramic frit. The tube is filled with the solution of silver triflate in the IL in which a 0.5 mm diameter silver wire is immersed. The standard potential of the Fc^+/Fc couple was measured to be −0.42 V against this reference.
- The auxiliary electrode is made out of a similar PTFE tube which is used as an anodic compartment comprising a 1 mm diameter platinum wire immersed in the neat ionic liquid.

Bulk electrolysis at SOLEIL was performed by a Metrohm Autolab PGSTAT 101 using chronoamperometry (Fig. 2).

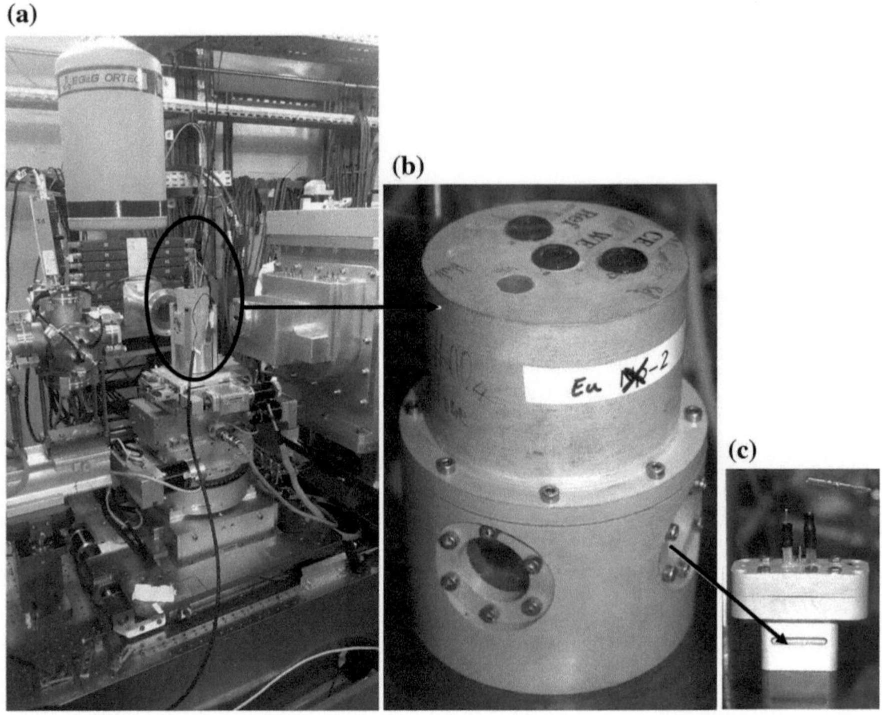

Fig. 2 XAS station of MARS beamline (**a**) with secondary containment (**b**) and XAS spectro-electrochemical inner cell (**c**). Points (1), (2), (3) and (4) on (**c**) are the different beam positions investigated during the experiment (c.f. Fig. 6)

Sample Preparation

The working solution was prepared by dissolving 0.05 M of $Eu(NTf_2)_3$ salt in $[EMIm][NTf_2]$. The sprectro-electrochemical cells were then prepared and sealed in the glove-box before transportation to SOLEIL Synchrotron.

X-Ray Absorption Spectroscopy

All XAS spectra were recorded at the Eu L_3 edge on the MARS beamline of SOLEIL Synchrotron (Saint-Aubin, France). The cell was installed on the Mars Beamline CX3 station dedicated to X-ray absorption spectroscopy. A 13-element HPGe solid state detector (ORTEC) was used to collect the data in fluorescence mode. The energy calibration of the monochromator was performed at the iron K-edge (first derivative at 7112 eV). All measurements were performed at room temperature.

Results and Discussion

Electrochemical Study

Thanks to cyclic voltammetry we could determine essential properties of the Eu(III)/Eu(II) couple and confront them to the literature. Three main criteria were checked in order to evaluate the reversibility of this couple. As shown in the insert of Fig. 3, the peak separation is larger than the 2.303 RT/nF (i.e., 62.1 mV at 40 °C for a monoelectronic transfer [20]) theoretically expected for a reversible couple and increases with increasing scan rate. By studying the evolution of the peak currents as a function of the scan rate, it was found that the peak currents varied linearly with the square-root of the scan rate (c.f. Fig. 3) which means that the reaction is diffusion-controlled. Finally, the ratio i_p^a/i_p^c was found to be constant and close to one. Thus, we can conclude that Eu(III)/Eu(II) in $[EMIm][NTf_2]$ is a quasi-reversible couple on platinum. The apparent standard potential was determined thanks to the following equation:

$$E_0^{'} \approx \frac{E_p^a - E_p^c}{2} = 0.18\,\mathrm{V\,vs\,Fc^+/Fc}$$

where E_p^a and E_p^c are the potentials of the anodic and cathodic peaks. This value is rather close to the value of +0.13 V found in the literature [21].

It is very likely that this reaction is limited both by slow diffusion of the ions in the viscous ionic liquid media and by the rearrangement of the first coordination

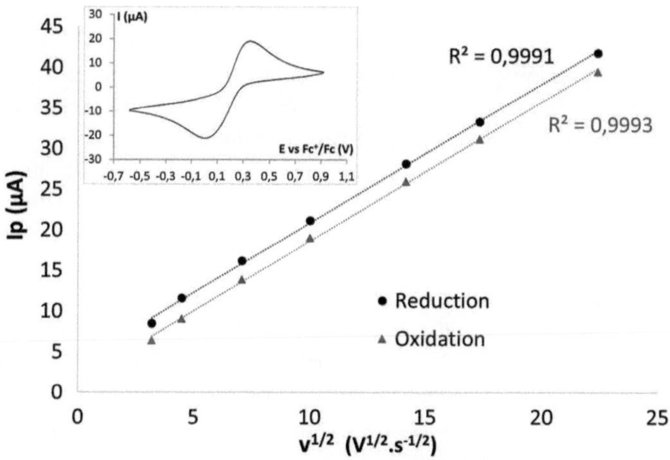

Fig. 3 Evolution of peak currents as a function of the square root of the scan rate. Insert: Voltamogram of $Eu(NTf_2)_3$ 0.05 M in $[EMIm][NTf_2]$ on on Pt-disc electrode (2 mm diam.) at 100 mV/s under natural convection regime, $[H_2O]$ = 35 ppm and T = 40 °C

sphere in the course of the reaction. Indeed, the first coordination spheres of the metal ion are obstructed with the bulky ions of the IL. Using the Cottrell equation for chrono-amperometry measurements:

$$j(t) = \frac{nFD^{1/2}C^{*}}{\pi^{1/2}t^{1/2}}$$

where j is the current density in $A.cm^{-2}$, n is the number of exchanged electrons, F the Faraday constant in $C{\cdot}mol^{-1}$, D the diffusion coefficient of the considered species in $cm^2{\cdot}s^{-1}$, C* the concentration in the bulk in $mol{\cdot}cm^{-3}$ and t the time, allowed us to determine a diffusion coefficient of $7.8.10^{-12}$ $m^2{\cdot}s^{-1}$ for Eu(III) in [EMIm][NTf_2] which correlates well with results from Yamagata et al. [21]. This value is nearly 120 times lower than the one in an aqueous $NaClO_4$ 1 M electrolyte [22]. Besides, Yamagata et al. [21] also showed that the peak separation is even larger in the more viscous 1-n-butyl-1-methylpyrrolidinium bis(trifluoromethane-sulfonyl)imide.

Knowing the potential for the reduction of Eu(III) to Eu(II), we are going to use spectro-electrochemical techniques in order to understand the unusual stability of Eu(II) in ionic liquid media.

UV-Vis Spectroscopy

Absorption spectra of $Eu(NTf_2)_3$ solutions in [EMIm][NTf_2] at different concentrations were recorded and are shown in Fig. 4a. They display an intense absorption band at 394 nm and a less intense one at 465 nm which are characteristic of Eu(III) [23]. Even though the prepared samples seem transparent to the naked eye, scanning toward UV wavelengths show an increase in the absorption signal consistent with a diffusion phenomenon. However, subtraction of the diffusion signal (c.f. Fig. 4b) enables the interpretation of the absorption spectra. The absorption at 394 nm varied linearly with the theoretical europium concentration in the ionic liquid. Thus, the observed diffusion is not due to an incomplete dissolution of the europium salt. It might be the consequence of a dimixtion phenomenon with residual water which is still under study. The limit of detection of Eu(III) in [EMIm] [NTf_2] with this set up is below 0.005 M.

The spectra recorded during electrolysis are presented on Fig. 5. It can be observed that the intensities of the absorption bands at 394 and 465 nm decrease with the electrolysis time at −0.28 V. In particular, it is noteworthy that after 15 h of electrolysis, Eu(III) can no longer be detected on the spectrum. Concentration of Eu(III) in the solution at this time is thus below 0.005 M, which means that at least 90% of the europium was electrolyzed.

When excited at 394 nm, europium is known to show intense emission bands at 592 and 615 nm [24]. A similar set up was developed for an in situ monitoring of

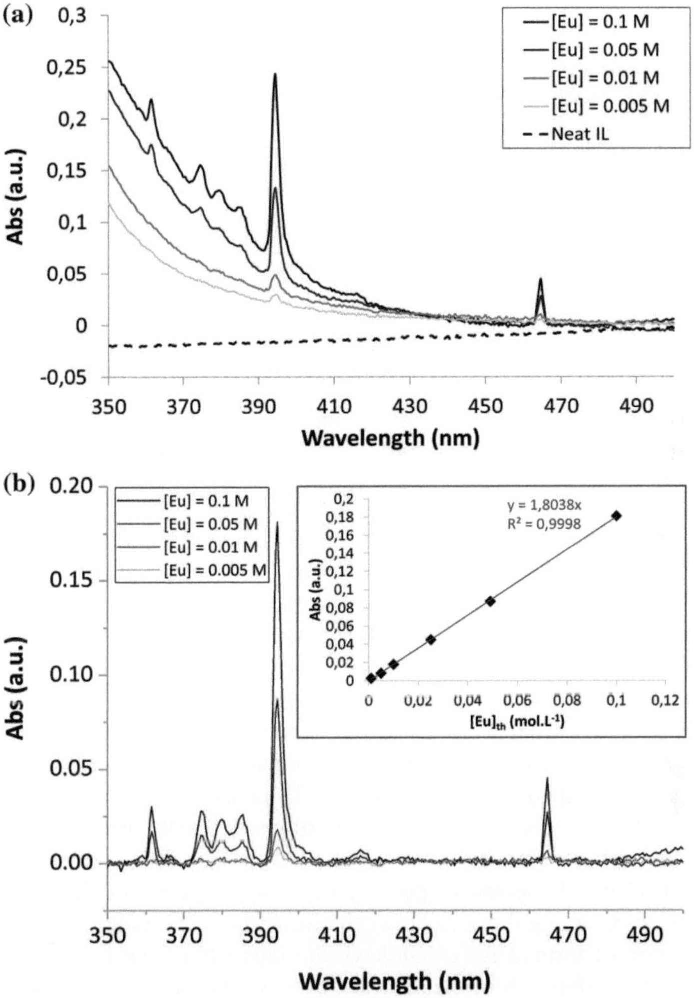

Fig. 4 Absorption spectra of $Eu(NTf_2)_3$ solutions in $[EMIm][NTf_2]$ at different concentrations before and after subtraction of the diffusion background. Insert on **b**: absorption at 394 nm as a function of the theoretical concentration of Eu in the ionic liquid

the electrolysis by fluorescence spectroscopy but unfortunately the coupling with the electrochemical cell did not give exploitable signals.

XANES Study of $Eu(NTf_2)_3$ in $[EMIm][NTf_2]$

Prior to electrolysis, the absorption spectrum of the $Eu(NTf_2)_3$ solution was recorded. The spectrum of the initial state is shown on Fig. 6. As expected, the L_3

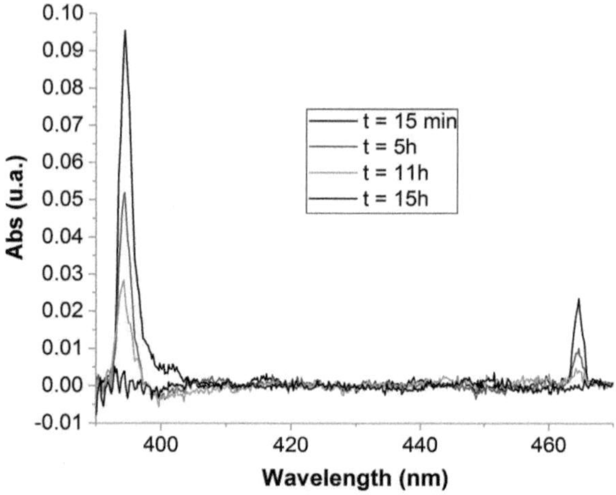

Fig. 5 Aborption spectra of an $Eu(NTf_2)_3$ 0.05 M solution in $[EMIm][NTf_2]$ recorded during constant-potential electrolysis at −0.28 V. The diffusion background was subtracted

absorption edge of Eu(III) can be seen at 6982.5 eV [15] followed by the first EXAFS oscillations. Electrolysis is then started with a set potential of −0.28 V vs Fc^+/Fc. It is to be noted here that due to the solution resistivity and to the potentiostat compliance, this potential is not reached immediately. The applied potential is in fact increased progressively from +0.14 to −0.28 V in 3 h. As can be seen in Fig. 6, another peak corresponding to the L_3 edge of Eu(II) appears at 6975 eV [15]. However, when the beam is let on the same position, the intensity of the Eu(II) absorption peak then decreases. This is an indication of photo-reoxidation of the Eu(II) previously formed during electrolysis. Thus, as shown on Fig. 6, by varying the beam position on the cell, higher proportions of Eu(II) in the ionic liquid are observed. The different positions are illustrated on Fig. 2. The porous voluminal carbon foam electrode seems to prevent rapid re-oxidation in the bulk solution. There seems to be an equilibrium between electrochemical reduction of Eu(III) and photo-oxidation of formed Eu(II). Indeed, the intensity of the Eu(II) peak did not increase anymore after 8 h of electrolysis, even when moving the cell.

Based on previous works [11, 15], a fitting method was developed in order to evaluate the Eu(III)/Eu(II) ratio. The fit consists in a linear combination of the Eu(III) reference spectrum, an arctangent function to fit the absorption threshold and a lorentzian function centred on 6975 eV to fit the Eu(II) contribution. This method is illustrated on Fig. 7. The proportion of Eu(III) is the x value that most closely fits the experimental curve.

This calculated proportion of the Eu(III) spectrum is then subtracted to the experimental data in order to isolate the XANES and EXAFS spectra of Eu(II). This work is carried out on all spectra containing at least 10% of Eu(II). A weighted average (Eu(II) proportions as weighting coefficients) of the resulting signals is calculated. This procedure allows to successfully extract the Eu(II) XANES spectra (c.f. Fig. 8). The absorption peak is observed at 6975 eV in agreement with Antonio et al. [15]. The maximum intensity ratio of XANES signal Eu(II)/Eu(III) is

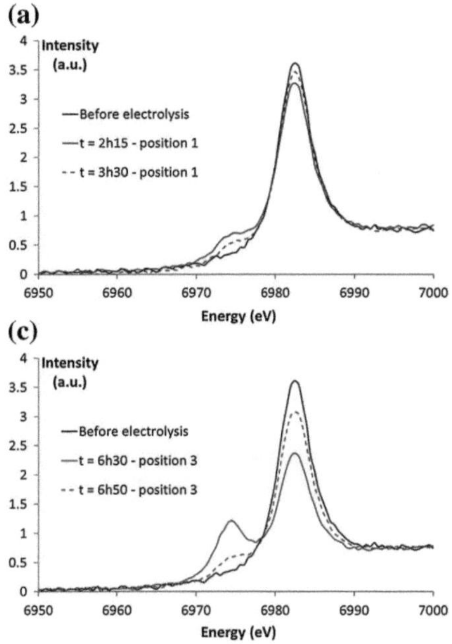

Fig. 6 Normalized Eu L_3 edge XANES spectra of $(Eu(NTf_2)_3$ 0,05 M in $[EMIm][NTf_2]$ before electrolysis (black solid line) and after different times of electrolysis (grey solid and dotted lines). Each figure corresponds to one of the beam positions indicated on Fig. 2

Fig. 7 Fitting method used to determine the ratio Eu(III)/Eu(II)

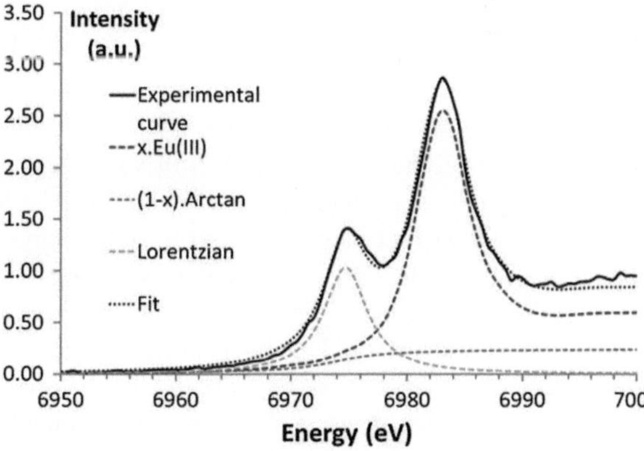

1.01 whereas in [15], authors report a ratio of 0.73 in sulfate media. This change in the intensity ratio indicates a very different geometry for Eu in IL system compared to sulfate media. This is in agreement with the authors' conclusion stating different Eu(II)/Eu(III) transition probabilities when varying the studied media [15].

The obtained Eu(II) spectrum is still noisy and a small contribution of Eu(III) seems to be observed at 6982.5 eV. However, the first EXAFS oscillation for Eu(II)

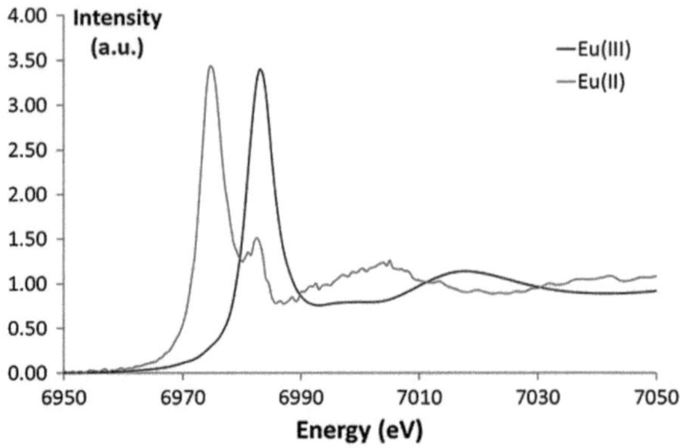

Fig. 8 Normalized XANES spectra of Eu(III) and Eu(II) in [EMIm][NTf_2]

is clearly shifted towards lower energies when compared with the one for Eu(III). This indicates elongated distances in the first Eu coordination sphere. This result is expected since the charge density on the metal ion is lower for Eu(II).

As explained previously, exhaustive electrolysis of Eu(III) could not be achieved and the maximum proportion of Eu(II) in solution determined by this method is 40% (Fig. 6, graph d, after 7.5 h, position 4). However, we believe that modifying the set up to include magnetic stirring and performing electrolysis prior to XAS-measurement could lead to higher proportions of Eu(II). An option would be to use an unfocused beam [25] to reduce the photo-oxidation effect.

Conclusions

Thanks to transient electrochemistry, Eu(II) was shown to be electrochemically stable in the ionic liquid [EMIm][NTf_2]. A formal potential of $E_0^{'} = 0.18\,\text{V}\ vs\ Fc^+/Fc$ was determined for this couple on a platinum electrode. The feasibility of a quasi-exhaustive electrolysis in a small-volume electrochemical cell was demonstrated by in situ UVS monitoring of the solution electrolysis at −0.28 V. XANES spectra of the solution before electrolysis were recorded and showed the characteristic L_3 absorption edge of Eu at 6982.5 eV. After 2 h of reductive electrolysis, the presence of Eu(II) was detected thanks to the apparition of a shifted absorption edge at 6975 eV. Unfortunately, photo-oxidation of Eu(II) to Eu(III) under X-ray irradiation did not allow us to detect more than 40% of Eu(II) in the cell. However, the feasibility of this kind of measurements in ionic liquid media is clearly demonstrated and ingenious modifications on the set up and protocol could lead to higher proportions of Eu(II) species and thus better resolved EXAFS spectra. Indeed, future EXAFS measurements would provide additional information on the electrolysis mechanism.

References

1. Moutiers G, Billard I (2005) Les liquides ioniques: des solvants pour l'industrie. Tech, Ing
2. Zhang S, Wang J, Lu X, Zhou Q (eds) (2014) Structures and interactions of ionic liquids. Springer, Verlag, Berlin, Heidelberg
3. Chavan SN, Tiwari A, Nagaiah TC, Mandal D (2016) Ether and siloxane functionalized ionic liquids and their mixtures as electrolyte for lithium-ion batteries. Phys Chem Chem Phys 18:16116–16126. https://doi.org/10.1039/C6CP01519K
4. Nockemann P, Thijs B, Parac-Vogt TN et al (2008) Carboxyl-functionalized task-specific ionic liquids for solubilizing metal oxides. Inorg Chem 47:9987–9999. https://doi.org/10.1021/ic801213z
5. Wang G, Shen S, Fang S et al (2016) New ether-functionalized pyrazolium ionic liquid electrolytes based on the bis(fluorosulfonyl)imide anion for lithium-ion batteries. RSC Adv 6:71489–71495. https://doi.org/10.1039/C6RA16348C
6. Bonhote P, Dias A-P, Papageorgiou N et al (1996) Hydrophobic, highly conductive ambient-temperature molten salts. Inorg Chem 35:1168–1178
7. Billard I, Moutiers G, Labet A et al (2003) Stability of divalent europium in an ionic liquid: spectroscopic investigations in 1-Methyl-3-butylimidazolium hexafluorophosphate. Inorg Chem 42:1726–1733. https://doi.org/10.1021/ic0260318
8. Binnemans K, Jones PT, Van Acker K et al (2013) Rare-earth economics: the balance problem. JOM 65:846–848. https://doi.org/10.1007/s11837-013-0639-7
9. Chen J (2016) Application of ionic liquids on rare earth green separation and utilization. Springer, Berlin, Heidelberg
10. Shearer CK, Papike JJ, Burger PV et al (2011) Direct determination of europium valence state by XANES in extraterrestrial merrillite: implications for REE crystal chemistry and martian magmatism. Am Mineral 96:1418–1421. https://doi.org/10.2138/am.2011.3860
11. Cicconi MR, Giuli G, Paris E et al (2012) Europium oxidation state and local structure in silicate glasses. Am Mineral 97:918–929. https://doi.org/10.2138/am.2012.4041
12. LLorens I, Solari PL, Sitaud B et al (2014) X-ray absorption spectroscopy investigations on radioactive matter using MARS beamline at SOLEIL synchrotron. Radiochim Acta. https://doi.org/10.1515/ract-2013-2241
13. Nockemann P, Thijs B, Lunstroot K et al (2009) Speciation of rare-earth metal complexes in ionic liquids: a multiple-technique approach. Chem Eur J 15:1449–1461. https://doi.org/10.1002/chem.200801418
14. Poineau F, Fattahi M, Auwer CD et al (2006) Speciation of technetium and rhenium complexes by in situ XAS-electrochemistry. Radiochim Acta. https://doi.org/10.1524/ract.2006.94.5.283
15. Antonio MR, Soderholm L, Song I (1997) Design of spectroelectrochemical cell for in situ X-ray absorption fine structure measurements of bulk solution species. J Appl Electrochem 27:784–792
16. Stumpf S, Billard I, Gaillard C et al (2008) TRLFS and EXAFS investigations of lanthanide and actinide complexation by triflate and perchlorate in an ionic liquid. Radiochim Acta 96:1–10. https://doi.org/10.1524/ract.2008.1461
17. Zhang S, Sun N, He X et al (2006) Physical properties of ionic liquids: database and evaluation. J Phys Chem Ref Data 35:1475. https://doi.org/10.1063/1.2204959
18. Randström S, Montanino M, Appetecchi GB et al (2008) Effect of water and oxygen traces on the cathodic stability of N-alkyl-N-methylpyrrolidinium bis(trifluoromethanesulfonyl)imide. Electrochim Acta 53:6397–6401. https://doi.org/10.1016/j.electacta.2008.04.058
19. Snook GA, Best AS, Pandolfo AG, Hollenkamp AF (2006) Evaluation of a Ag| Ag + reference electrode for use in room temperature ionic liquids. Electrochem Commun 8:1405–1411. https://doi.org/10.1016/j.elecom.2006.07.004
20. Bard AJ, Faulkner LR (2001) Electrochemical methods—fundamentals and applications, 2nd edn. Wiley, New York

21. Yamagata M, Katayama Y, Miura T (2006) Electrochemical behavior of samarium, europium, and ytterbium in hydrophobic room-temperature molten salt systems. J Electrochem Soc 153: E5. https://doi.org/10.1149/1.2136088
22. Rabockai T (1977) Influence of water on the diffusion coefficient of Eu in aqueous formamide solutions.pdf. Electrochim Acta 22:489–490
23. Binnemans K (2015) Interpretation of europium(III) spectra. Coord Chem Rev 295:1–45. https://doi.org/10.1016/j.ccr.2015.02.015
24. Gupta R, Gupta SK, Gamre JS et al (2015) Understanding the dynamics of Eu^{3+} ions in room-temperature ionic liquids—electrochemical and time-resolved fluorescence spectroscopy studies: the dynamics of Eu^{3+} ions in room-temperature ionic liquids. Eur J Inorg Chem 2015:104–111. https://doi.org/10.1002/ejic.201402713
25. Fieser ME, Ferrier MG, Su J et al (2017) Evaluating the electronic structure of formal Ln^{II} ions in $Ln^{II}(C_5\ H_4\ SiMe_3)_3^{1-}$ using XANES spectroscopy and DFT calculations. Chem Sci 8:6076–6091. https://doi.org/10.1039/C7SC00825B

The Electrolytic Production of Rare Earths from Their Oxides

James C. Withers

Abstract There is no production of rare earth metals in the U.S. in spite that the U.S. is the largest consumer of rare earths in some form that includes consumer and Department of Defense (DoD) devices. The one U.S. mine has been purchased by the Chinese that even enhances their monopoly on rare earths. A research effort has demonstrated rare earth oxides can be extracted from bastnaesite ore as well as coal ash at efficiencies in the upper 90s percent. The rare earth oxide can be carbothermically treated to produce an oxycarbide which is highly electrically conductive. The rare earth oxycarbide can be used as an anode in anhydrous fused salts electrolysis to produce a highly purified rare earth powdered metal. It is also possible to electrofractionate the individual metals which can be combined with co-deposition of iron and boron to produce the highly magnetic alloy $Nd_2Fe_{14}B$.

Keywords Rare earth · Extraction · Bastnaesite ore · Coal ash
Electrolysis · Electrofractionate · Froth flotation

Introduction

Rare earths are utilized in some form in most high technology consumer products and rare earth use have become enabling throughout defense systems. The U.S. is the largest consumer of rare earth applications, yet no rare earths are produced in the U.S. Until the latter quarter of the twentieth century the U.S. was the largest producer of rare earths then China became the low cost supplier of rare earths. With environmental concerns in the U.S. of mining and extracting rare earths, all productions of rare earths ceased in the U.S. By the turn of the 21st century China had a monopoly of 97% of the world's supply of rare earths. By the middle of the first decade of the 21st century the price of rare earths had escalated by a factor of 10 in some cases as well as allocations in supply. By the middle of the second decade of

J. C. Withers (✉)
ATS-MER, LLC, Tucson, USA
e-mail: jwithers@atsmer.com

H. Kim et al. (eds.), *Rare Metal Technology 2018*, The Minerals, Metals & Materials Series, https://doi.org/10.1007/978-3-319-72350-1_10

the 21st century the prices of rare earths had settled to 2–5 times the original price with China maintaining a 97% monopoly of supply and the only rare earth mine in the U.S. had closed and was subsequently purchased by the Chinese-led Consortium, MP Mine Operations, earlier this year [1].

During the early 2000s ATS-MER developed processing to carbothermically reduce many metal oxides to a corresponding oxycarbide which has relatively good to high electrical conductivity. Initially, TiO_2 was the focus that produced Ti_2OC that can be used as an anode to electrowin titanium metal as well as the Ti_2OC chlorinates to $TiCl_4$ in the 280–400 °C range. The titanium electrowinning process has been scaled to production demonstration. More recently, ATS-MER demonstrated rare earth oxides (e.g. Nd_2O_3) could be carbothermically reduced to the oxycarbide (e.g. $Nd_2O_2C_2$) as illustrated in the XRD pattern shown in Fig. 1 which can be used as an anode to electrowin the rare earths (e.g. Nd). The rare earth oxycarbide (e.g. $Nd_2O_2C_2$) chlorinates in the same temperature range as for the titanium oxycarbide (e.g. $NdCl_3$).

Concurrently, ATS-MER in coordination with the University of Arizona (U of A) demonstrated most rare earth ores as well as coal fly ash and particulate bastnaesite could be froth flotation processed with select additives that extracted up to 97% of the rare earth at virtually ambient temperatures as compared to typically around 70% extraction at near boiling point of the water suspension [2].

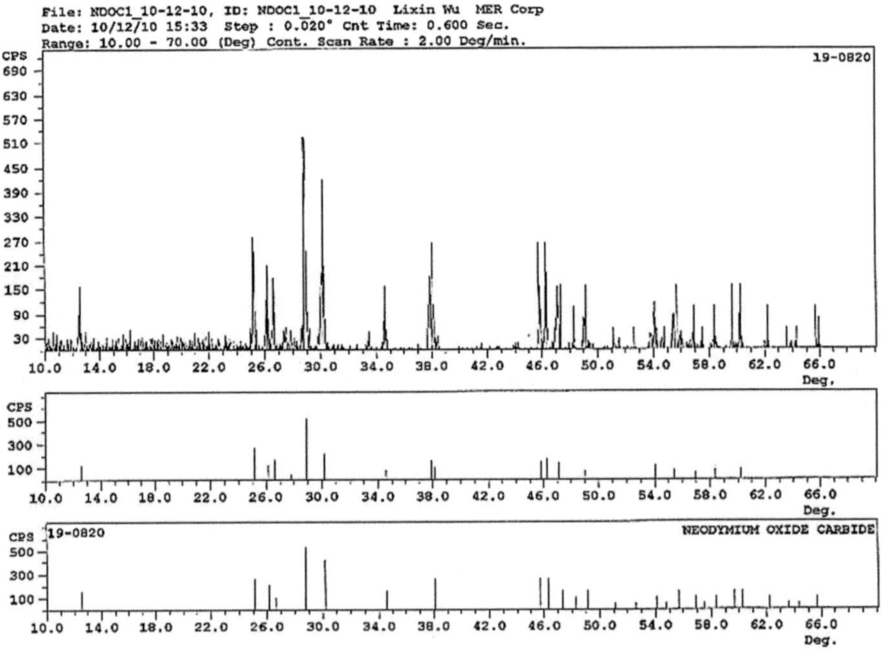

Fig. 1 Conformation Nd_2O_3 is carbothermically reduced to Nd_2O_{22}, XRD performed at University of Arizona

Experimental

ATS-MER obtained Mountain Pass ore samples and some select coal fly ash samples. Both have been subjected to the advanced froth flotation processing to demonstrate the high efficiency extraction of the rare earths. The rare earth extracts were subjected to carbothermic reduction to produce the respective oxycarbide. The oxycarbide was utilized as an anode in atmospherically controlled fused salt electrolytic cells such as shown in Fig. 2. The fused salt electrolyte can be any combination of the alkali and/or alkaline earth chlorides. Typically, ATS-MER used the eutectic of KCl–LiCl and operated in the 500–600 °C range. In the cathode current density range of 0.5–1.5 amps/cm^2 range a powder is produced at the cathode and primarily CO at the anode with traces of CO_2.

As an alternate to the use of the oxycarbide, it was chlorinated to produce the rare earth chloride. Since permanent magnets are a primary use of rare earths and particularly neodymium with or without dysprosium, a goal was to produce the composition that is most often used in permanent magnets, $Nd_2Fe_{14}B$. In this case neodymium oxycarbide ($Nd_2O_2C_2$) was chlorinated at the low temperature of 300–600 °C to produce $NdCl_3$ which was vacuum sublimed and collected without exposure to the atmosphere. The $NdCl_3$ was added to the fused salt electrolyte KCl–LiCl. Iron chloride ($FeCl_2$) was also added to the KCl–LiCl–$NdCl_3$ electrolyte. Boron trichloride gas (BCl_3) was then meter bubbled into the electrolyte in an electrolytic cell shown in Fig. 3. Adjusting the cathode current density to 0.25–1.5 amps/cm^2, the $FeCl_2$ concentration and the BCl_3 gas flow, it was possible

Fig. 2 Bench scale pilot cell that has electrolytically produced titanium from titanium oxycarbide

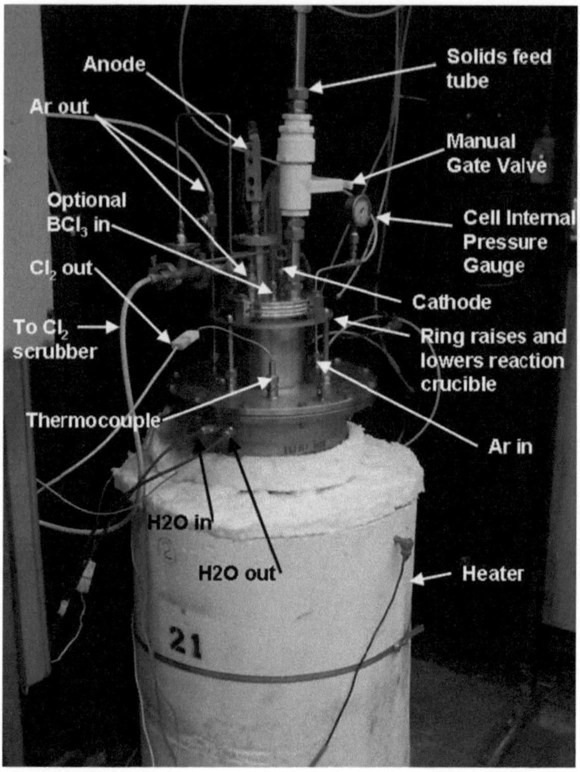

Fig. 3 Cell to electrolytically produce $Nd_2Fe_{14}B$ powder

to electrolytically produce the stoichiometry $Nd_2Fe_{14}B$ as a powder as shown in Fig. 4 along with the XRD analysis shown in Table 1. The $Nd_2Fe_{14}B$ was separated from the fused chloride electrolyte by vacuum sublimation to less than 10 ppm residual chlorine. Without exposure to the atmosphere, the $Nd_2Fe_{14}B$ could be cold pressed and sintered or hot pressed to high density for subsequent use as permanent magnets. It was also possible to utilize additive manufacturing to consolidate the $Nd_2Fe_{14}B$ powders to a net shape for magnets.

An alternate electrolytic process to produce $Nd_2Fe_{14}B$ powder was to use iron-iron boron anodes in the proper ratios in the electrolyte of KCl–LiCl–$NdCl_3$. Anodes in a ratio of 7:1 Fe–FeB commercial powder were cold isopressed at high pressures that produced a compact that could be used as an anode containing iron and boron. That Fe–FeB ration 7:1 anode in the KCl–LiCl–$NdCl_3$ electrolyte also produced $Nd_2Fe_{14}B$ powder which could be utilized by pressing or additive manufacturing to produce material for permanent magnets.

With the advanced froth flotation process to efficiently extract the rare earths from the Mountain Pass ore and carbothermic processed, it is possible to electrowin all the mixed rare earths as powder at the cathode. The mixed rare earth powder is then isopressed at $\sim \geq 20{,}000$ psi to produce a consolidate that can then be used as an anode to electrofractionate in a halide fused salt electrolyte. Using very

Fig. 4 Example of electrolytically produced $Nd_2Fe_{14}B$ powder

Table 1 EDX analysis of electrolytically produced Nd–Fe-B powder compared to Molycorp and Chinese supplied Nd–Fe-B powder

Typical alloy composition[a]			
wt%	ATS-MER electrolytically produced Nd–Fe-B powder	Molycorp supplied Nd–Fe-B powder	Chinese Commercial Nd–Fe-B powder
Nd	26.3	29.7	17.1
Pr	–	–	2.9
Zr	–	–	4.0
Fe	68.3	65.9	69.2
O	1.5–3.5	4.0–6.5	6.7
B[b]	1.0	1.0	0.2

[a]EDS analysis on ATS-MER's SEM
[b]Curcumin method using UV-vis absorbance spectroscopy

precise voltage control to the third decimal place, it is possible to electrorefine each separate rare earth. With up to three electrorefining steps it is possible to produce 99.9+% fractionation of each rare earth.

References

1. Topf, Andrew, "Mountain Pass sells for $20.5 million", Mining.com, 16 June 2017, http://www.mining.com/mountain-pass-sells-20-5-million/
2. ATS-MER, LLC Improved Flotation Separation of Rare Earth Ore, Contract N00014-13-P-1010, Tucson, AZ

Commercial Processes for the Extraction of Platinum Group Metals (PGMs)

Rekha Panda, Manis Kumar Jha and D. D. Pathak

Abstract Platinum Group Metals (PGMs) play a significant role in the manufacturing of catalysts, super alloys, electronics, space materials, biomedical equipments, jewellery, etc. due to their excellent electrical and thermal conductivity as well as chemical resistivity. The rising demand of PGMs in industrial applications and their limited natural resources have laid emphasis on the development of feasible and eco-friendly processes for the extraction of these metals from different sources to meet their future requirements. Present review reports commercial processes based on pyro-/hydro- and hybrid techniques to recover PGMs from various resources. The salient findings on different processes used for recovery of PGMs have been reviewed with respect to various methodologies and objectives.

Keywords Platinum group metals · Primary source · Secondary source
Spent catalysts · Waste recycling

Introduction

Platinum group metals (PGMs) comprises of six noble metals namely Ruthenium (Ru), Rhodium (Rh), Palladium (Pd), Osmium (Os), Iridium (Ir), Platinum (Pt) which are found together in the d-block of periodic table. These transition metals possess similar physical and chemical properties. The unique properties of PGMs such as catalytic activity, chemical inertness, resistance towards corrosion, thermal as well as electrical stability make them a vital component of many industrial applications, thus, they are also called 'Vitamin of modern industry' [1–4].

R. Panda (✉) · M. K. Jha
Metal Extraction and Recycling Division, CSIR-National Metallurgical Laboratory, Jamshedpur 831007, India
e-mail: rekhapanda1608@gmail.com

R. Panda · D. D. Pathak
Department of Applied Chemistry, Indian Institute of Technology (ISM), Dhanbad 826004, India

H. Kim et al. (eds.), *Rare Metal Technology 2018*, The Minerals, Metals & Materials Series, https://doi.org/10.1007/978-3-319-72350-1_11

Recent technological modernization involving advanced chemistry has commercially accelerated the use of PGMs in the field of vehicle and equipment construction, chemical industry, oil refining, medical practices, jewellery making, etc. [5]. Pt and Pd are of major commercial significance followed by Rh, Ir and Ru whereas Os has rare viable applications [6, 7]. The global demand of PGMs (Pt, Pd and Rh) is over 590 tons while their natural resources are only 66,000 tons all over the world. South Africa is the leading producer of PGMs in the world followed by Russia, Canada, Zimbabwe, USA and Colombia. Extensive deposits of PGMs are located in the norite belt of the Bushveld Igneous Complex covering the Transvaal Basin in South Africa, the Stillwater Complex in Montana, United States, the Thunder Bay District of Ontario, Canada, and the Norilsk Complex in Russia. PGMs are also found associated with base metal (Cu, Ni) sulfide minerals where their content is almost 2–10 g/t [8]. Nowadays, reefs like Merensky, Upper Group Two (UG-2) and Plat reefs are also mined due to presence of significant quantity of PGMs in them [3, 9]. They are also recovered as by-products depending on their concentration in the ore [2, 10].

It has been observed that high value of PGMs coupled with their increasing demand has fuelled its processing from low-grade resources using elaborated techniques [3]. Despite expensive multi-step processes, different innovative methods are continuously being explored to extract PGMs from primary resources. But the depletion of high grade PGMs resources due to continuous mining, has laid emphasis on their production of PGMs from secondary resources viz. automobile catalysts, e-waste, industrial waste (solid/liquid), etc. Hence, in order to explore improved possibilities for the recovery of PGMs, an attempt has been made to provide a general overview on prevailing commercial technologies. The present paper gives an overview of the commercial processes developed for the recovery of PGMs from various resources using pyro-/hydro- or hybrid techniques. The paper will be helpful for researchers in future to develop new process flow-sheet for extraction of PGMs keeping in view the drawbacks of the existing commercial process.

Processing of Primary Resources to Recover PGMs

Although extensive deposits of PGMs are available but deposits for their economical extraction are inadequate [11]. Primary ores of PGMs are broadly divided into four types: (i) Stratiform deposits ($\sim$10–1000 MT) containing 3–10 g/t PGMs; (ii) Norite intrusions ($\sim$10–1000 MT) having 1–3 g/t PGMs (iii) Ni-Cu bearing sills ($\sim$10–1000 MT) with 2–15 g/t PGMs and (iv) Placer deposits containing coarse PGMs (mainly Pt). PGMs ores are mined through conventional underground or open cut techniques followed by grinding. Further, gravity-based separation and flotation is generally used to produce a PGM-rich concentrates [12].

PGMs from high grade chromite ores (containing 200–2000 g/t PGM along with 0.4–2.8% Cr_2O_3) are conventionally recuperated by matte-smelting-refining process as shown in Fig. 1 [3, 11]. The chromite rich ores are generally processed by conventional flotation in a mill-float-mill-float (MF2) open circuit prior to recovery of PGMs, in order to prevent the accumulation of chromite fines generated during over grinding of ores. Initially, the ore is ground by crushing and ball milling in several stages or by SAG (semi-autogenous grinding) and then followed by smelting. The matte obtained undergoes hydrometallurgical treatment for removal of impurities such as Fe, Co, Ni and Cu leaving 10–50% of PGMs in the slime [13]. Emission of SO_2, accumulation of highly refractory chromite spinel layers in the furnace and environmental pollution as a result of high temperature smelting are some of the major limitation of the conventional smelting and converting processes [14–16].

Several researchers have also reported hydrometallurgical or the combination pyro-hydrometallurgical processes as a pre-treatment step for the recovery of PGM from ores/concentrates [17–22]. Hydrometallurgical processes consisting of leaching operation for enrichment PGMs from base metal (Cu, Ni, Co, Fe) sulfide minerals have also been employed. During this process, the base metals are dissolved leaving behind the PGM concentrate for further refining. The same process has also been used in Ni and Cu-Ni refineries for the enrichment of PGMs [23]. Apart from high grade ores, hydrometallurgical processes have also been developed for the commercial extraction of PGMs from low grade refractory sulfide ores and concentrate as presented in Table 1.

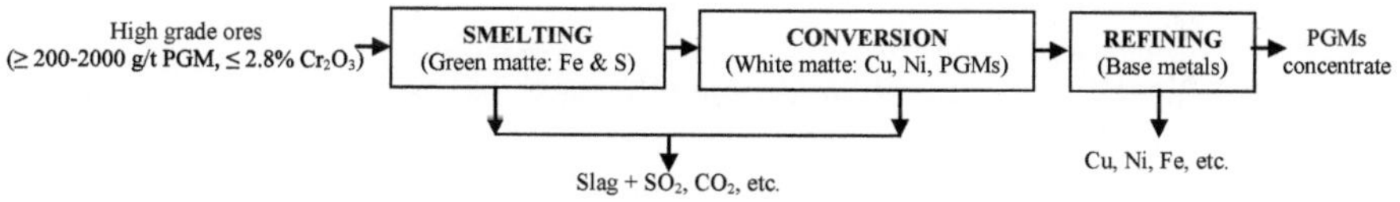

Fig. 1 Conventional matte-smelting-refining technology to get PGM concentrates [3]

Table 1 Technologies for commercial extraction of PGMs from low grade ores

Processes	Medium	Scale	References
Anglo American Corporation/University of British Columbia Process	Sulfate	Pilot	[24, 25]
Albion process	Sulfate	Commercial	[26–32]
Galvanox™ process	Sulfate	Pilot	[32–34]
Total pressure oxidation process	Sulfate	Commercial	[24, 25, 35]
Hydro copper process	Chloride	Commercial	[36–39]
Kell process	Sulfate + Chloride	Pilot	[40]
Nitrogen species catalyzed process	Sulfate + Nitrogen	Commercial	[25, 35]

Processing of Secondary Resources to Recover PGMs

PGMs are vital component of several products like mobile phones, industrial catalysts, ceramic glazes, hard disks, aircraft turbines, etc. Catalysts and electronic wastes are two imperative secondary resources containing significant amount of PGMs due to their remarkable resistance to high temperature corrosion and oxidation. Among these, PGMs are widely used as a catalyst in various chemical reactions like reduction, reformation, hydrogenation, isomerisation, conversion, etc. [41–53]. Automobile industry is the largest consumer of PGMs. Almost 34% Pt, 55% Pd and 95% Rh out of their total demand is used for the manufacturing of catalytic filters-neutralizers of exhaust gases in automobile industries [54–57]. Varying amount of Pt, Pd and Rh are used in auto catalysts depending upon the type of vehicle, manufacturer, country, etc. which helps in regulating the harmful emission of CO, NO_x and hydrocarbons. During the catalytic conversion, Pt helps in converting hydrocarbons and CO to H_2O and CO_2, while Rh is highly efficient in reducing NO_x to N_2 whereas Pd alone can handle all three pollutants, but less efficiently compared to Pt and Rh [58, 59]. Thus, spent automobile catalysts are important supplementary source for the recycling and recovery of PGMs. Processing 2 mg of spent automobile catalysts to recover PGMs can prevent the mining of 150 kg of their ores [58]. The rise in demand of PGMs in automobile industries and strict environmental regulations make their recycling indispensable. The spent automobile catalyst contains an average of ~4 g PGMs which is quite high in comparison to primary resources of PGMs [58, 60, 61]. Moreover, the high price of Pt, Pd and Rh, makes their recovery from used catalysts profitable. Based on their chemical composition and nature, the recycling of these PGM-bearing catalysts is carried out through hydrometallurgical as well as pyrometallurgical processes. Several corporations and industries like Umicore, Belgium; Hereaus, Germany; BASF/Engelhard, USA; Johnson Matthey, UK; Nippon/Mitsubishi, Japan, etc. have already developed successful commercial processes for the recycling of PGMs from secondary resources [2].

Pyrometallurgical process is usually employed to concentrate the PGMs followed by refining technology to recover them. Pyrometallurgical process including crushing, batching, granulation, smelting, separation, has become a traditional method to recover PGMs from spent catalyst [41, 58]. The spent catalyst is initially mixed with fluxes, collector and reductant during smelting and a PGM-collector alloy is obtained which further undergoes purification [41]. The choice of collector plays an important role during smelting. The selection between collector and PGMs is based on their mutual solubility, melting point and chemical properties. Metals like Cu, Ni, Pb and Fe are generally considered good collectors [1]. PGM-Pb collection is one of the oldest methods [62, 63] that has been used to process secondary resources in Inco, Johnson Matthey, Impala, etc. before 1980s. The process is simple to operate, require low smelting temperature, followed by simple refining process, and needs less investment but the major disadvantages are low recovery of Rh and generation of hazardous lead oxide. PGM-Cu collection is another method for treating the spent auto-exhaust catalyst in an electric arc furnace with addition of fluxes (SiO_2, CaO, etc.), collector ($CuCO_3$ or CuO) and reductant [64, 65]. PGMs are collected at low temperature and weak reduction atmosphere. A semi-industrial process combining pyrometallurgy and electrolytic refining for the recovery of PGMs from spent auto-exhaust catalyst using metal copper collection has been performed by the Institute for Mining and Metallurgy Bor, Serbia as shown in Fig. 2 [66]. The Nippon PGM Co. Japan works on the well-known Rose Process (Fig. 2) which is basically copper collection process. The final concentrates contain almost 30% of PGMs whereas the Cu_xO produced can be reused during the smelting process [1, 58].

The Umicore operated at Hoboken, Belgium is an integrated metal smelter and refinery, which also recovers PGMs along with other metals from auto-catalysts/printed circuit boards/electronic components [67]. PGM-Fe collection process mainly involves the method of plasma arc smelting and mineral phase reconstruction. The plasma arc smelting technology was very popular during 1980s to recover PGMs from spent auto-exhaust catalyst [68–70]. High energy density, high temperatures and flexibility in the plasma gases are the vital advantages of this technology [71]. The Plasma arc smelting process for recovery of PGMs from the spent auto-exhaust catalyst has also been commercialized in Texasgulf, USA and Safina, Czech Republic [41, 70]. On the other hand, the short lifetime of plasma gun accessory restricted its practical industrial application. Based on the findings,

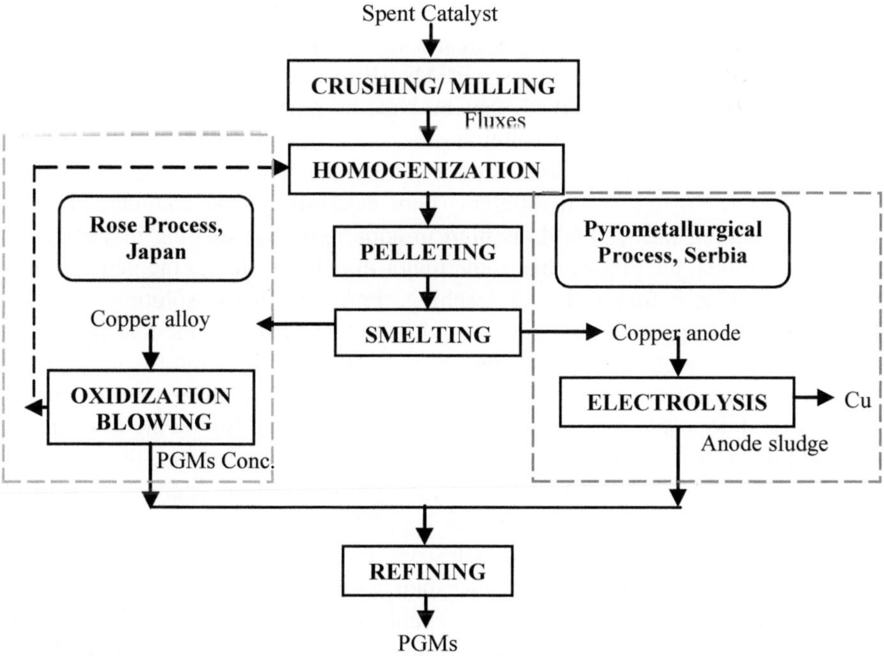

Fig. 2 Industrial process for the recovery of PGMs from spent automobile catalyst [1]

a mineral phase reconstruction process was proposed to recover PGMs from spent auto-exhaust catalyst [72]. The Johnson Matthey process involves smelting of crushed catalyst with flux materials in a crucible containing molten collector metal (Fe or Cu), using a plasma torch [62]. The operation is carried out at temperatures between 1500–1650 °C where the alloy of collector metal is tapped off and ~95% PGMs are recovered by conventional refining methods. From the above studies, it can be concluded that metal smelting collection process is appropriate for processing various secondary materials containing PGMs. The affinity of PGM particles towards the collector metal is an important factor behind the success of this process whereas other factors like fluxes, collector, smelting equipment, operating system, etc. should also be considered. From industrial point of view, PGM-Cu collection technology has wide applications due to high efficiency, low smelting temperature, less pollution and easy industrialization.

Hydrometallurgical processing for the recovery of PGMs mainly involves dissolution using suitable acidic and alkaline solutions in the presence of additives like O_2, I_2, Br_2, Cl_2, H_2O_2, etc. [7, 73–82]. The spent catalyst containing PGMs is pre-treated before hydrometallurgical processing. The PGMs present in the catalysts are encapsulated by specific substances which lead to decrease in their leaching efficiency. Thus, pre-treatment steps such as fine grinding, roasting, reduction, pressure leaching, etc. are necessary prior to leaching of PGMs. Several researchers have reported various pre-treatment methods (oxidization roasting, reduction roasting, pre-leaching, etc.) to destroy organic substances on the surface of spent catalysts or change the supporter forms which hinders the leaching of PGMs present in spent catalysts [41, 58]. After pre-treatment, leaching of the spent catalysts are carried out for maximum dissolution of PGMs. HCl is the most common complexing agent, while HNO_3, Cl_2, or H_2O_2 can be used as oxidant [41, 83]. Aqua regia, commonly used leachant for dissolution precious metals, but not all PGMs can be dissolved with it. Several researchers have studied the use of aqua regia [84–90] to recover Pt, Pd, Rh, etc. from different spent catalysts on the commercial scale. Although leaching rate of PGMs is high in aqua regia but keeping in view environmental aspects, the process has some major drawbacks due to the generation of NO_x, Cl_2 and acid fumes during leaching. From the leach solution obtained, purification and separation of PGMs could be achieved by the method of cementation, solvent extraction, ion exchange, etc. An alternative method is to leach the ceramic material of the catalyst with NaOH or hot H_2SO_4 under pressure where PGMs (Pt, Pd, Rh) are insoluble and remains in the residue. But owing to relatively low yield of PGMs and generation of huge quantities of wastewater, the process is infeasible. The general flow-sheet for extraction of PGMs from spent catalyst using hydrometallurgical technique is shown in Fig. 3 [1].

The chemical and metallurgical industries have used the method of cyanide leaching to recover PGMs [13, 91] due their ability to form stable complexes in alkaline medium. PGMs extraction by cyanidation show poor kinetics at room temperature and atmospheric pressure, thus, effective leaching of the PGMs is

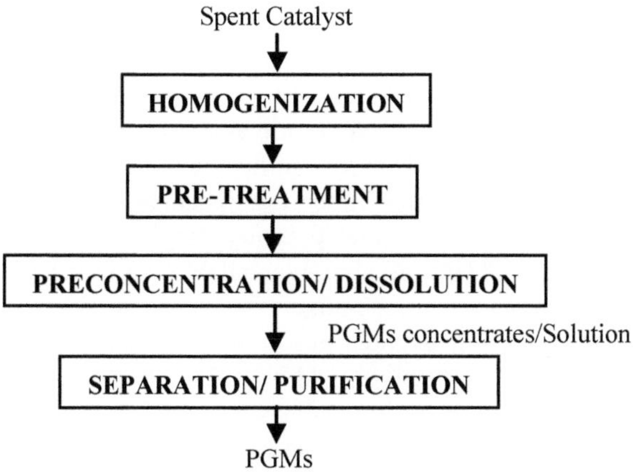

Fig. 3 Hydrometallurgical processing of PGMs [1]

carried out at high temperature and pressure. In cyanide leaching process, the rate of reaction rate is proposed to be controlled by a surface chemical reaction, which is similar to gold cyanidation mechanism [92]. Cyanide leaching of PGMs requires special equipment as well as proper management of the toxic waste generated which may lead to severe environmental problem. Thus, the industrial application of this process is difficult and it is still in exploratory stage. Several processes for the commercially extracting PGMs from secondaries have been successfully developed. Platinum Lake Technology Inc., Canada [93] has successfully developed a hydrometallurgical process for the recovery PGMs (95% Pt and 98% Pd) from spent automotive catalysts. Nippon PGM Co. has reported the production of PGMs from different resources on commercial-scale [94]. Heraeus, Germany has reported the recovery of PGMs from spent materials using hydrometallurgical processes consisting of leaching in HCl in presence of oxidant followed by selective precipitation and ion exchange [2]. BASF Catalysts LLC, USA also developed a novel process for the recovery of PGMs from membrane electrode assemblies (MEAs) eradicating the release of HF (highly toxic gas) generated during the current combustion recycling process [2]. Studies have also been carried out at the Mining and Materials Processing Institute of Japan to recycle PGMs from the residue of automobile catalyst with high leaching efficiency [7]. Hydrometallurgical route offers a faster rate of metal recovery at low capital costs. In addition, the energy consumption is low compared to pyrometallurgical process, which requires high temperatures to melt the raw material. Moreover, the wastewater generated at the industrial scale could be treated in effluent treatment plant and possibilities could be explored for further recovery of value added products.

Conclusion

Based on the review it can be concluded that the recovery and recycling of PGMs from both primary and secondary resources are essential due to their rising demand in various industrial applications. Due to significant conflict between availability of natural resources of PGMs and their increasing demands, it is necessary to exploit indigenous resources of PGMs. Several pyrometallurgical and hydrometallurgical processes for recovery of PGMs are already available but development of modern and productive technologies to utilize indigenous resources as well as improvement in prevailing technologies, will be helpful to meet the future demand of PGMs in various applications. More emphasis should be laid on the recycling of PGMs from secondary sources (spent automobile catalysts, e-waste, industrial wastes, etc.) in order to economise the natural resources and to minimise the environmental pollution in connection to production of PGMs. Thus, R & D efforts should be made to develop hybrid processes consisting of physical beneficiation/pyro-/hydro-/electro metallurgy for efficient recovery of PGMs from various resources.

Acknowledgements The authors are thankful to the Director, CSIR-National Metallurgical Laboratory, Jamshedpur for the permission to publish this paper. One of the authors Ms. Rekha Panda would like to extend her sincere gratitude to CSIR, New Delhi (Grant: 31/10 (64)/ 2017-EMR-I) for providing Senior Research Fellowship to carry out this research work.

References

1. Dong H, Zhao J, Chen J, Wu Y, Li B (2015) Recovery of platinum group metals from spent catalysts: a review. Int J Miner Process 145:108–113
2. Jha MK, Lee JC, Kim MS, Jeong J, Kim BS, Kumar V (2013) Hydrometallurgical recovery/ recycling of platinum by the leaching of spent catalysts: a review. Hydrometallurgy 133:23–32
3. Mpinga CN, Eksteen JJ, Aldrich C, Dyer L (2015) Direct leach approaches to platinum group metal (PGM) ores and concentrates: a review. Miner Eng 78:93–113
4. Afolabi AS, Nkobane MP, Abdulkareem AS (2012) Development of PGMs and chrome extraction circuit from UG-2 ore. In: Proceedings of the world congress on engineering 3
5. Kononova ON, Melnikov AM, Borisova TV (2012) Simultaneous sorption recovery of platinum and rhodium from sulfate–chloride solutions. Hydrometallurgy 117–118:101–107
6. Fact Sheet (2012) Platinum group metals. Polinares working paper 35 March 2012
7. Harjanto S, Cao YC, Shibayama A, Naitoh I, Nanami T, Kasahara K, Okumura Y, Liu KJ, Fujita T (2006) Leaching of Pt, Pd and Rh from automotive catalyst residue in various chloride based solutions. Mater Trans 47(1):129–135
8. Kyriakakis G (2005) Extraction of gold from platinum group metal (PGM) ores. Dev Miner Process 15:897–917
9. Cramer LA (2001) The extractive metallurgy of South Africa's platinum ores. J S Afr Inst Min Metall
10. Liu SJ (2013) Metallurgy of platinum group metals. Central South University Press, Changsha

11. Cole S, Joe Ferron C (2002) A review of the beneficiation and extractive metallurgy of the platinum group elements, highlighting recent process innovations. SGS minerals services technical paper-03
12. Vermaak CF (1995) The platinum-group metals: a global perspective. Mintek, Randburg, South Africa
13. Chen J, Huang K (2006) A new technique for extraction of platinum group metals by pressure cyanidation. Hydrometallurgy 82:164–171
14. Eksteen JJ (2011) A mechanistic model to predict matte temperatures during the smelting of UG2-rich blends of platinum group metal concentrate. Miner Eng 24(7):676–687
15. Eksteen JJ, Bezuidenhout GA, Van Beek B (2011) Cracking a hard nut: an overview of Lonmin's operations directed at smelting of UG2-rich concentrate blends. J South Afr Inst Min Metall 111(10):681–690
16. Ritchie S, Eksteen JJ (2011) Investigating the effect of slag bath conditions on the existence of "mushy" zones in PGM smelting furnaces using computational fluid dynamics. Miner Eng 24(7):66–675
17. Li Y, Kawashima N, Li J, Chandra AP, Gerson AR (2013) A review of the structure, and fundamental mechanisms and kinetics of the leaching of chalcopyrite. Adv Colloid Interface Sci 197–198:1–32
18. Li J, Miller JD (2006) A review of gold leaching in acid thiourea solutions. Miner Process Extr Metall Rev 27:177–214
19. Wang S (2005) Copper leaching from chalcopyrite concentrates. JOM
20. Xiao Z, Laplante AR (2004) Characterizing and recovering the platinum group Minerals-a review. Miner Eng 17:961–979
21. Fleming CA (2002) Platsol™ process provides a viable alternative to smelting. SGS minerals services technical paper 2002–01
22. Maurice D, Hawk JA (1999) Ferric chloride leaching of a mechanically activated pentlandite-chalcopyrite concentrate. Hydrometallurgy 52:289–312
23. Kerfoot DGE (1986) Review of hydrometallurgical nickel refining operations: Responses to questionnaire. In: Ozberk E, Marcusson SW (eds), Nickel metallurgy, Volume I, extraction and refining of nickel, 25th annual conference of metallurgists, CIM, March, p 426441
24. Dreisinger D (2006) Copper leaching from primary sulfides: options for biological and chemical extraction of copper. Hydrometallurgy 83:10–20
25. Gok O, Anderson CG (2013) Dissolution of low-grade chalcopyrite concentrate in acidified nitrite electrolyte. Hydrometallurgy 134–135:40–46
26. Nazari G, Dixon DG, Dreisinger DB (2012) The role of galena associated with silver-enhanced pyrite in the kinetics of chalcopyrite leaching during the Galvanox™ process. Hydrometallurgy 113–114:177–184
27. Nazari G, Dixon DG, Dreisinger DB (2012) The role of silver-enhanced pyrite in enhancing the electrical conductivity of sulfur product layer during chalcopyrite leaching in the Galvanox™ process. Hydrometallurgy 111–112:35–45
28. Nazari G, Dixon DG, Dreisinger DB (2012) The mechanism of chalcopyrite leaching in the presence of silver-enhanced pyrite in the Galvanox™ process. Hydrometallurgy 113–114:122–130
29. Kowalczuk P, Chmielewski T (2008) Search for a new technology producing copper from chalcopyrite. Scientific papers of the institute of mining, No 122, conferences No 51, 7th PhD studies scientific conference, interdisciplinary topics in mining and geology, Ofic Wyd PWr, Wroclaw, pp 94–100
30. Ellis S, Mines KCG, Kalgoorlie WA (2008) Ultra fine grinding- a practical alternative to oxidative treatment of refractory gold ores
31. Turner DW, Hourn M (2013) Albion process simplicity in leaching: copper applications. Xstrata technology
32. Nazari GT (2012) Enhancing the kinetics of pyrite catalyzed leaching of chalcopyrite. Ph.D thesis, University of British Columbia

33. Dixon DG, Tshilombo AF (2005) Leaching process for copper concentrates. United States Patent Application 2005(0269208):A1
34. Dreisinger D (2012) Hydrometallurgical extraction of base, rare, and precious metals from complex and low grade resources. In: XXVI International Mineral Processing Congress (IMPC) 2012 Proceedings/New Delhi, India/24–28 September 2012
35. Milbourne J, Tomlinson M, Gormely L (2003) Use of hydrometallurgy in direct processing of base metal/PGM concentrates. Hydrometallurgy 625
36. Lundström M, Aromaa J, Forsen O, Hyvarinen O, Barker MH (2005) Leaching of chalcopyrite in cupric chloride solution. Hydrometallurgy 77:89–95
37. Lundström M, Aromaa J, Forsen O (2009) Redox potential characteristics of cupric chloride solutions. Hydrometallurgy 95:285–289
38. Senanayake G (2009) A review of chloride assisted copper sulfide leaching by oxygenated sulfuric acid and mechanistic considerations. Hydrometallurgy 98:21–32
39. Turkmen Y, Kaya E (2009) Acidified ferric chloride leaching of a chalcopyrite concentrate. J Ore Dressing 11(22)
40. Liddell KS, Adams MD (2012) Kell hydrometallurgical process for extraction of platinum group metals and base metals from flotation concentrates. J South Afr Inst Min Metall 112
41. Aberasturi DJD, Pinedo R, Larramendi IRD, Larramendi JIRD, Rojo T (2011) Recovery by hydrometallurgical extraction of the platinum-group metals from car catalytic convertors. Miner Eng 24:505–513
42. Matjie RH, Scurrel MS, Bunt J (2005) The selective dissolution of alumina, cobalt and platinum from a calcined spent catalyst using different lixiviants. Miner Eng 18:801–810
43. Pinheiro AA, Lima TS, Campos PC, Afonso JC (2004) Recovery of platinum from spent catalysts in a fluoride-containing medium. Hydrometallurgy 74:77–84
44. Avci AK, Trimm DL, Erhan Aksoylu A, İlsen Önsan Z (2004) Hydrogen production by steam reforming of n-butane over supported Ni and Pt–Ni catalysts. Appl Catal A 258:235–240
45. Ghosh SK, Mandal M, Kundu S, Nath S, Pal T (2004) Bimetallic Pt–Ni nanoparticles can catalyze reduction of aromatic nitro compounds by sodium borohydride in aqueous solution. Appl Catal A 268:61–66
46. Malyala RV, Rode CV, Arai M, Hegde SG, Chaudhari RV (2000) Activity, selectivity and stability of Ni and bimetallic Ni-Pt supported on zeolite Y catalysts for hydrogenation of acetophenone and its substituted derivatives. Appl Catal A 193:71–86
47. Mahmoud MHH (2003) Leaching platinum-group metals in sulfuric acid/chloride solution. J Miner Met Mater Soc 55(4):37–40
48. Epron F, Gauthard F, Barbier J (2002) Influence of oxidizing and reducing treatments on the metal-metal interactions and on the activity for nitrate reduction of a Pt-Cu bimetallic catalyst. Appl Catal A 237:253–261
49. Saberi MA, Mao RLV, Martin M, Mak AWH (2001) Effect of Zn loading of the Pt-Zn-Hy trifunctional catalysts on the hydroisomerization of n-heptane. Appl Catal A 214:229–236
50. Grzelczuk S, Popowicz M, Berak JM, Schimmelpfennig Z, Gora M, Staszak S (1985) PL 131079 B1. Instytut Chemii Przemyslowej, Poland
51. Jeliyaskova M, Sariev I, Koralska S, Aneva S, Pankova M, Yancheva M (1982) Recovery of platinum from spent catalysts. Chem. Techniek 34(12):651–653
52. Kluksdhal HE (1971) New bimetallic catalyst with high activity for petroleum naphtas reforming. US Patent: 3558477
53. McCoy CS, Munk P (1971) Symposium on catalytic reforming. 68th National meeting of AIChE, Houston, Tex., paper 42a
54. Buslaeva TM (1999) Platinum group metals and their role in contemporary society. Sorosovskiy Obrazovatelny Zhurnal 11:45–4
55. Taran OP, Polyanskaya EM, Ogorodnikova OL, Descorme C, Besson M (2010) Ruthenium carbon-based catalysts for catalytic wet air oxidation of phenol. J Siberian Fed Univ Chem 3(3):245–252
56. Zolotov YA, Varshal GM, Ivanov VM (2003) Analytical Chemistry of Platinum Group Metals. Editorial URSS, Moscow

57. Kim CH, Woo SI, Jeon SH (2000) Recovery of platinum-group metals from recycled automotive catalytic converters by carbochlorination. Ind Eng Chem Res 39:1185–1192
58. Fornalczyk A, Saternus M (2009) Removal of platinum group metals from the used auto catalytic converter. Metalurgija 48(2):133–136
59. Votsmeier M, Kreuzer T, Gieshoff J, Lepperhoff G (2009) Automobile exhaust control. In: Ullmann's encyclopedia of industrial chemistry, Wiley-VCH, Weinheim, Germany
60. Rumpold R, Antrekowitsch J (2012) Recycling of platinum group metals from automotive catalysts by an acidic leaching process. In: The Southern African Institute of mining and metallurgy platinum, pp 695–714
61. Chevalier P (2004) PGM, Canadian minerals yearbook, pp 41.1–41.16
62. Benson M, Bennett CR, Harry JE, Patel MK, Cross M (2000) The recovery mechanism of platinum group metals from catalytic convertors in spent automotive exhaust systems. Resour Conserv Recy 31:1–7
63. Keyworth B (1982) The role of pyrometallurgy in the recovery of precious metals from secondary materials. In: 6th international precious metals institute, California, pp 509–538
64. Ezawa N (1993). Process of recovering platinum group metal. US Patent 5252305
65. Hoffmann JE (1988) Recovering platinum-group metals from auto catalyst. JOM 40(6):40–44
66. Ivanović SZ, Trujuć VK, Gorgievski MD, Mišić LD, Božić DS (2011) Removal of platinum group metals from the spent automobile catalyst by the pyrometallurgical process. In: 15th international research/expert conference "Trends in the development of machinery and associated technology", pp 12–18
67. Hagelüken C (2004) Umicore precious refining-the power of integration. Precious Metals Market Report. Umicore AG & Co.KG, Belgium, pp 10–11
68. Mishra RK, Reddy RG (1986) Pyrometallurgical processing and recovery of precious metals from auto-exhaust catalyst using plasma arc smelting. In: 10th International Precious Metals Institute, Florida, pp 217–231
69. Saville J (1985) Recovery of PGM's by plasma arc smelting. In: 9th International precious metals institute conference, New York, pp 157–167
70. Zhao HZ (1998) Analysis of phases in metallurgical products of plasma treating PGMs secondary resources. Chin J Nonferr Met 8(2):314–317
71. Burkhard R, Hoffelner W, Eschenbach RC (1994) Recycling of metals from waste with thermal plasma. Resour Conserv Recycl 10(2):11–16
72. Wang YH, Wu XF, Tong WF, Zhao C, Zan LH, Fan XX, Li BY, Li N (2009) A method for extraction of Pt, Pd and Rh from automotive catalyst by mineral phase restruction. Chinese Patent, 200910094112.7
73. Angelidis TN (2001) Development of a laboratory scale hydrometallurgical procedure for the recovery of Pt and Rh from spent automotive catalysts. Top Catal 1–4:419–423
74. Bolinski L, Distin PA (1992) Platinum group metals recovery from recycled auto catalyst by aqueous processing. Publ Australas Inst Min Metall 9/92
75. Duyvesteyn S, Liu H, Duyvesteyn WPC (1994) Recovery of platinum group metals from oxide ore-TML process. In: Proceedings of international symposium on hydrometallurgy. Chapman & Hall, London, pp 887–912
76. Han KN, Meng X (1996) Redox leaching of ores and spent catalysts using halide-containing solutions for recovery of platinum-group metals and rhenium. US Patent: 5542957
77. Meng X, Han KN (1995) Recovery of platinum and palladium from spent automobile catalytic converters by leaching with solutions containing halogen salts, ammonium and oxidants. In: Queneau PB, Peterson RD (eds), Minerals, 3rd ed. International symposium of recycling of metals engineering materials, metals and materials society, Warrendale, PA, pp 501–513
78. Mishra RK (1988) Recovery of platinum group metals from automobile catalytic converters-a review. The Minerals, Metals & Materials Society: Precious Metals 89:483–501
79. Mishra RK (1993) A review of platinum group metals recovery from automobile catalytic converters. In: Mishra RK (ed) Proceedings of International Precious Metals Ins Conf, Newport, RI: Precious Metals, pp 449–474

80. Yoo S (1998) Metal recovery and rejuvenation of metal-loaded spent catalysts. Catal Today 44:27–46
81. Eugenia G, Petru GS, Constanta G, Florica Z, Sergiu P, Dumitru G, Nedelcu A, Costel S, Dino S, Florica MB, Nicolae SV (1983) Recovery of platinum and other components from spent catalysts of the platinum. Romanian Patent, RO 71056 B 19801205
82. Ezawa N (1989) Recovery of platinum-group metals from spent catalysts. Japanese Patent: JP 01108390 A2 19890425
83. Barakat MA, Mahmoud MHH (2004) Recovery of platinum from spent catalyst. Hydrometallurgy 72:179–184
84. Baghalha M (2012) The leaching kinetics of an oxide gold ore with iodide/iodine solutions. Hydrometallurgy 113–114:42–50
85. Jafarifar D, Daryanavard MR, Sheibani S (2005) Ultrafast microwave-assisted leaching for recovery of platinum from spent catalyst. Hydrometallurgy 78:166–171
86. Schreier G, Edtmaier C (2003) Separation of Ir, Pd and Rh from secondary Pt scrap by precipitation and calcinations. Hydrometallurgy 68:69–75
87. Horner BT, McGrath RB (1994) European patent: 94–300378 (Johnson Matthey PLC)
88. Perte E, Ghiara C, Marc M, Ceuca O, Crucin O (1988) Romania patent: 94014 B1(Institutul de Chimie)
89. Muraki M, Mitsui Y (1986) Method for collecting platinum and palladium from platinum catalyst. Japanese patent: 61110731 A2 (Nippon Magnetic Dressing)
90. Grzelczuk S, Popowicz M, Berak JM, Schimmelpfennig Z, Gora M, Staszak S (1985) PL 131079 B1. Instytut Chemii Przemyslowej, Poland
91. Atkinson GB, Kuczynski RJ, Desmond DP (1992) Cyanide leachingmethod for recovering platinum group metals from a catalytic converter catalyst. U S Patent 5(160):711
92. Wadsworth ME, Zhu X, Thompson JS, Pereira CJ (2000) Gold dissolution and activation in cyanide solution: kinetics and mechanism. Hydrometallurgy 57:1–11
93. Ryder JM, Dymock K (1990) Platinum lake technology process for recovery of precious metals from scrap automotive catalysts. In: Proceedings of recycling of metalliferous materials conference. The Institution of Mining and Metallurgy, Birmingham, England, pp 255–258
94. Suzuki S, Ogino M, Matsumoto T (2007) Recovery of platinum group metals at Nippon PGM Co., Ltd. J MMIJ 123:734–736

Recovery of Valuable Metals from Waste Printed Circuit Boards by Using Iodine-Iodide Leaching and Precipitation

Altansukh Batnasan, Kazutoshi Haga and Atsushi Shibayama

Abstract This study presents a viable approach for recovery of precious metals such as gold (Au), silver (Ag), palladium (Pd), and base metals, including copper (Cu), nickel (Ni), cobalt (Co), lead (Pb) and zinc (Zn) from waste printed circuit boards (WPCBs) via iodine-iodide leaching and precipitation. The behaviours of dissolution and precipitation of precious and base metals during iodine-iodide leaching and precipitation processes were discussed. Sodium hydroxide (NaOH) was used to remove base metal impurities exist in the pregnant leach solution under alkaline conditions. Precious metals remained in the resulting solution from NaOH precipitation were recovered by reduction using ascorbic acid (L-AA) solution. Results show that under optimum leaching conditions, almost all (> 99%) of Au was dissolved in an iodine-iodide solution when the dissolution efficiencies of other precious metals (Ag, Pd) and base metals, besides calcium (leaching of 25%) were less than 1 and 6%, respectively. The study revealed that more than 95% of Cu, Ni, Pb, Zn, Fe and Mn were initially removed from the pregnant leach solution at pH of 9.3 with addition of 0.1 M NaOH. Then 99.8% Au, 81.7% Ag and 74% Pd were precipitated from the obtained solution after NaOH precipitation while L-AA dose was 0.6 ml/ml at the condition. It can be concluded that the precious and base metals could be recovered selectively and economically from WPCBs via iodine-iodide leaching followed by precipitation using NaOH and L-AA.

Keywords Valuable metals · Precious and base metals · Iodine-iodide Leaching · Precipitation · Reduction

---

A. Batnasan (✉) · A. Shibayama
Graduate School of International Resource Sciences, Akita University,
1-1 Tegata Gakuen Machi, Akita 010-8502, Japan
e-mail: altansux@gipc.akita-u.ac.jp

K. Haga
Graduate School of Engineering Science, Akita University,
1-1 Tegata - Gakuen Machi, Akita 010-8502, Japan

H. Kim et al. (eds.), *Rare Metal Technology 2018*, The Minerals, Metals & Materials Series, https://doi.org/10.1007/978-3-319-72350-1_12

Introduction

With the rapid increase in population, economic growth, and the development of mobile computing devices, the use of handheld devices such as cell phones, tablets, electronic games and laptops is rapidly increasing. Since the relatively short life span of these devices, a huge amount of the spent handheld devices is discarded, which cause harm to the environment in various ways [1–3]. However, a main component like printed circuit board of these devices contains the considerable amount of valuable metals such as gold (Au), silver (Ag), palladium (Pd), platinum (Pt), copper (Cu), nickel (Ni), cobalt (Co), zinc (Zn) and aluminum (Al), which are occurred in limited quantities in nature [4, 5]. Hence recycling of electronic wastes generated from spent electronic devices is very important to maximize the sustainable utilization of natural resource and to minimize the environmental impact. To satisfy this objective, a main challenging task is to recover precious metals from waste printed circuit boards (WPCBs) via available approaches. Numerous processes based on pyrometallurgy and hydrometallurgy are utilized to recover metals from WPCBs [6–8]. Due to the higher operating cost associated with energy consumption and toxic gas emissions from WPCBs, which contain 24 wt% ceramics and 13 wt% polymers besides the valuable metals, many studies have been conducted on the leaching of precious metals especially Au from WPCBs in various aqueous media using cyanide, aqua regia, thiosulphate, thiourea, ammonia, bromine, and iodine [6–12].

In hydrometallurgical route, considerable attention has been given to the pursuit of alternative reagents to cyanide for the processing of precious metals. Iodine-iodide solution is one of the alternative reagents for the extraction of gold from gold ores and WPCBs because of its good stability, selectivity and effectiveness for gold leaching. A major disadvantage of this leaching process is a high cost of reagents [6, 7, 13–16]. Therefore there is still an obvious necessity for further development of the iodine-iodide leaching process in order to improve the effectiveness and to reduce the consumption of reagents through leaching process. Despite the dissolution of gold in iodine-iodide solution, the recovery of gold from pregnant gold-iodide solutions by subsequent hydrometallurgical techniques such as extraction, adsorption, precipitation and purification is still in progress.

In this paper, a complete hydrometallurgical method consisting of iodine-iodide leaching and sequential precipitation technique for recovery of precious and base metals from WPCBs is presented. The precipitation of precious metals and base metal impurities exist in the pregnant leach solution with sodium hydroxide and ascorbic acid are discussed. Effect of base metal impurities contained in WPCBs on the processing of precious metals is also considered.

Experimental Section

Materials

Waste printed circuit boards collected from discarded computers were used in this study. Reagent grade chemicals used were: iodine, potassium iodide (KI), sodium hydroxide (NaOH) and L-ascorbic acid ($C_6H_8O_6$). Distilled water was used for the preparation of all aqueous solutions: iodine-iodide, 0.1 M NaOH and 0.1 M L-AA. Compositions of the solid samples from high pressure oxidative leaching (HPOL), iodine-iodide leaching and precipitations were determined by X-ray Diffraction (XRD), RINT-2200/PC (Rigaku) and X-ray Fluorescence (XRF), ZSX Primus II (Rigaku). Concentrations of metal ions exist in the aqueous phases from leaching and precipitations were determined using inductively coupled plasma optical emission spectrometer (ICP-OES), SPS-5500 (Seiko Instruments Inc.). pH and oxidation-reduction potential (ORP) of all aqueous solutions after leaching and during precipitation were measured by pH/ORP controller (Laqua, D-74). Particles precipitated under different pH conditions were investigated Field Emission–Scanning Electron Microscope (FE-SEM/EDS), JSM-7800F.

Iodine-Iodide Leaching

Before iodine-iodide leaching, WPCBs collected was treated by sequential pre-treatment processes, namely incineration at 800 °C in a furnace equipped with exhaust gas cleaning systems (DOWA HOLDINGS Co., Ltd, JAPAN) and high pressure oxidative leaching (HPOL) with a dilute sulfuric acid solution in an autoclave at 120 °C for 30 min in the presence of oxygen gas, in order to remove impurities such as volatile organic materials and high concentrations of some base metals (Cu, Zn, Al, Ni and Fe) due to their reverse effect on the leaching of precious metals in aqueous iodine-iodide solutions [17, 18]. Chemical content and constituent of the solid residue of WPCBs derived from HPOL are shown in Table 1 and Fig. 1, respectively.

The solid residue of WPCBs obtained from HPOL was dissolved in a mixture solution of iodine-iodide in a baffled volumetric flask immersed in a thermostatically controlled water bath under the selected conditions described below [18]: iodine-iodide mass ratio, 1:6; pulp density, 10%; agitation speed, 500 rpm; leaching

Table 1 The chemical content of the solid residue of WPCBs from HPOL

Metals											
Au	Ag	Pd	Cu	Zn	Al	Ni	Fe	Pb	Mn	Co	SiO_2
g/t			wt%								
458	2504	209	0.24	0.2	0.7	0.5	0.7	2.5	0.02	0.004	37.5

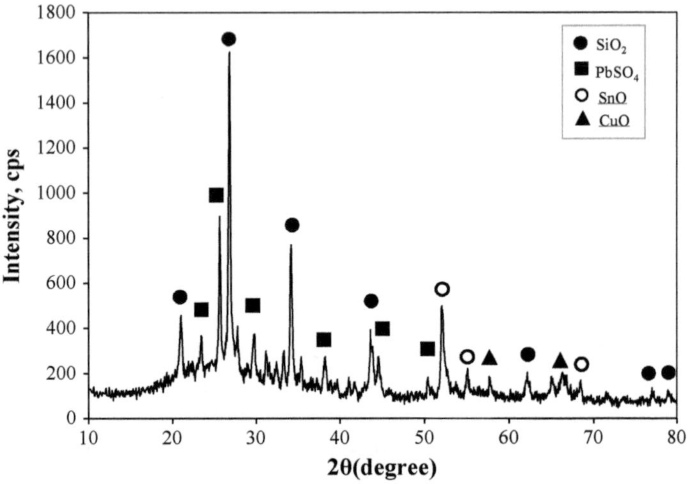

Fig. 1 XRD pattern of the solid residue of WPCBs after HPOL

temperature, 40 °C and leaching time, 24 h. After leaching for a certain period of time under the conditions, the slurry sample was cooled to room temperature and separated solids from leachate by filtration. The leachate/pregnant leach solution derived from the iodine-iodide leaching of the residue of WPCBs was used for further study.

Recovery of Precious Metals

In order to recover precious metals such as Au, Ag and Pd, base metal impurities (Cu, Ni, Co, Zn, Pb, Ca, Fe and Mn) present in the pregnant leach solution (PLS) were first removed by selective precipitation using 0.1 M sodium hydroxide solution under different pH conditions. After removal of some metal impurities, precious metals remained in the resulting solution were recovered by reduction using 0.1 M ascorbic acid as a reducing agent. During mixing the pregnant leach solution and NaOH or L-AA in the appropriate ratio in a flask, the solution was stirred at room temperature with 500 rpm agitation speed and continued the stirring for 10 min. The effects of pH, oxidation/reduction potential (Eh) of the solution and precipitation media on the removal of the base metal impurities and the recovery of precious metals were investigated under various conditions. At the end of precipitation, the solution was centrifuged at 3500 rpm for 10 min and the precipitate obtained was washed with 10 ml ethanol and 10 ml distilled water and dried at 70 ° C for 24 h. Characteristics and composition of precipitates obtained from different precipitating conditions in the presence of NaOH and L-AA were measured by SEM-EDS. The oxidation-reduction potential, ORP (E_{meas}, V) values obtained from precipitation by using the commercially available electrode is converted to standard electrode potential (Eh, V) by providing the specific conversion factor (E_{ref}, V), which is determined using Quinhydrone, MSDS (CAS:106-34-3) solution [19, 20]:

$$Eh = E_{meas} + E_{ref} \quad (1)$$

The ORP value (0.262 V) of the reference electrode calibrated is consistent with the data in reference textbooks from manufacturer [19, 20]. Recovery efficiency (*R%*) of metals was calculated by a mass balance method using the concentrations of metals in the PLS (C_{in}) and in the aqueous phases (C_{aq}) after precipitations determined by ICP-OES, as follows:

$$R\% = \frac{C_{in} - C_{aq}}{C_{in}} * 100\% \quad (2)$$

Results and Discussion

Iodine-Iodide Leaching

The pregnant leach solution with an initial pH of 2.84 and Eh of 0.58, V derived from the iodine-iodide leaching under the previously specified conditions contains precious metals such as Au, Ag and Pd in appropriate amounts, and some other base metal impurities like Ca, Cu, Al, Pb, Mg, Zn, Fe etc. in measurable amounts. The concentration of metals in the pregnant leach solution is shown in Table 2.

Table 2 shows that gold content in the pregnant leach solution is 81–285 fold higher than other precious metals. It can be referred that iodine-iodide leaching is a selective method for dissolution of gold from the sample due to its preferential characteristics for gold leaching. The dissolution mechanism of gold in aqueous iodine-iodide solutions has been described in the literature [13–16, 18]:

$$2Au + I_3^- + I^- \rightarrow 2Au\,I_2^- \quad (3)$$

$$3Au + 4I_3^- \rightarrow 3Au\,I_4^- \quad (4)$$

It is apparent from Table 2 that the precious and base metals are dissolved simultaneously in the iodine-iodide solution. For this reason, the base metal impurities need to be removed from the pregnant leach solution in order to recover precious metals.

Table 2 Concentration of metals in the pregnant leach solution, mg/L

Au	Ag	Pd	Ca	Cu	Al	Pb	Mg	Zn	Fe	Ni	Sr	Mn	Co
114	1.4	0.4	1181	38.5	17.4	13.6	9.6	7.4	3.1	2.5	1.5	0.9	0.14

Precipitation of Metals Under Alkaline Conditions

The metal precipitation and variation in Eh value as a function of pH of the pregnant leach solution were exhibited in Fig. 2. The initial pH and Eh values of the leach solution were varied from pH 2.84 and 0.58 V to pH 13 and −0.01 V with the addition of NaOH due to more reduction of the solution, providing the increase in the precipitation of metals from the solution. It is shown that at a pH below 9.3 and Eh value of 0.21 V, more than 95% of the most base metal impurities like Fe, Al, Pb, Zn, Co, Cu, Mn and Ni were precipitated from the pregnant leach solution, while Ca, Mg and Sr precipitation were less than 7%. However the precipitation efficiency of these metals (Ca, Mg and Sr) increased with increasing of pH and maximum precipitation efficiency reached 61.8% for Ca, 100% for Mg and 33.4% for Sr, respectively at a pH value of 13 and Eh of −0.01 V. For precious metals, about 0.2% of Au and 15% of Ag were precipitated at pH 9.3, when the precipitation of Pd did not take place. A further increase of the solution pH resulted in an increase of Au, Ag and Pd precipitation and their maximum precipitation reached 98.8% for Au, 41.7% for Ag and 68.4% for Pd at pH 13, respectively (Fig. 2).

As a result of aforementioned preliminary experiments, it is inferred that precious metals and base metal impurities could be separated under different pH conditions (around pH 9.0), on the other hand most of the metals present in the pregnant leach solution can be simultaneously recovered at pH over 12, converting metal ions in the aqueous solution into their insoluble forms as hydroxide and metallic state. Based on the above result, precipitation experiment was carried out with 0.1 NaOH at pH 9.3, in order to remove base metal impurities from the

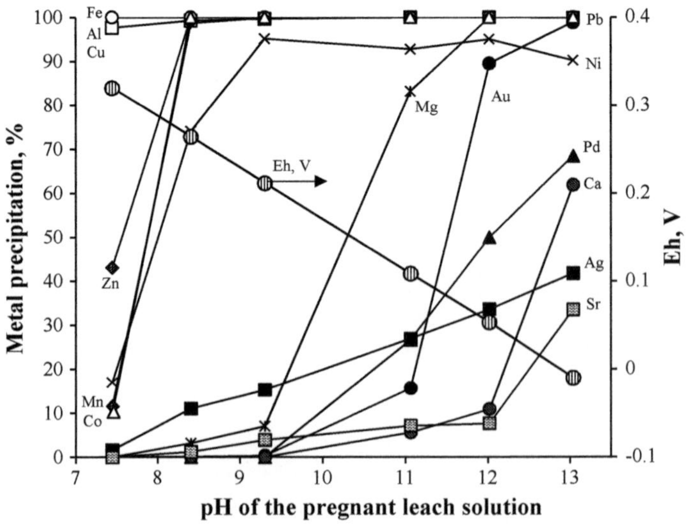

Fig. 2 Metal precipitation and variation in Eh value as a function of solution pH

pregnant leach solution. Under the conditions, main base metal impurities like Cu, Ni, Pb and Zn were removed with high efficiency (>95%). The resulting solution containing precious metals and some base metal impurities especially Mg, Ca and Sr was used in a subsequent study.

Precipitation of Precious Metals

The resulting solution from NaOH precipitation was treated with different dosages of 0.1 M L-AA at room temperature and 500 rpm for 10 min. Figure 3 shows the precipitation efficiencies of metals with various dosages of L-AA. Results indicate that all of metal ions in the solution were precipitated under the conditions and their precipitation efficiencies increased with increasing the dosage of L-AA added. Nevertheless, the precipitation efficiencies of metal impurities (Zn, Ca, Ni, Sr, Pb and Fe) apart from Cu were relatively lower though the precipitation. The highest precipitation efficiency was achieved with 0.2 ml/ml L-AA dose: 99.0% for Au, 94.0% for Cu, 76.8% for Ag respectively, while the Pd precipitation was 26.7%. Further increasing the amount of L-AA, precipitation efficiencies of Au, Ag and Cu were slightly increased, whereas Pd precipitation intensely increased up to 74.0%, respectively (Fig. 3). It was observed from the variations in solution pH and Eh that the Eh value increased rapidly with increasing the L-AA dose up to 0.2 ml/ml in the solution, and then become almost constant, whereas pH value of the solution decreased below pH 2.0 due to the enhanced reactivity of L-AA during the reaction. It means that the profile towards metals precipitation in the presence of L-AA is

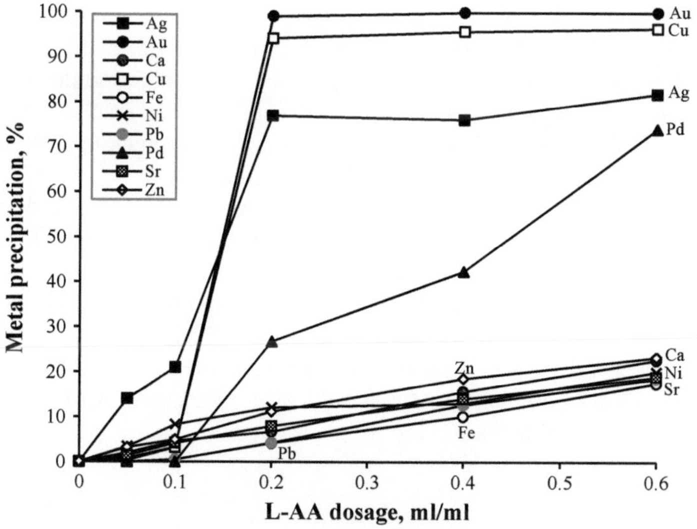

Fig. 3 Metal precipitation efficiency as a function of dosage of L-AA in the solution

attributed to the changes in solution pH and Eh values. On the other hand the precipitation efficiency of metals is associated with the molar ratio of L-AA/metal, and reduction property of the metal with L-AA. For instance, the molar ratio of L-AA and gold (L-AA/Au) was varied from 0.08 to 1.0 by adding the different doses of L-AA. The maximum precipitation efficiency reached 99.8% for Au when the L-AA/Au molar ratio was 1.0. It was identified that trace amounts of precious metals especially Ag and Pd, and base metal impurities like Cu, Ca, Ni and Sr still remained in the aqueous solution after reduction with L-AA. Therefore, the solution derived from the pregnant leach solution via sequential precipitation with NaOH and L-AA was purified in the subsequent stage by addition of NaOH to precipitate metals completely at pH 13.

The metal distribution in each phase after three-stage precipitation in aqueous solutions are summarized in Table 3. It can be shown that apart for Pd, Ca and Sr, high recovery yields were obtained for Au, Ag and other base metal impurities under sequential precipitations. The selective and complete removal of Al, Fe and Zn from the pregnant leach solution was achieved by one-stage precipitation at pH 9.3 using NaOH, whereas other metal impurities such as Cu, Ni, Pb and Co were required a two- and three-stage precipitation. For precious metals, about 99.8% Au recovery was accomplished by a one-stage precipitation at pH 2 in the presence of L-AA, but a three-stage precipitation was required for complete recovery of Ag and Pd.

Based on these results, a viable process for recovery of precious and base metals from WPCBs is proposed. A schematic representation of the developed process for recycling of WPCBs is shown in Fig. 4. Generally, the proposed flowchart consists of several sequential treatment processes as follows: (1) reduction of collected WPCBs size to 2 cm by dismantling and shredding; (2) pre-treatment processes, which include incineration at 800 °C, grinding the WPCB ash sample to smaller than 106 μm and high-pressure oxidative leaching (HPOL) with a dilute sulfuric acid solution in an autoclave at 120°C for 30 min in the presence of oxygen, for removal of volatiles and base metal impurities; (3) iodine-iodide leaching for extraction of metals present in the solid residue of WPCBs resulted from HPOL; (4) removal of base metal impurities and recovery of precious metals via the sequential precipitation technique using NaOH and L-AA.

The final precipitates obtained via a three-step precipitation process by the use of NaOH and L-AA were characterized by FE-SEM (data not shown). It was found that the selective separation of precious metals and base metal impurities is possibly by sequential precipitations, but it is no feasible to individually separate metal from the other metals. It means that the precipitate at pH 9.3 contains Cu, Ni, Zn, Fe and Pb, whereas at pH 13, main contents in the precipitate are Ca, Mg and Sr. The precipitate at pH 1.9 contains Au, Ag and Pd with trace amounts of impurities like Cu, Ca, Ni and Pb (Table 3).

Table 3 Distribution of metals in each phase through sequential precipitation

Precipitation	Reagent	pH	Phase	Metal distribution (%)											
				Au	Ag	Pd	Cu	Al	Ca	Fe	Ni	Pb	Co	Sr	Zn
Stage 1	NaOH	9.3	Aqueous	99.8	84.7	100	0.4	0	98.4	0	11	12.6	19.3	93.8	0
			solid	0.2	15.3	0	99.6	100	1.6	100	89	87.4	80.7	6.2	100
Stage 2	L-AA	2.0	Aqueous	0	8.7	57.7	0.2	0	82.8	0	0.7	0	0	79.8	0
			solid	99.8	76.0	42.3	0.2	0	15.6	0	10.3	12.6	19.3	14	0
Stage 3	NaOH	13	Aqueous	0	0	18.5	0	0	24	0	0	0	0	29.2	0
			solid	0	8.7	39.2	0.2	0	58.8	0	0.7	0	0	50.6	0

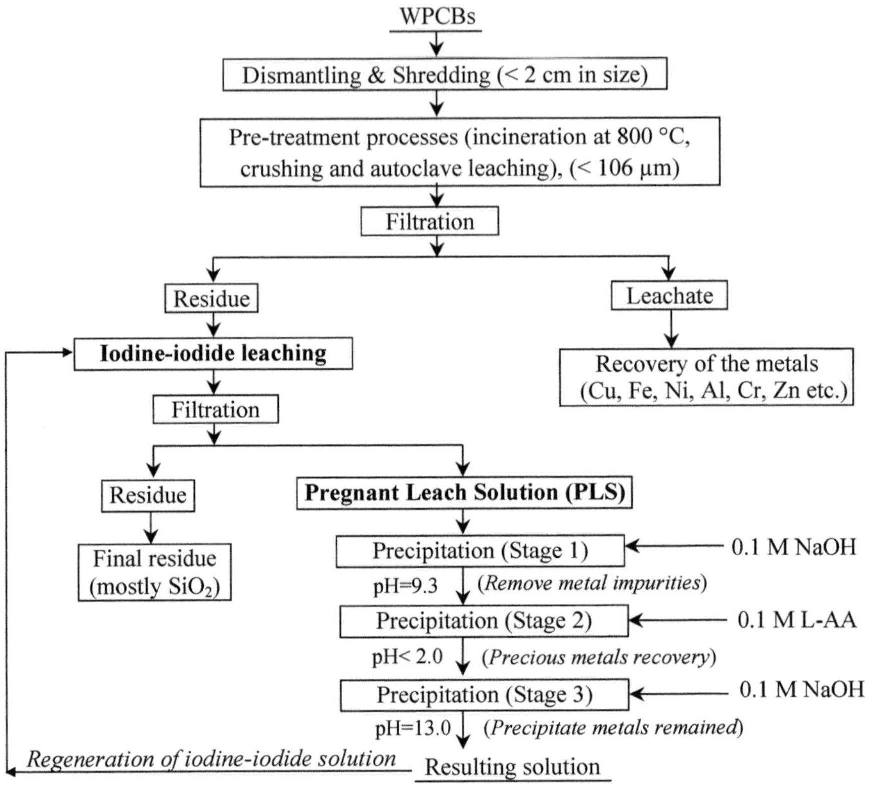

Fig. 4 The proposed flowchart for recycling of WPCBs

Conclusions

This paper aims to discuss the developed recycling approach for recovery of precious and base metals from WPCBs by using iodine-iodide leaching and sequential precipitation technique. The results obtained from this study are summarized as follows:

- The vast majority (>99%) of gold was dissolved in the iodine-iodide solution from WPCBs, whereas less than 1% of Ag and Pd were leached at the conditions. The concentration of Au, Ag and Pd in the pregnant leach solution were 114, 1.4 and 0.4 mg/L, respectively. The lower dissolution of Ag and Pd in the iodine-iodide solution may be related to the leaching conditions and ratio of iodine species to Ag and Pd, respectively. Leaching efficiency of Ca, Cu, Ni, Co and Zn were around 1–25%, while those of other metal impurities did not exceed 1%.

- Apart from Ca and Sr, all metal impurities were completely removed from the pregnant leach solution through sequential precipitation using NaOH at pH 9.3 and pH 13, respectively.
- The metal recovery were 99.8% for Au, 81.7% for Ag and 74% for Pd from the resulting solution after NaOH precipitation with a 0.6 ml/ml dosage of L-AA at pH of 1.9. It was revealed that recovery of Ag and Pd remained in the aqueous phase from L-AA reduction were improved by precipitation with NaOH at pH 13.
- Results indicate that precious and base metals could be effectively recovered from WPCBs using iodine-iodide leaching followed by sequential precipitation with NaOH and L-AA.

Acknowledgements The authors gratefully acknowledge the financial support from the Japan Society for the Promotion of Science (JSPS) through "New Frontier Leader Program for Rare-Metals and Resources" and grant KAKENHI-16H04182.

References

1. Abhishek KA, Gabriel IZ, Xianlai Z, Jinhui L (2017) Evaluating waste printed circuit boards recycling: opportunities and challenges, a mini review. Waste Manage Res 35(4):1–11
2. Kui H, Jie G, Zhenming X (2009) Recycling of waste printed circuit boards: a review of current technologies and treatment status in China. J Hazard Mater 164:399–408
3. Ravi V (2012) Evaluating overall quality of recycling of e-waste from end-of-life computers. J Cleaner Prod 20:145–151
4. Ning C, Lin CSK, Hui DCW, McKay G (2017) Waste printed circuit board (PCB) recycling techniques. Top. Curr. Chem (Z) 375(2):43. https://doi.org/10.1007/s41061-017-0118-7
5. Kaya M (2016) Recovery of metals and nonmetals from electronic waste by physical and chemical recycling processes. Waste Manag 57:64–90
6. Syed S (2012) Recovery of gold from secondary sources-A review. Hydrometallurgy 115–116:30–51
7. Jirang C, Lifeng Zh (2008) Metallurgical recovery of metals from electronic waste: a review. J Hazard Mater 158:228–256
8. Zhang L, Xu Z (2016) A review of current progress of recycling technologies for metals from waste electrical and electronic equipment. J Clean Prod 127:19–36
9. Cayumil R, Khanna R, Rajarao R, Mukherjee PS, Sahajwalla V (2016) Concentration of precious metals during their recovery from electronic waste. Waste Manag 57:121–130
10. Hilson G, Monhemius AJ (2006) Alternatives to cyanide in the gold mining industry: what prospects for the future? J Clean Prod 14:1158–1167
11. Fleming CA, McMullen J, Thomas KG, Wells JA (2003) Recent advances in the development of an alternative to the cyanidation process: thiosulfate leaching and resin in pulp. Miner Metall Process 20(1):1–9
12. Aylmore MG (2005) Alternative lixiviants to cyanide for leaching gold ores. In: Adams MD (ed) Advances in gold ore processing 2005. Elsevier, Western Australia, pp 501–541
13. Homick RP, Sloan H (1976) Gold reclamation process, US3957505
14. Qi PH, Hiskey JB (1993) Electrochemical behaviour of gold in iodide solutions. Hydrometallurgy 32:161–179
15. Davis A, Tran T, Young DR (1993) Solution chemistry of iodide leaching of gold. Hydrometallurgy 32:143–159

16. Angelidis TN, Kydros KA, Matis KA (1993) A fundamental rotating disk study of gold dissolution in iodine-iodide solutions. Hydrometallurgy 34:49–64
17. Shibayama A, Tongamp V, Altansukh B, Haga K, Hosoi A (2013) Electronic waste treatment: Part 1. Autoclave oxidation-leaching using pyrite waste from mine tailing. Hydrometallurgy 137:92–100
18. Altansukh B, Haga K, Ariunbolor N, Kawamura S, Shibayama A (2016) Leaching and adsorption of gold from waste printed circuit boards using iodine-iodide solution and activated carbon. Eng J 20(4):29–40
19. Process Probe™ ORP Sensors, Broadley James, (2013), Doc. Nr. P2000–2013, http://www.broadleyjames.com/wp-content/uploads/2015/02/ProcessProbe-ORP-sensorsP2000.pdf. Accessed 8 Mar 2017
20. Bier AW, Hach-Lange (2009) Introduction to oxidation reduction potential measurement, Hach Company, https://www.hach.com/hqdguide-ORP. Accessed 26 Feb 2016

Cyclone Electrowinning of Antimony from Antimonic Gold Concentrate Ores

Weijiao Yang, Liugen Sun, Yihang Hu, Yongqiang Yang, Xingming Jiang and Hua Wang

Abstract During the past decades, cyclone electrowinning (CE) has been regarded as a novel separation technology, which can be used for direct electrowinning of metals from low concentration solution. Recently, CE was proven effective for the treatment of antimonic gold concentrates. The experimental results of cyclone electrowinning of antimony show the current efficiency increases from 43.8 to 72.2% with the increasing current density (from 254.8 to 647.3 A/m^2), and on the other hand, power consumption per ton antimony decreases from 6100 to 4200 kWh with the aforementioned increasing current density. The grade of the obtained cathode antimony could reach 98.6%. Compared with the traditional electrowinning using parallel-board electrodes, the economic profit of CE was 33.1% higher attributed to the high current efficiency of CE. In addition, CE could adapt to high current density. All these characteristics indicate the CE method can be implemented commercially for processing antimonic gold concentrates.

Keywords Cyclone electrowinning · Antimonic gold concentrate ores
Antimony · Gold

W. Yang · X. Jiang (✉) · H. Wang (✉)
State Key Laboratory of Complex Nonferrous Metal Resources
Clean Utilization, Faculty of Metallurgical and Energy Engineering,
Kunming University of Science and Technology, Kunming 650093, China
e-mail: 2008k321@sina.com

W. Yang
e-mail: weijiao1110@163.com

H. Wang
e-mail: wanghua65@163.com

W. Yang · L. Sun · Y. Hu · Y. Yang
Beijing General Research Institute of Mining and Metallurgy,
Beijing 100160, China

H. Kim et al. (eds.), *Rare Metal Technology 2018*, The Minerals,
Metals & Materials Series, https://doi.org/10.1007/978-3-319-72350-1_13

Introduction

China is the world's largest antimony resource country and the largest producer. Current China owns approximately 47.5% of world's antimony resource and produces 78.1% of world's antimony. Its antimony deposits are found in Hunan, Guangxi, Guizhou, Tibet and Yunnan provinces, which (except Tibet) also produce mostly antimony ore in China. Antimony-gold ore is one of the antimony resource in China, whose proved reserve is up to 1000 tons [1, 2].

The commercial route to extraction antimony from the antimony-gold ore to further separating antimony and gold involves alkaline leaching—electrowinning process [3–6]. In this process, the ore is leached in sodium hydroxide solution for antimony and gold separation, the resultant antimony leaching solution (antimony electrolyte) is extracted via electrowinning to obtain cathode antimony. However, antimony electrowinning in a plate cell problems with environmental pollution, high power consumption and high antimony concentration solution. For instance, considerable quantities of alkaline fog due to sodium hydroxide volatilization [5]. The current efficiency in such solution is about 70%, even to 50%. Additionally, sodium sulphide, sodium sulfate and other sodium salt accumulation lead to huge and complicated waste liquor [6]. Therefore, several methods have been proposed to improve the antimony electrowinning technology, including slurry electrolysis, ferric chloride leaching- electrowinning [7, 8]. However, these processes with difficulties in realizing commercial production efficiently.

Cyclone electrowinning (CE) is a relatively recent multi-metals separation technology which has been under continuous developed during past decades [9–15] (Fig. 1). It is widely used in acidic solutions, including sulphuric acid, hydrochloric acid, nitric acid and cyanide, et al., to extract low concentration copper, cobalt, nickel, zinc, gold and silver. Compared with the traditional plate electrowinning, CE not only raises the electrolyte flow speeds to strengthen the contact between the electrolyte and the electrode, but also is more suitable for metal extraction from low metal concentration solution. These seem that CE can improve the antimony electrowinning technology to obtain high grade cathode antimony with low current efficiency.

In this study, an improved alkaline leaching—electrowinning process is used for the separation and recovery of antimony. The general process flow of this process is shown in Fig. 2. As mentioned above, one of the advantages of CE is suitable for extraction metals from low concentration metals electrolytes through intensive current density. Therefore, this paper tried to discuss the effects of current density on antimony electrowinning from the low concentration antimony electrolytes (5–20 g/L), in order to provide a clear relationship between the current density and the CE technique.

Fig. 1 Photograph of cyclone electrowinning workshop

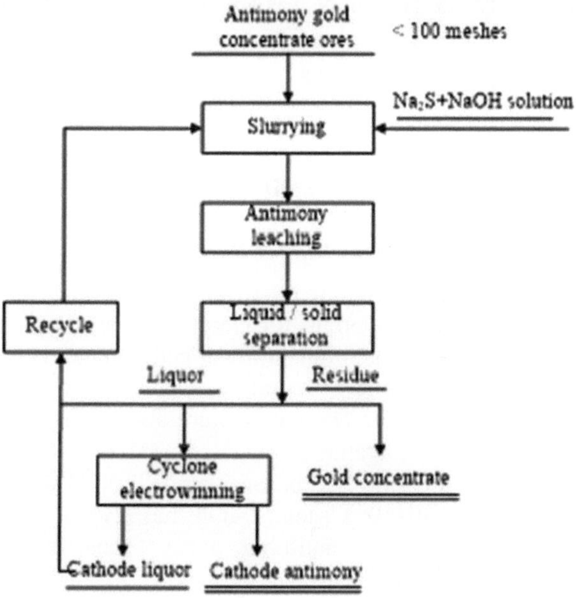

Fig. 2 The general flow sheet of cyclone electrowinning of antimony gold concentrate ores

Experimental

The raw antimony gold concentrate used for the tests was analyzed and the details were summarized in Table 1. This ore was leached in a Φ2000 × 2500 leaching tank and the resultant slurry was filter using a 250 m^2 pressure filter to obtain both solid (gold concentrate) and liquor (antimony leach solution, typical components seeing Table 2). Table 2 shows the typical leach solution, named the electrolyte, was performed using an 20 L cylcone electrowinning cylinder equipped with a

Table 1 Chemical components of the raw ore samples (%)

Component	Sb	Au (g/t)	Ag (g/t)	As	S	C	Fe
Content	4.52	48.65	12.30	4.32	12.62	2.27	10.62

Table 2 Typical components of the electrolyte (g/L)

Component	Sb	Au	As	Fe, Ag etc.
Content	10 ~ 16	~1	0.39	<0.05

Note Fixed conditions: the L/S = 1:1, NaOH 20 g/L, Na_2S 40 g/L, 40 °C

WYK-5400 stabilized current power supply. Stainless steel and titanium based lead dioxide were utilized as the cathode cylinder and anode, respectively. After electrowinning, both the resultant cathode antimony and cathode liquor were sampled and analyzed. The liquor was analyzed via chemical determination to obtain the Sb, NaOH and Na_2S content. The solid sample was washed three times, dried, and then analyzed via fire assaying.

Results and Discussion

Electrowinning Temperature

Temperature is important for electrowinning. Higher temperature is favorable for lowering resistance to obtain lower cell voltage. However, higher temperature also increases electrolyte evaporation, which leads to more serious alkaline fog by NaOH evaporation. The temperature above 50 °C leads to much serious alkaline fog than the temperature below 50 °C [5]. Considering the operation environment, temperature below 50 °C is considered. Table 3 shows that when the temperature in the range of 30–50 °C, temperature has few effects on the current efficiency, cell voltage and cathode antimony purity. Thus, 30–50 °C is suitable for practical antimony electrowinning.

Table 3 Effect of temperature on electrowinning parameters

Temperature (°C)	Current efficiency (%)	Cell voltage (V)	Cathode Sb purity (%)
30	73.42	3.0	97.83
40	72.29	2.8	97.20
50	71.88	2.6	97.12

Current Density at 5 g/L Sb Electrolyte

An antimony electrolyte with 5 g/L antimony was chosen to investigate the influences of current density on CE for antimony extraction. The CE tests were carried out in an alkaline solution with 45 g/L NAOH and 25 g/L Na_2S for electrowinning 5 h at 40 °C.

The results for 5 g/L Sb electrolyte are shown in Figs. 3, 4, and 5. Figure 3 indicates that increasing the current density is beneficial to current efficiency. However, 250 A/m^2 is proposed the best current density to reach the highest current efficiency of 47.23%. As experiment, such antimony electrolyte was treated in the same electrowinning conditions, and the largest current density was less than 100 A/m^2 with the current efficiency of 33–35%. This suggests that CE can reach higher current efficiency via higher current density.

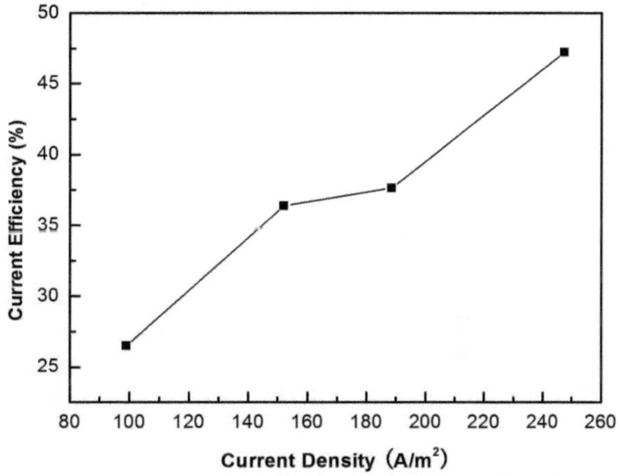

Fig. 3 Effect of current density on current efficiency (fixed conditions: 5 g/L Sb advanced electrolyte, NaOH 45 g/L, Na_2S 20 g/L, 40 °C and 5 h)

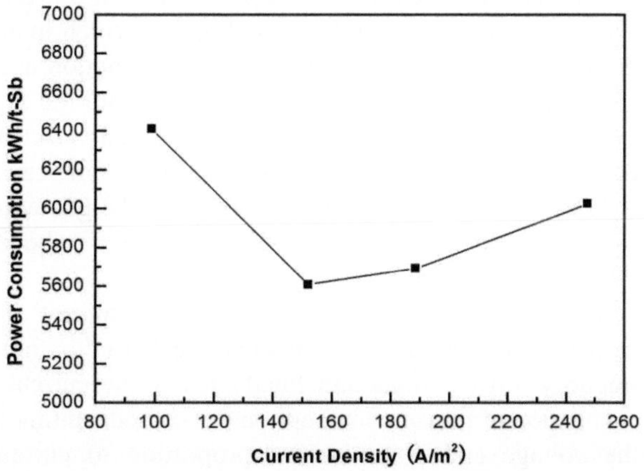

Fig. 4 Effect of current density on power consumption (fixed conditions: 5 g/L Sb advanced electrolyte, NaOH 45 g/L, Na_2S 20 g/L, 40 °C and 5 h)

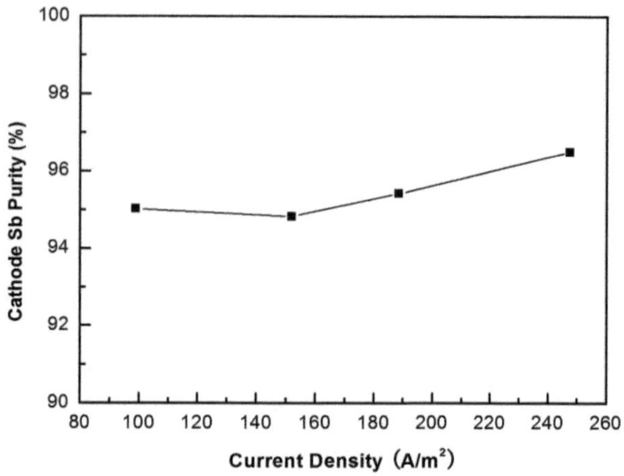

Fig. 5 Effect of current density on cathode Sb purity and Au-in-cathode Sb (fixed conditions: 5 g/L Sb advanced electrolyte, NaOH 45 g/L, Na_2S 20 g/L, 40 °C and 5 h)

Figure 4 shows the tendency of power consumption per cathode antimony changing. When the current density in the range of 150–200 A/m^2, the consumption is in a relative low level, approximate 5600–5700 kWh/t-Sb. Thus, it suggests that current density in the range of 150–200 A/m^2 is the most appropriate for antimony electrolyte containing 5 g/L antimony.

From Fig. 5, it can be seen that increasing current density is in favor of cathode antimony grade. However, the effects of current density are not much, only approximate 1.5% cathode antimony purity change.

Current Density at 12 g/L Sb Electrolyte

An antimony electrolyte with 12 g/L antimony was also chosen to investigate the influences of current density on CE for antimony extraction. The other CE tests conditions were the same for the conditions of 5 g/L Sb electrolyte as stated above. The results are shown in Figs. 6, 7 and 8, which indicate that the effects of current density on current efficiency, power consumption and cathode antimony grade for 12 g/L antimony electrolyte are different from the 5 g/L antimony electrolyte CE. Firstly, the current density for 12 g/L antimony electrolyte is higher than the current density for 5 g/L antimony electrolyte, the former can reach 475 A/m^2 and the latter is 250 A/m^2, 1.9 times higher. And, for 12 g/L antimony electrolyte CE, it has higher current efficiency (68.32%) and better cathode antimony with the grade of 99.79%.

Figure 6 shows that the increasing of current efficiency is a W-shaped with increasing current density. This may be caused by higher current density leading to impurity ion reactions and finally lower the current efficiency. According to the calculation of power consumption per cathode antimony, this data is proportional to the voltage and is in inverse proportion to current density. Thus, the power

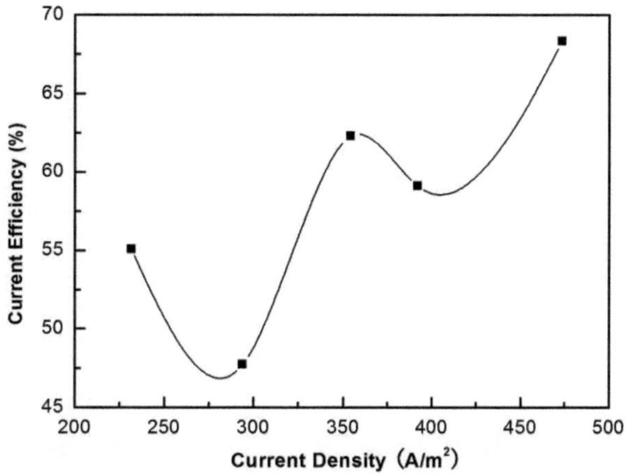

Fig. 6 Effect of current density on current efficiency (fixed conditions: 12 g/L Sb advanced electrolyte, NaOH 45 g/L, Na_2S 20 g/L, 40 °C and 5 h)

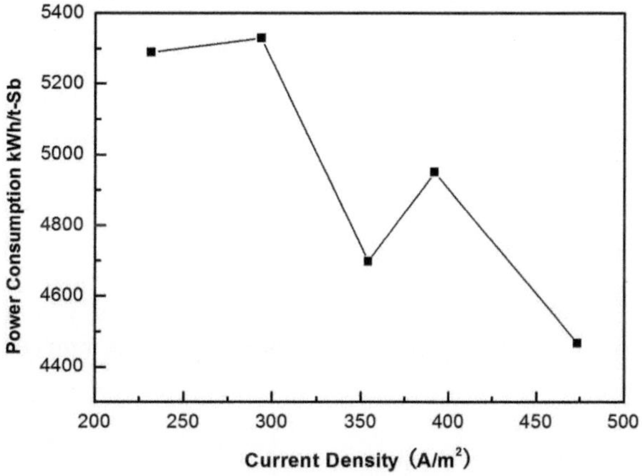

Fig. 7 Effect of current density on power consumption (fixed conditions: 12 g/L Sb advanced electrolyte, NaOH 45 g/L, Na_2S 20 g/L, 40 °C and 5 h)

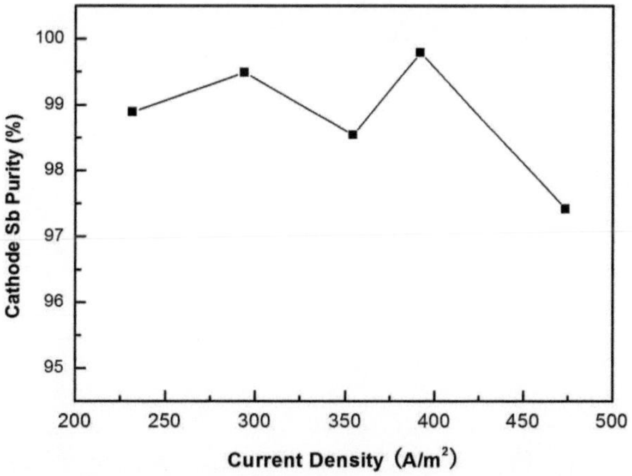

Fig. 8 Effect of current density on cathode Sb purity and Au-in-cathode Sb (fixed conditions: 12 g/L Sb advanced electrolyte, NaOH 45 g/L, Na_2S 20 g/L, 40 °C and 5 h)

consumption per cathode antimony declines with current density increasing, seeing Fig. 7. Figure 8 indicates that the purity of cathode antimony almost gradually increases with increasing current density under the experiment conditions. However, beyond 400 A/m^2 the influence of current density is inverse, thus 400 A/m^2 is deemed sufficient. It is well known that higher current density is more easily to heat up the electrolyte leading to a series of by-products, this may one of the reasons why the cathode antimony grade obtained from 475 A/m^2 is lower than which obtained from 400 A/m^2.

Considering Figs. 6, 7 and 8, 400 A/m^2 is an appropriate current density for antimony extraction from a 12 g/L antimony electrolyte by CE.

Current Density at 20 g/L Sb Electrolyte

An antimony electrolyte with 20 g/L antimony was used to investigate the influences of current density on antimony CE. The reason for choosing 20 g/L antimony electrolyte is that such concentration seems too high to antimony extraction by CE, while is enough by traditional plate cell electrowinning. As mentioned in the references, CE is much more suitable for low metal concentration electrolyte. The other conditions for 20 g/L antimony electrolyte CE are also the same as mentioned above, except of the temperature (50 °C). It should be noted that the temperature of the electrolyte can easily reach 50 °C, due to higher current density. These experimental results are shown in Figs. 9, 10 and 11.

Figure 9 demonstrates that the current density of 567.4 A/m^2 reaches the highest current efficiency of 70.24%, which is much higher than the antimony electrolyte

Fig. 9 Effect of current density on current efficiency (fixed conditions: 20 g/L Sb advanced electrolyte, NaOH 45 g/L, Na_2S 20 g/L, 50 °C and 5 h)

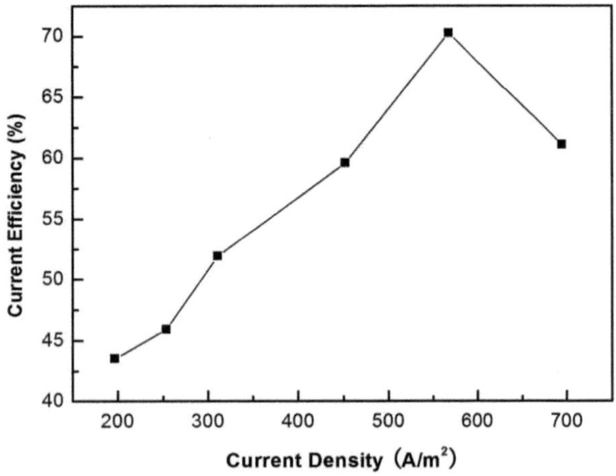

Fig. 10 Effect of current density on power consumption (fixed conditions: 20 g/L Sb advanced electrolyte, NaOH 45 g/L, Na_2S 20 g/L, 50 °C and 5 h)

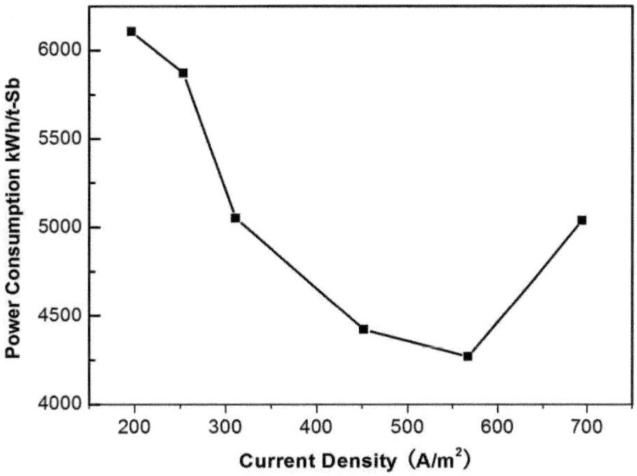

Fig. 11 Effect of current density on cathode Sb purity and Au-in-cathode Sb (fixed conditions: 20 g/L Sb advanced electrolyte, NaOH 45 g/L, Na_2S 20 g/L, 50 °C and 5 h)

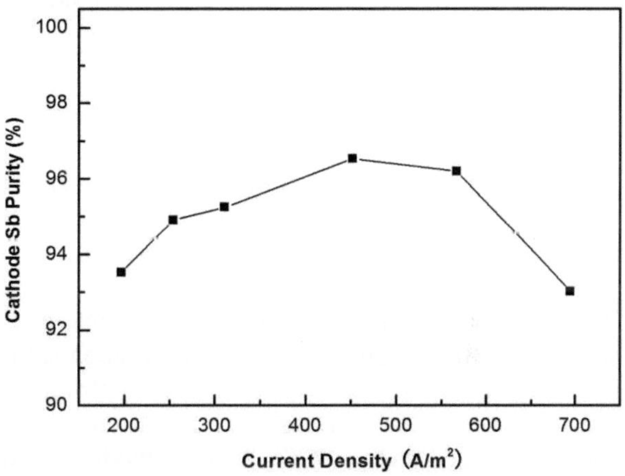

containing 12 g/L antimony. During the experiment, we found that the current density of 567.4 A/m^2 caused the temperature of the electrolyte up to 60 °C. This may explain why the current efficiency is lower in the current density of 694.7 A/m^2 than of 567.4 A/m^2. The range of the current density is much wider for the 20 g/L antimony electrolyte than for 12 g/L antimony electrolyte. Thus, the power consumption per ton antimony is also wider (Figs. 7 and 10). However, the minimum power consumption of these two batches of electrolyte is not much different, 4269 kWh/t-Sb for 20 g/L antimony electrolyte, and 4467 kWh/t-Sb for 12 g/L antimony electrolyte. Figure 11 is given a parabola curve for the relationship between the current density and the cathode antimony grade. The grades in Fig. 11 are lower than in Fig. 8. These results indicate that high antimony concentration of the electrolyte is not very necessary for antimony CE.

Table 4 Technical indicators for CE and PCE

Indicators	CE	PCE
Sb concentration in the advanced electrolytes (g/L)	10.83	16.75
Sb concentration in the spent electrolytes (g/L)	4.85	16
Time (h)	5	24
Cathode antimony grade (%)	97.42	96.40
Current density (A/m^2)	210	470
Current efficiency (%)	49.82	68.32
Power consumption (kWh/t-Sb)	3579	4467

Table 5 Economical indicators for CE and PCE

Indicators	CE	PCE
Annual earning (1000 RMB)	2140.5	1577.4
Annual power cost (1000 RMB)	84	107.5
Annual profit (1000 RMB)	1956.5	1469.9

Comparison of CE with Plate Cell Electrowinning (PCE)

To clear the technical-economic indicators, comparison tests between CE and PCE were carried out under the conditions as follow: NaOH 45 g/L, Na_2S 20 g/L, 40 °C and 5 h. It is worth noting that the antimony advanced electrolyte for PCE is 16.75 g/L, and is 10.83 g/L for CE, due to high antimony concentration need for PCE. These results are shown in Table 4. Table 4 says that CE has advantages for antimony extraction from the low concentration antimony advanced electrolyte. To know the economic indicators of these two electrowinning techniques, the benefits and costs were calculated based on the assumption that the annual operation day is 300 days, total current is 7000 A, the cathode antimony sale price is 40,000 RMB/t-Sb, and the electricity price is 0.79 yuan per kilowatt. The results are shown Table 5. From Table 5, we can find CE has better benefits than PCE. Both considering Tables 4 and 5, we can say that CE is a more promising technique in extracting metal from low concentration metal solution in the scale of industrial production.

Conclusions

The cyclone electrowinning technique has been developed to treat antimony gold concentrate ores. A series of tests was conducted to investigate the relationship between the current density and antimony CE technical indicators, such as current efficiency, power consumption per ton antimony, and cathode antimony grade.

For relative low antimony concentration (5 g/L) in the electrolyte, the current efficiency and antimony grade increase with current density increasing, the minimum power consumption per ton antimony is 5600 kWh/t-Sb.

In the moderate antimony concentration (12 g/L) in the electrolyte, 400 A/m^2 is an appropriate current density for antimony electrolyte by CE. And the resultant cathode antimony purity is 99.79%, together with current efficiency of 59.11% and 4950 kWh/t-Sb. When the current density is 475 A/m^2, the cathode antimony purity is 97.42%, and the current efficiency is 68.32%, together with the power consumption of 4467 kWh/t-Sb.

The relative high antimony concentration (20 g/L) in the electrolyte, the current density is 567.4 A/m^2, the minimum power consumption is obtained of 4269 kWh/t-Sb, and the cathode antimony purity is 96.20%.

Although the cathode antimony purity is not so good, and the power consumption of per ton antimony is not minimum enough, the technical and economical indicators for CE is better than PCE under the same experiment conditions, except the electrowinning equipment. Thus, it is worth mentioning that the CE is promising results addressed in the current work assisting in further development promising technique for extracting antimony from low antimony concentration electrolyte for commercial application after further study.

References

1. U.S. Geological survey, Mineral commodity summaries (Washington, U.S. Geological Survey, 2016)
2. Ministry of land and resources of the People's Republic of China, China's mineral resources report (Beijing, Geology Publishing House, 2015) (in Chinese)
3. Multani RS, Feldmann T (2016) Demopoulos, antimony in the metallurgical industry: a review of its chemistry and environmental stabilization options 164:141–153
4. Tang MT, Zhao TC, Lu JL et al (1992) Principle of chlorination hydrometallurgy process and its application. J Cent-South Inst. Min Metall 4:405–411
5. Zhang LZ, Jin RG, Qi XC, et al (2015) Study on antimony electrowinning from sodium sulphide alkaline leaching solution of antimony-bearing gold concentrates. Nonferrous Metal (Metall) 10:22–24,33. (in Chinese)
6. Wang CY, Qiu DF, Jiang PH (2002) Status and development of antimony metallurgy technology in China. Nonferrous Metals (Metall) 5:6–10. (in Chinese)
7. Yang JG, Wu YT (2014) A hydrometallurgical process for the separation and recovery of antimony. Hydrometallurgy 143:68–74
8. Ubaldini S, Veglio F, Fornari P, Abbruzzese C (2000) Process flow-sheet for gold and antimony recovery from stibnite. Hydrometallurgy 57:187–199
9. Barr N (1996) Metal recovery apparatus (US Patent: No. 5529672), June 25
10. Treasure PA (2000) Electrolytic zinc recovery in the EMEW Cell. In: Paper presented at the proceedings of the TMS Fall Extraction and Processing Conference, Missouri, USA, pp 185–191
11. Hu ZY, Xia XW (2011) The vortex electrolytic technology from copper recovery of copper and zinc in zinc ash study. J Chem Eng Equip 11:68–71 (in Chinese)

12. Deng T (2009) Rotational-flow electrolysis and device, (CN Patent: No. 200910136782.0), May 15, 2009
13. Peng FC, Xu Z, Ji ZG et al (2017) Removal of copper and arsenic from contaminated acid by cyclone electrowinning technology. Chin J Rare Metal 41(4):410–415 (in Chinese)
14. Escobar V, Treasure T, Dixon RE (2003) High current density EMEW copper electrowinning. In: Paper presented at the proceedings of the TMS Fall Extraction and Processing Conference, Chicago, USA, pp 1369–1380
15. Guo XY, Shi WT, Li D, Tian QH (2010) Recovery of copper and nickel from electroplating sludge by cyclone electrowinning. Chin J Nonferrous Metal 20(12):2425–2430 (in Chinese)

Part III
Base and Rare Metals

Thermodynamic Study of Ga Extraction for Trace Element Analysis by ICP-MS

Kyungjean Min, David Johnson and Kevin Trumble

Abstract Inductively Coupled Plasma Mass Spectrometry (ICP-MS) has a detection limit of sub-parts-per-trillion (ppt) level. However, it was found the detection limit increased to ~10 ppb for the analysis of impurities in ultra-pure Ga of 7 N (99.99999%) and 8 N (99.999999%) by ICP-MS due to matrix-induced interference. By extracting Ga during sample preparation, matrix-induced interference can be reduced and a lower detection limit can be achieved in the ICP-MS measurement. The dextran-based resin Sephadex G-25 can chemically separate the impurity Ge from Ga. The hydroxide complexes of Ga and Ge are adsorbed on Sephadex G-25 and desorbed into acid depending on pH. Viability of Ge separation from Ga by pH change was evaluated by thermodynamic calculation of the hydrolysis reaction of Ga and Ge. The distributions of Ga-hydroxide species and Ge-hydroxide species in 0.64 M HNO_3 were derived by numerical calculation for thermodynamic equilibrium. The effect of presence of medium and ionic strength in aqueous solutions on hydrolysis reaction was evaluated. From the derived speciation diagram for Ga and Ge hydrolysis reactions, the optimal pH range to separate impurity Ge from Ga for ICP-MS sample preparation was investigated.

Keywords Gallium · Separation · ICP-MS · Speciation diagram
Hydrolysis

Introduction

The purity of the source Ga has a proportional correlation with electron mobility in the semiconductor thin film of GaAs–AlGaAs [1, 2]. As pure Ga is demanded in semiconductor processing, the ultra-pure Ga of 6 N (99.9999%) with 1 ppm total nominal impurity concentration, 7 N (99.99999%) with 100 parts per billion (ppb) impurity concentration and 8 N (99.999999%, 10 ppb total) are commercially

K. Min (✉) · D. Johnson · K. Trumble
School of Materials Engineering, Purdue University, West Lafayette IN 47907, USA
e-mail: kjmin@purdue.edu

H. Kim et al. (eds.), *Rare Metal Technology 2018*, The Minerals, Metals & Materials Series, https://doi.org/10.1007/978-3-319-72350-1_14

available. To measure each impurity concentrations of ultra-pure Ga of 6 N, 7 N and 8 N grades, elemental analysis technique with the detection limit of ~1 ppb is required. However, elemental techniques with such high detection limit are limited.

Inductively Coupled Plasma Mass Spectrometry (ICP-MS) is known to be the state-of-the-art technique in elemental analysis with sub-parts-per-trillion (sub-ppt) detection limit. However, the detection limit of ICP-MS is degraded to 10 ppb in the metal analysis such as Ga due to cross contamination, spectral interference and matrix-induced interference (matrix effect) [3, 4]. Particularly, the high total dissolved solids (TDS) amount required in liquid ICP-MS samples to measure ultra-low concentration causes significant signal depression due to a matrix effect, resulting in the degradation of detection limit. Therefore, removing matrix Ga in the ICP-MS sample could be one of the methods to improve the detection limit.

Germanium is one of the main impurities which disturbs electron mobility in GaAs–AlGaAs heterostructures and its presence is often found in Ga relative to other elements because Ga and Ge are chemically similar. Hence only a few methods have been reported to separate Ga and Ge. Rafaeloff used organic solvent extraction by methyl ethyl ketone (MEK) to separate Ga from Ge and other group III elements [5]. The extracted percentage of Ga was from 84.3 to 99.99% depending on the concentration of acids such as HCl and H_2SO_4. The Ge of 0.08% was coextracted with Ga.

Fitzsimmons and Mausner used a Sephadex G-25 column to separate Ge from Ga [6]. The objective solution for separation containing 100 μg of Ga, Ge and Zn in HCl was adjusted to pH 13.0 with NaOH and added on Sephadex G-25. Gallium was eluted with 0.1 M NaOH and Ge was eluted with 0.1 M HCl. The column was rinsed with water between the elution of Ga and Ge. The extracted percentage of Ga was 98% with coextraction of 6.2% Ge in the 0.1 M NaOH elution. The Ge of 92% was extracted with coelution of 2.3% Ga.

In a different study by Fitzsimmons and Mausner [7], the sample solution containing Ga, Ge, Co and Zn in HCl was adjusted to pH 12.5 with $Na_3C_6H_5O_7$ and NaOH. The eluents for Ge and Ga extraction were the same as that in their previous study, but the column was rinsed using both $Na_3C_6H_5O_7$ and water between elution. The percentage of eluted Ge was reported as a function of pH from 11 to 13.5 of the sample solution. The maximum percentage of eluted Ge was 96% at pH 13. At pH 12, 90.8% of Ge was eluted.

Harada and Tarutani extracted Ge in rocks using Sephadex G-25 [8]. Germanium was preconcentrated from carbonate solution and desorbed into 0.1 M HNO_3, maintaining the ionic strength to 0.1 M with NaCl. The pH of sample solution was adjusted to pH 12 by NaOH and HCl was used to elute Ge. The degree of Ge extracted was evaluated by comparing the concentration of Ge adsorbed with that found in a standard rock sample. The degree of Ge adsorbed was smaller than 50% of Ge found in all samples.

As found in the previous studies, the separation of Ge and Ga by Sephadex G-25 is influenced by pH. This is because Ga and Ge species with different solubility are produced in different percentages from hydrolysis reaction depending on pH [9]. Consequently, an understanding of the hydrolysis reaction of Ga and Ge, the produced Ga and Ge species distribution, and the associated separation mechanism are required to establish experimental parameters for Ga extraction. Although experimental methodology and results of Ga extracted are informed by previous studies, the experimental parameters and conditions in the previous studies are not applicable to the ICP-MS sample due to spectral interference. For example, the chloride ions in a sample solution can cause interference in the measurement of Ge by ICP-MS. In this study, the effect of ionic strength in the sample solution and medium concentration was evaluated. The Ga and Ge hydrolyses distribution for an ideal ICP-MS sample condition for pure-Ga analysis were derived by thermodynamic calculations and pH range over which Ge can be separated from Ga was determined.

Background

The general hydrolysis reaction for metal cation M^{z+} is defined by [9, 10],

$$xM^{z+} + yH_2O \leftrightarrow M_x(OH)_y^{(xz-y)^+} + yH^+ \tag{1}$$

The corresponding equilibrium quotient of Q_{xy}, which is the concentration ratio of products to reactant, is defined by Eq. (2).

$$\mathrm{Log}Q_{xy} = \mathrm{Log}K_{xy} + a\frac{\sqrt{I}}{1+\sqrt{I}} + bm_x \tag{2}$$

Here, K_{xy} is the equilibrium constant of hydrolysis reaction, I is the ionic strength in aqueous solution (HNO_3 in this study) and m_x is the medium concentration. The medium (e.g. NaCl) does not participate as a reactant in the hydrolysis reaction but it interacts with ions in the aqueous solution and contributes to the equilibrium quotient [10]. The coefficient a is the product of the change in charges squared for the reaction and the Debye-Hückel limiting slope and the coefficient b is related to the interaction coefficient summed over the formation reaction for a hydrolysis product.

The speciation diagram which shows the distribution of hydrolysis species as a function of pH is derived from thermodynamics calculation using the relation between free energy of formation and the equilibrium quotient in Eq. (2). There are four hydrolyses mononuclear species formed from Ga^{3+}. Germanium(IV) with the +4-oxidation state occurs in $Ge(OH)_4$, resulting in three hydrolysis reactions. The equilibrium constant and coefficients to obtain the equilibrium quotient for each reaction of Ga and Ge hydrolysis possible are listed in Tables 1 and 2. The species $Ge_8O_{16}(OH)_3^{3-}$ from $Ge(OH)_4$ was excluded in the analysis of species distribution for Ge hydrolysis because it is found at the initial Ge concentration greater than 0.005 M [9]. The initial Ge concentration before the hydrolysis reaction in this study would be smaller than 1 ppm, considering the 1 ppm total nominal impurity concentration of 6 N Ga.

Table 1 Gallium hydrolysis reaction at 25 °C [9]

Reaction	LogK_{xy}	*a*	*b*
$Ga^{3+} + H_2O \leftrightarrow GaOH^{2+} + H^+$	−2.6	−2.044	0.4
$Ga^{3+} + 2H_2O \leftrightarrow Ga(OH)_2^+ + 2H^+$	−5.9	−3.066	0.4
$Ga^{3+} + 3H_2O \leftrightarrow Ga(OH)_3 + 3H^+$	−10.3	−3.066	0.2
$Ga^{3+} + 4H_2O \leftrightarrow Ga(OH)_4^- + 4H^+$	−16.6	−2.044	0.1

Table 2 Germanium hydrolysis reaction at 25 °C [9]

Reaction	LogK_{xy}	*a*	*b*
$Ge(OH)_4 \leftrightarrow GeO(OH)_3^- + H^+$	−9.31	1.022	−0.2
$Ge(OH)_4 \leftrightarrow GeO_2(OH)_2^{2-} + 2H^+$	−21.9	3.066	−0.4
$Ge(OH)_4 \leftrightarrow Ge_8O_{16}(OH)_3^{3-} + 3H^+$	−14.24	6.132	0.2

The ICP-MS sample were prepared to 4% HNO_3 [3]. The concentration of HNO_3 is directly connected to the ionic strength. The condition of target for hydrolysis reaction to derive speciation diagram for Ga and Ge is the sample solution containing 10,000 ppm Ga and 100 ppb Ge in 4% (~0.64 M) HNO_3. The pH for separation is adjusted with NaOH. The speciation diagrams for Ga and Ge hydrolysis were derived using parameters listed in Tables 1 and 2. The series of nonlinear equations obtained from the relation of free energy of formation and equilibrium constant used to calculate the fraction of hydrolysis species for Ga and Ge as a function of pH were solved by Newton's method.

Results and Discussion

The relative activity of hydrolysis reaction for each species can be estimated by the hydrolysis constant. The hydrolysis constants for each reaction of Ga and Ge are plotted as a function of ionic strength in Figs. 1 and 2. Among four different mononuclear hydrolyses of $GaOH^{2+}$, $Ga(OH)_2^+$, $Ga(OH)_3$, $Ga(OH)_4^-$ produced from Ga^{3+}, two species of interest are derived. Cases both with the presence and absence of 0.1 M medium were represented in all reactions.

Overall, the medium addition (0.1 M in the aqueous solution) does not significantly influence the hydrolysis constants, and thus does not increase the formation of hydrolyses. The effect of ionic strength change on the hydrolysis reaction is much greater than that of medium addition. While the Ga hydrolysis reaction increases, Ge hydrolysis reaction decreases as ionic strength increase. This result shows the hydrolysis of Ga reacts opposite to the hydrolysis of Ge for mole concentration increase in the aqueous solution.

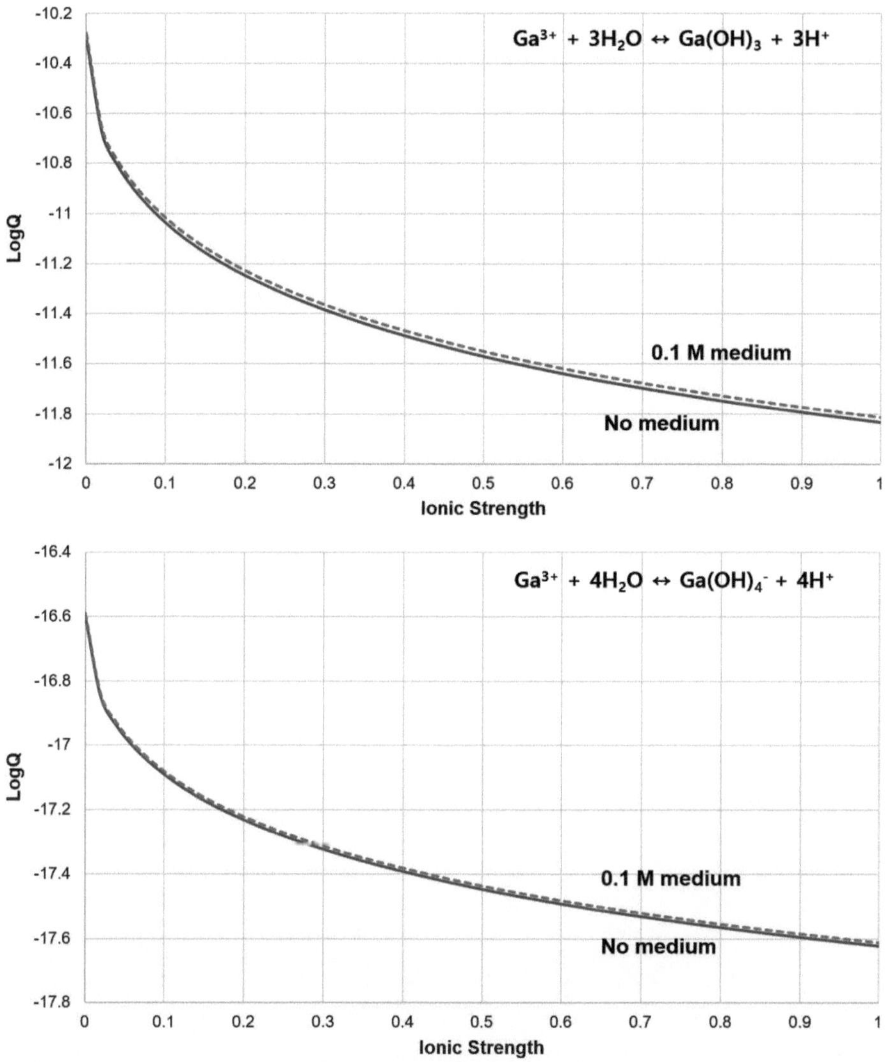

Fig. 1 Hydrolysis constant (quotient) of Ga-hydroxide species $Ga(OH)_3$ (top) and $Ga(OH)_4^-$ (bottom) as a function of ionic strength with and without 0.1 M medium

The speciation distributions for Ga and Ge hydrolysis are shown in Figs. 3 and 4. Among Ga species, $Ga(OH)_3$ is insoluble in the aqueous solution [6, 9]. On the other hand, Ge hydrolyses of $GeO(OH)_3^-$ and $GeO_2(OH)_2^{2-}$ are both soluble in the aqueous solution. In addition, Ge forms chelates with polyols, which is the hydroxyl functional group of Sephadex G-25, and the polyols are dissociated by HCl. If Ga is mostly present in the form of insoluble species of $Ga(OH)_3$ and Ge is present in the form of soluble species, then the Ga stays in Sephadex G-25 and Ge

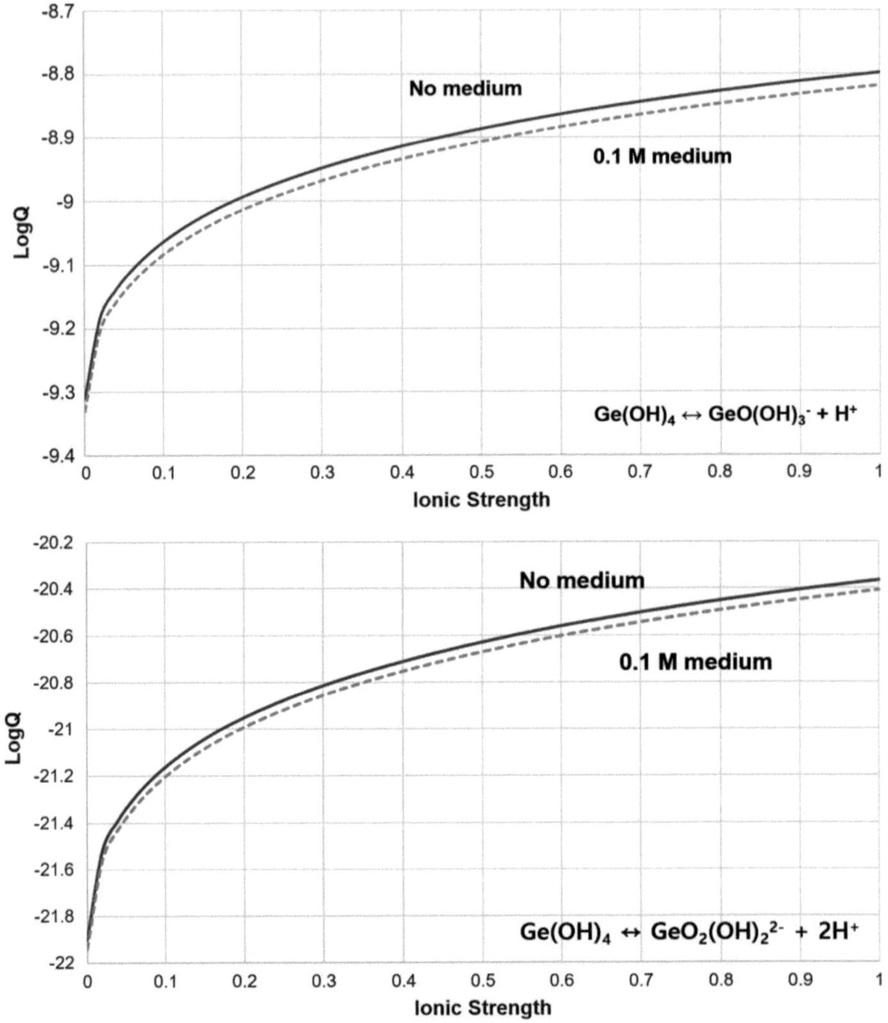

Fig. 2 Hydrolysis constant (quotient) of Ge hydroxide species $GeO(OH)_3^-$ (top) and $GeO_2(OH)_2^{2-}$ (bottom) as a function of ionic strength with and without 0.1 M medium

can be separated when it reacts with HCl and loses the chelate form. Consequently, the pH range that insoluble Ga species of $Ga(OH)_3$ and Ge can be separated is pH 10–11 in this study, considering the pH change in the Sephadex G-25 column during Ge elution by HCl addition.

The effect of 0.1 M medium addition was also studied. However, the difference of speciation distribution was so small that speciation diagrams cannot be differentiated. In the case of increased ionic strength, all speciation diagrams showed the tendency to move toward increasing pH. This indicates high concentration of acid

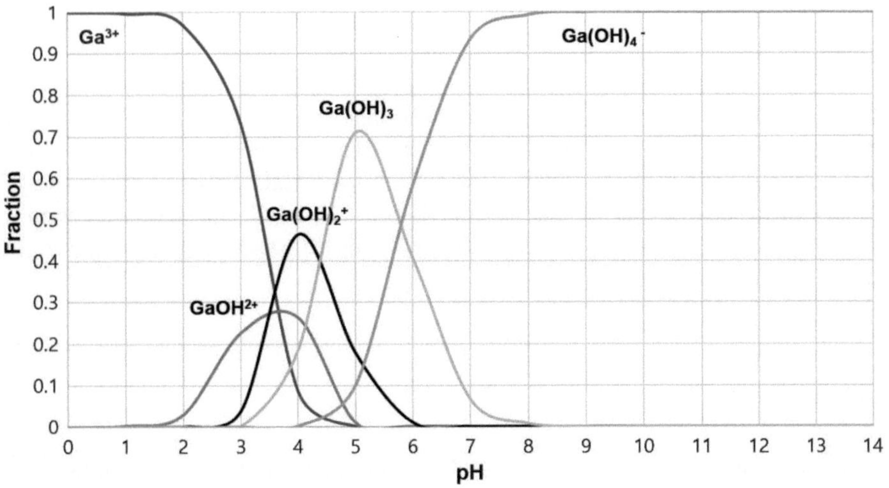

Fig. 3 Speciation diagram of Ga-hydroxide distribution as a function of pH, I = 0.64 M in the absence of medium

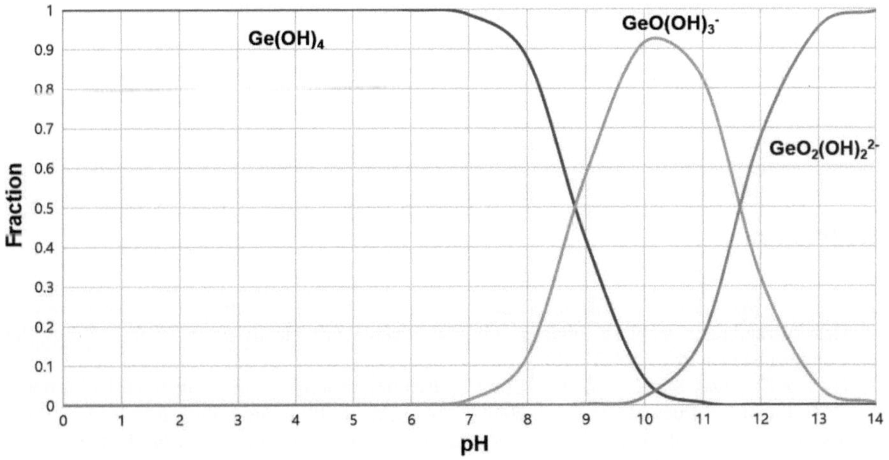

Fig. 4 Speciation diagram of Ge-hydroxide distribution as a function of pH, I = 0.64 M in the absence of medium

in the starting sample solution needs to be adjusted to higher value of pH to obtain enough fraction of soluble species of Ge to be separated. That would be a disadvantage in ICP-MS samples, where high amount of concentrated Ge is preferred to reach the detectable concentration, because it requires additional NaOH.

Harada and Tarutani [8] used pH 12 to separate Ge by Sephadex G-25 despite using lower concentration of 0.1 M HNO_3 where Ge was desorbed than in this study [8]. This is because the ionic strength could be greater than our condition due to high concentration of aqueous solutions to preconcentrate Ge from the rock sample.

The Ga speciation diagram also can be derived using thermodynamic data obtained from the solubility measurement of α-GaOOH. However, a different value of the hydrolysis constant from that in Table 1 was obtained [11]. Since the thermodynamic data from the solubility measurement of α-GaOOH was based on 150 °C and extrapolated for 25 °C, the hydrolysis constant for Ga speciation diagram in this study, which targeted to the column experiment at room temperature, used the data in Table 1 based on 25 °C.

Conclusion

For the separation of Ge from Ga by Sephadex G-25, the hydrolysis reactions of Ga and Ge in 4% HNO_3 at 25 °C were studied. The speciation diagrams of Ga and Ge hydrolysis were derived by thermodynamic calculation. The pH range to separate Ga and Ge by solubility difference was 10–11. The low concentration of acid in the starting sample has an advantage on the detection limit of ICP-MS. The effect of medium addition was negligible in the speciation distribution.

Acknowledgements Support for this research from the W. M. Keck Foundation is gratefully acknowledged (Grant No. 52445). The authors also would like to thank the ultra-pure GaAs Keck Project team members at Purdue University.

References

1. Manfra MJ (2014) Molecular beam epitaxy of ultra-high-quality AlGaAs/GaAs Heterostructures: enabling physics in low-dimensional electronic systems. Annu Rev Condens Matter Phys 5:347–373
2. Hwang EH, Das Sarma S (2008) Limit to two-dimensional mobility in modulation-doped GaAs quantum structures: how to achieve a mobility of 100 million. Phys Rev B. 77:235437
3. Min K (2014) Analysis of high-purity of Ga by ICP-MS, MS thesis, Purdue University
4. Min K, Johnson D, Trumble R, unpublished work
5. Rafaeloff R (1971) Separation of gallium from group III elements, Germanium, Copper, Arsenic, and Iron. Anal Chem 43(2):272–274
6. Fitzsimmons J, Mausner L (2015) Evaluation of materials for separation of germanium from gallium, zinc and cobalt. J Chem Chem Eng 9:462–467
7. Fitzsimmons J, Mausner L (2015) Development of a production scale purification of Ge-68 from irradiated Ga metal. Radiochim Acta 103(2):117–123
8. Harada A, Tarutani T (1988) Spectrophotometric determination of germanium in rocks after selective adsorption on Sephadex gel. Anal Chim Acta 209:333–338
9. Baes Jr. CF, Mesmer RE, (1976) The hydrolysis of cations. Wiley
10. Brown PL, Ekberg C, (2016) Hydrolysis of metal ions. Wiley-VCH
11. Wood SA, Samson IM (2006) The aqueous geochemistry of gallium, germanium, indium and scandium. Ore Geol Rev 28(1):57–102

Electrodeposition of γ-MnO_2 from Manganese Nodule Leach Liquor: Surface Modification and Electrochemical Applications

A. Baral, B. C. Tripathy and M. K. Ghosh

Abstract The electrolytic manganese dioxide (EMD) production through electrodeposition from manganese nodules by reduction leaching in H_2SO_4 medium using sucrose as reductant followed by purification through 2-stage precipitation methods is described. The purified leach liquor was the starting solution for electrodeposition. Influence of in situ addition of a cationic surfactant Cetyltri-methylammonium bromide (CTAB) (0–500 mg dm^{-3}) on the surface morphology and electrochemical properties of the EMD was examined. The XRD patterns of the deposited material were indexed to the γ-MnO_2 phase. FESEM results demonstrated that in the presence of CTAB uniform nano-fibrous type grains were formed. Higher concentration of CTAB in the electrolytic bath however, drastically decreased the current efficiency (CE) and increased the energy consumption (EC) during electrodeposition. Under the optimum CTAB concentration of 100 mg dm^{-3} the obtained CE and EC values were 87.64% and 1.70 kWh/Kg respectively.

Keywords Electrolytic manganese dioxide (EMD) · Nanomaterials
CTAB · Electrodeposition · Discharge capacity

Introduction

Electrolytic manganese dioxide (EMD) is considered as an integral part of energy storage devices like electrochemical capacitors (EC) and primary batteries. To meet the future demand of MnO_2 in energy storage application, it is beneficiary to exploit Mn bearing natural resources. Polymetallic manganese nodules found on the ocean floors are enriched with many metals such as Mn, Fe, Co, Ni, Cu, Zn, Pb, Mo, and Ti [1]. The major component of manganese nodule is manganese (20–30%) as manganese dioxide. The mineralogy of manganese oxides is more varied than other

A. Baral · B. C. Tripathy · M. K. Ghosh (✉)
CSIR-Institute of Minerals and Materials Technology, Bhubaneswar 751013, India
e-mail: mkghosh@immt.res.in

H. Kim et al. (eds.), *Rare Metal Technology 2018*, The Minerals,
Metals & Materials Series, https://doi.org/10.1007/978-3-319-72350-1_15

oxides such as aluminium or iron oxides, because besides Mn^{4+} they may contain Mn^{3+} and Mn^{2+} and for the charge balancing foreign ions as Li, Na, K, Ca, Ba, Al and Fe are easily accommodated in the structure. Numerous approaches were taken for the extraction of valuable metals from polymetallic nodules where as there are very few reports on the production of high pure manganese dioxide from these natural Mn-bearing resources. Manganese oxide is one of the most attractive inorganic materials because of its variable oxidation state and several polymorphic forms. Amongst the oxides MnO_2 is most widely applied in catalysis, ion exchange, molecular adsorption, biosensors, and energy storage areas owing to its unique physical and chemical properties. Presently tuning the shape and size of nanostructures has been one of the challenging issues for material scientists. One-dimensional (1D) nanoscale building blocks, such as nanotubes, nanobelts, nanowires, and nanorods, have attracted and increasing interest due to their importance in fundamental research and potential wide-ranging applications [2].

The electrochemical performance of MnO_2 strongly depends on its particle size, morphology, crystalline structure and bulk density. Thus, the study of manganese oxide electrode materials focuses on how to change its morphology to improve its electrochemical performance. Surfactant was widely used for preparation of various electrode materials due to its special properties [3].

Herein, we focused on the preparation of MnO_2 through electrolytic deposition process from purified polymetallic nodule leach liquor. In one of our previous studies we had reported the dissolution of metals through reductive leaching in presence of sulfuric acid and sucrose as a reductant from Indian Ocean manganese nodule [4]. The aim here is to prepare pure $MnSO_4$ solution from the nodule leach liquor by incorporation of purification steps. The purified solution has been taken as a source of electrolytic γ-MnO_2 production. The possible growth mechanism of nano γ-MnO_2 fibres was established and the detailed electrochemical properties of the electrodes were investigated. Mainly we discussed on the role of cationic surfactant (CTAB) on the changes in morphology and chemical properties of the as obtained EMD samples.

Experimental

Materials

Polymetallic Indian Ocean nodules collected from National Institute of Oceanography (NIO) Goa, India was used for this work. The nodules were crushed and ground to −100 μm size. Chemical composition of the nodule obtained through standard wet chemical analysis method is: Cu 0.93%, Ni 1.17%, Co 0.09%, Mn 24.4%, Fe 6.5%. The other chemicals used in this study were of analytical grade and purchased from Thermo Fischer Scientific, India.

Methods

Leaching

Leaching was carried out in a double-walled glass reactor (Model KGW-Isotherm, Germany, Capacity 500 mL) with provision for circulation of hot water through outside shell of the reactor in order to achieve the required temperature. The reactor lid has ports for reflux condenser, sample addition cum sampling, stirrer shaft. The hot water from Julabo® constant temperature water bath at predetermined temperature was circulated between the walls of the reactor to maintain a constant temperature. The stirring was accomplished through a Teflon anchor agitator. For each experiment 400 mL leach solution containing required amount of H_2SO_4, sucrose and 40 g of powdered nodule samples were used.

Purification of Mn Bearing Solutions

Two-stage purification process was carried out for the removal of impurities from leach liquor. Iron removal was done by addition of $Ca(OH)_2$ for raising the solution pH to 4 followed by second stage purification for the removal of other impurities like Cu, Ni, Co and Zn where sodium sulphide (Na_2S) was used as precipitating reagent. In 1st stage purification process the alkali reagent was added slowly to increase the solution pH up to 2.5. The content was agitated for an hour with proper mixing and stabilization. Then with further addition of alkali and simultaneous agitation pH was raised to $\sim$4. After filtration the 2.5 times stoichiometric amount of Na_2S was added to the filtrate, mildly agitated for 30 min and then filtered. The Mn concentration of the final purified solution was measured through volumetric analysis using EDTA.

Preparation of the MnO_2 Samples

Electrodeposition of MnO_2 was carried out in a 500 mL glass beaker placed in a thermostatic water bath in which pure lead (Pb) and stainless steel (SS) 304 grade electrodes were used as anode and cathode respectively. A regulated power supplier [0–30 V, 10A, DC Power supply, APLAB, INDIA] was used to regulate the direct current. The cell voltage and current at regular intervals were mapped through a digital multimeter. The electrodeposition process was carried out for 6 h and the depletion of Mn metal was calculated from the Mn concentration of the electrolyte determined volumetrically at each 1 h interval. Anodic current efficiency (CE) was calculated from the weight gained by the anode after electrolysis.

Characterization of the as-Prepared MnO_2 Nanofibers

X-ray powder diffraction patterns were obtained using X'PERT PANanalytical diffractometer (Model PW 1830) using Cu cathode (Cu–Kα; λ = 1.54060A°). Field Emission Scanning Electron Microscopy (FESEM) was performed on (Zeiss—Supra 55 Model). RENISHAW In via Raman Spectrometer was used to collect the Raman spectra in which an argon ion laser beam was used as the excitation source at 540 nm.

Electrochemical Activity of as Prepared Nanofiber

The electrochemical activity of the EMD samples were measured in the two electrode system where MnO_2 was used as cathode and Zn was taken as anode. The MnO_2 electrode was prepared by mixing 20% graphite powder with 80% EMD powder and grinding for 1 h followed by dropwise addition of poly vinyl alcohol (PVA). The mixture was placed in a die containing stainless steel mesh and compressed into a pellet by giving 10 ton load. The discharge profile was recorded at room temp with constant discharge current in 9 M KOH electrolyte solution. The applied discharge current was 10 mA with a cut-off voltage of 0.9 V. The galvanostatic measurements were carried out in a WBCS Battery Cycler by WonATech, Korea, (Model WBCS3000M2).

Results and Discussions

Leaching

Manganese (as MnO_2) is the major metal content in manganese nodule along with other valuable metals in the matrix and on the surface of the nodule structure. The dissolution of tetravalent manganese ions into divalent form is the main key point of releasing other locked metals especially Co from the MnO_2 matrix. The detailed leaching optimization on nodule using sucrose as reducing agent has been previously reported in one of our studies [4]. In the present study standard leaching conditions used were: 10% (v/v) H_2SO_4, 7% (w/w) sucrose, 90 °C, 2 h and S:L ratio 1:10. Under this conditions recovery achieved were: Cu 90.3%, Ni 97.4%, Co 84.4%, Mn 98% and Fe 92.3%. The leach liquor composition is given in Table 1.

Purification

The leach liquor obtained from nodule leaching was first subjected to adsorption with activated charcoal for 1 h stirring to remove organic contaminations followed by two stage purification. In the 1st stage $Ca(OH)_2$ was used to remove iron as Fe $(OH)_3$ by sequentially raising the pH first by 2.5 and then to 4.0. The concentration of different metal ions after Fe removal is given in Table 1. It is observed some small amount of metal ions losses due to mostly by adsorption of freshly precipitated $Fe(OH)_3$. In the 2nd stage purification, sodium sulphide was used as precipitating agents to separate Co^{2+}, Ni^{2+}, Cu^{2+}, Zn^{2+} as their respective insoluble sulphides. The sodium sulphide requirement for this purpose is 3 times stoichiometric requirement for Cu, Ni and Co. Due to Na_2S treatment solution pH after sulphide precipitation was increased to ~6.0. To the purified solution H_2SO_4 was added to maintain the molar ratio of Mn and the acid at 2:1 before electrolysis.

Electrolysis

The electrodeposition of MnO_2 from an acidic sulphate solution can be described by the following electrochemical reactions [5].

$$\textbf{At Anode}: Mn^{2+} + 2H_2O \leftrightarrow MnO_2 + 4H^+ + 2e^- \quad (1)$$

$$\textbf{At Cathode}: H^+ + 2e^- \leftrightarrow H_2 \quad (2)$$

However, formation of MnO_2 in aqueous sulfate solution is not a straightforward one and takes place through two steps process.

$$Mn^{2+}_{sol} \rightarrow Mn^{3+}_{ads} + e^- \quad (3)$$

$$2Mn^{3+}_{ads} \rightarrow Mn^{2+}_{sol} + Mn^{4+}_{ads} \quad (4)$$

In the 1st step reaction Mn^{2+} oxidation to Mn^{3+} occurs at the growing MnO_2 surface with some related solid intermediates (like MnOOH, Mn_2O_3, etc.). As the Mn^{3+} is unstable in hot acidic solution it readily disproportionates into Mn^{2+} and Mn^{4+}. Then fast hydrolysis reaction of Mn^{4+} forms MnO_2 deposit which traps Mn^{3+} ions [6–8].

Table 1 Concentration of various metal ions in leach liquor and after 2 stages of purification

	Cu (mg dm^{-3})	Ni (mg dm^{-3})	Co (mg dm^{-3})	Fe (mg dm^{-3})	Mn (g dm^{-3})
Leach liquor	840	1140	76	5400	23.9
After 1st stage	800	1110	71	60	23.5
After 2nd stage	1	25	5	1	23.5

The Effect of CTAB Concentration on Electrochemical Parameters

Investigations were carried out to study the effect of CTAB on the electrodeposition of MnO_2. The variation in current efficiency (CE) and energy consumption (EC) in the presence of different concentrations of cetyl tri-methyl ammonium bromide (CTAB) are given in Table 2.

In the absence of additive in the electrolytic bath maximum 71.19% of CE was achieved. In presence of 100 mg dm^{-3} CTAB in the bath solution, the CE increased to 87.64% and the EC decreased from 2.23 to 1.70 KWh kg^{-1}. Further increase in CTAB concentration reduced the CE and increased the EC. This might have happened due to the adsorption of surfactant on the electrode surface. At lower concentration moderate adsorption of surfactant molecule on the electrode surface helped in the uniform nucleation of MnO_2 and rapid growth which increases the CE, however, higher surfactant concentration accelerates the adsorption rate which tends to non-uniform motion of the particles towards the substrate reducing the CE [7].

Characterisation of EMD

Figure 1 shows the x-ray diffraction patterns of the EMD with and without CTAB. For both the samples the peak positions were same and indexed to γ phase of MnO_2 (JCPDS 01-082-2169) with space group *Pnam/60* orthorhombic crystal plane. In presence of CTAB deposited EMD demonstrates slightly broader peaks indicating decrease in particle size.

The Raman scattering spectra for EMD are shown in Fig. 2. It is observed that the Raman shifts of both the materials are approximately same. The main characteristic peaks for γ-MnO_2 are basically found in the region of 500–700 cm^{-1} [9–11]. The Raman band at 634 cm^{-1} is shifted to 637^{-1} in case of MnO_2 prepared in presence of CTAB. Fibrous γ-MnO_2 consists of the combination of two different crystal planes of MnO_2 so the structure of γ is developed by the random intergrowth of Ramsdellite (MnO_2, Ram) and Pyrolusite (β-MnO_2) structures. Peaks at lower band are attributed to the stretching vibrations of MnO_6 octahedra [12].

The surface morphologies of deposited EMD samples are shown in Fig. 3. It is observed that in the absence of CTAB in the electrolytic bath (Figs. 3a, b), the

Table 2 Effect of CTAB concentrations on current efficiency and energy consumption

CTAB (mg dm^{-3})	Current efficiency (%)	Energy consumption (kWh kg^{-1})	Cell voltage (V)
0	71.19	2.23	2.58
100	87.64	1.70	2.42
250	80.28	1.97	2.57
500	70.64	2.32	2.66

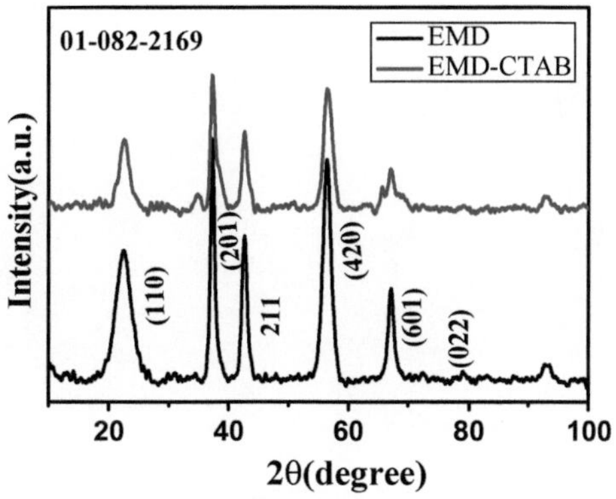

Fig. 1 X-ray diffraction patterns of EMD without CTAB and with 100 mg dm^{-3} CTAB

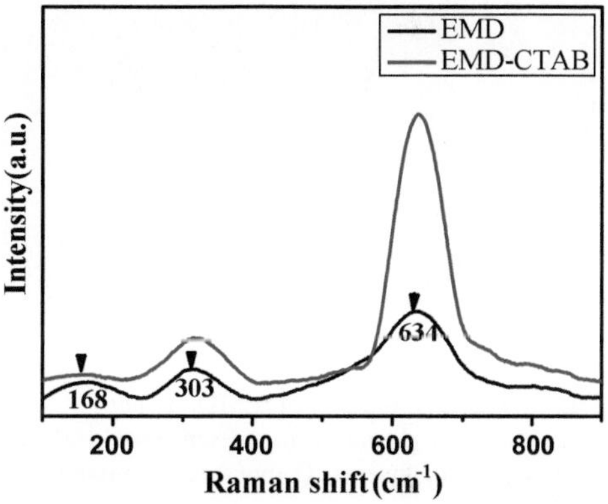

Fig. 2 Raman scattering (RS) spectra of EMD without CTAB and with 100 mg dm^{-3} CTAB

growth of nanoparticles are represented by both spherical layered aggregates along with rod like structure development (Fig. 3b) without distinct interparticle boundaries. This may due the formation mechanism of manganese dioxide crystal nucleus. In the absence of surfactant, $MnSO_4$ reacts with H_2SO_4 first and MnO_2 nuclei grows as the reaction continues which ultimately forms irregular coarse surface [3]. From Fig. 3c, it can be predicted that in presence of 100 mg dm^{-3} CTAB MnO_2 forms regular fibre like growth with clear and well defined grain boundaries. It can be observed from higher magnification image (Fig. 3d) that EMD deposit in presence of CTAB emboldens the nano-particles to grow in a way similar to the orientation of dwarf hairgrass type patterns (moss carpet) where each single grass is rooted from the flat base (Inset of Fig. 3d). The high resolution images of both the powdered samples (Figs. 3b, d) depict that in the presence of CTAB the surface morphology changes. The presence of CTAB in the electrolytic bath reduces the agglomeration formation on the electrode surface and distributes the particles evenly to get uniform smooth deposit (Figs. 3c, d).

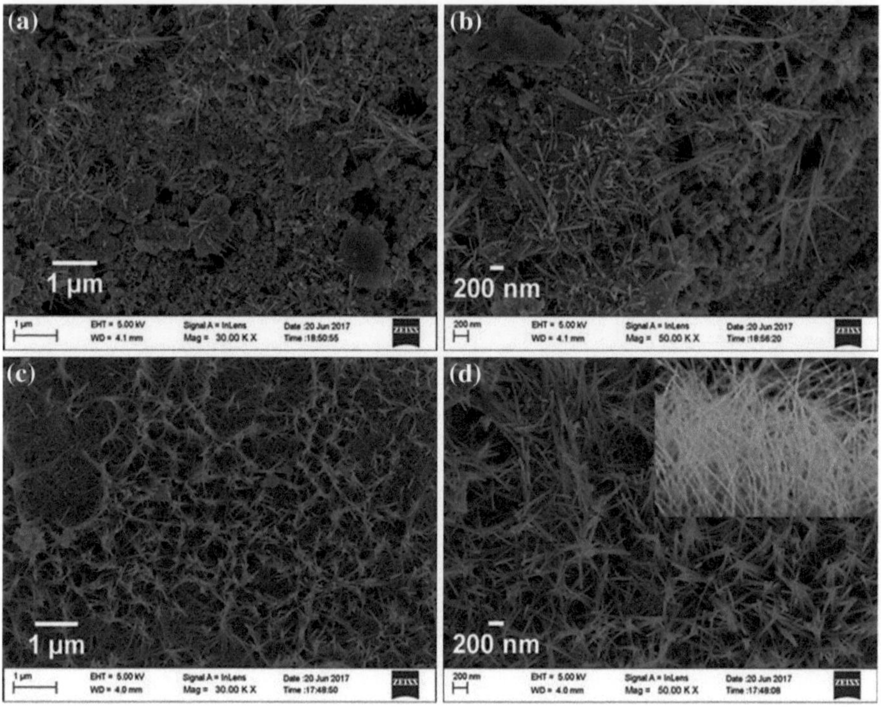

Fig. 3 FESEM images of EMD: **a** and **b** without surfactant (**a**: low magnification, **b**: high magnification) **c** and **d** in presence of 100 mg dm^{-3} CTAB (**c**: low magnification, **d**: high magnification). Inset in D moss carpet image

Electrochemical Applications

The electrochemical performance tests of EMD both in presence and absence of surfactant were carried out in galvanostatic mode. The self-discharge profiles are the characteristic measurements of primary battery activity. During discharge of the Zn–MnO_2 alkaline batteries the following possible reactions take place [13].

The MnO_2 reduction occurs at cathode in two steps as:

$$MnO_2 + H_2O + e^- \leftrightarrow MnOOH + OH^-$$

$$MnOOH + H_2O + e^- \leftrightarrow Mn(OH)_2 + OH^-$$

In the 1st step MnO_2 loses one electron and forms MnOOH where, the oxidation state of Mn is +3. Then in the 2nd step it is further reduced to Mn^{2+} by losing one more electron and is converted to $Mn(OH)_2$. The reaction of Zn electrode proceeds through dissolution-precipitation process [13]. The reaction profile progresses

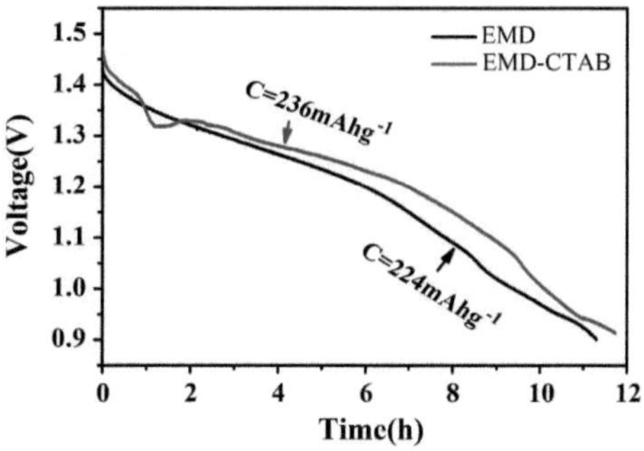

Fig. 4 Discharge performance of the as-synthesized γ-MnO_2 and γ-MnO_2@CTAB

through formation of super saturated zincate solution along with electron transformation reactions such as:

$$Zn + 4OH^- \leftrightarrow Zn(OH)_4^{2-} + 2e^-$$

Porous ZnO is produced on the Zn surface by the following chemical reaction:

$$Zn(OH)_4^{2-} \leftrightarrow ZnO + OH^- + H_2O$$

The capacity of alkaline battery is limited because of Zn electrode potential is reduced [13, 14] at the end of the reaction whereas the potential of MnO_2 remains constant throughout the discharge reaction.

The discharge capacities of MnO_2 and MnO_2 @ 100 mg dm^{-3} CTAB at the current density of 10 mA were examined with a cut-off voltage (COV) of 0.9 V. It can be observed from the discharge profile graph of the sample (Fig. 4) that by addition of CTAB in the electrolytic bath the discharge capacity of EMD is enhanced. From the potential–time plot it is noticed that a significant voltage drop at the very beginning of discharge process is obtained due to the system resistance.

The discharge profiles of EMD deposits reveal that in absence of surfactant the discharge capacity of MnO_2 is less i.e. 224 mA hg^{-1} but by adding 100 mg dm^{-3}CTAB discharge capacity is increased to 236 mA hg^{-1}. The presence of surfactant accelerates the passage of electrons because the primary batteries release all the electrochemical energy in one step (self-discharge capacity) [13].

Conclusion

Electrolytic manganese dioxide can be prepared from Mn bearing natural resources such as polymetallic nodule containing other metal besides manganese through leaching and 2-stage purification steps. The surface modification through cationic surfactant CTAB can greatly improve the electrochemical properties. Electrodeposition from purified solution produces γ-MnO_2 whether or not CTAB is added during electrodeposition. Concentration of added CTAB of more than 100 mg dm^{-3} during electrodeposition adversely affects both CE and EC. Presence of 100 mg dm^{-3} surfactant decreases the size of the particles and results in uniform deposits with clear grain boundaries. Coarse deposits are formed at higher concentrations and in absence of CTAB in electrolytic bath. The battery activity of as prepared EMD is intensified in presence of 100 mg dm^{-3} CTAB as compared to without CTAB and the increase in self-discharge capacity is found to be 236 mA hg^{-1} from 224 mA hg^{-1}.

Acknowledgements The authors are thankful to the Director, CSIR-IMMT for his kind consent to publish this work. Authors are also thankful to Subrat K Padhi, Senior Research Fellow, CSIR-IMMT for his help and suggestions.

References

1. Post JE (1999) Manganese oxide minerals: crystal structures and economic and environmental significance. Proc Natl Acad Sci 96:3447–3454
2. Song XC, Zhao Y, Zheng YF (2007) Synthesis of MnO_2 nanostructures with sea urchin shapes by a sodium dodecyl sulfate-assisted hydrothermal process. Cryst Growth Des 7 (1):159–162
3. Zhang H, Gu J, Jiang Y, Zhao J, Zhang X, Wang C (2014) Effects of sodium dodecyl sulfate on the electrochemical behavior of super capacitor electrode MnO_2. J Solid State Electrochem 18:235–247
4. Baral A, Dash B, Ghosh MK, Subbaiah T, Minakshi M (2015) Pathway of sucrose oxidation in manganese (pyrolusite) nodule. J Ind Eng Chem 54 (49):12233–12241
5. Zhang W, Cheng CY (2007) Manganese metallurgy review. Part I: leaching of ores/secondary materials and recovery of electrolytic/chemical manganese dioxide. Hydrometallurgy 89:137–159
6. Biswal A, Dash B, Tripathy BC, Subbaiah T, Shin SM, Sanjay K, Mishra BK (2013) Influence of alternative alkali reagents on Fe removal during recovery of Mn as Electrolytic Manganese Dioxide (EMD) from Mn sludge. Hydrometallurgy 140:151–162
7. Ghaemi M, Khosravi-Fard L, Neshati J (2005) Improved performance of rechargeable alkaline batteries via surfactant-mediated electro synthesis of MnO_2. J Power Sources 141:340–350
8. Biswal A, Sanjay K, Ghosh MK, Subbaiah T, Mishra BK (2011) Preparation and characterization of EMD from manganese cake—a by product of manganese nodule processing. Hydrometallurgy 110:44–49
9. Julien C, Massot M, Rangan S, Lemal M, Guyomard D (2002) Study of structural defects in γ-MnO_2 by Raman spectroscopy. J Raman Spectrosc 33:223–228
10. Wei M, Konishi Y, Zhou H, Sugihara H, Arakawa H (2005) Synthesis of single-crystal manganese dioxide nanowires by a soft chemical process. Nanotechnology 16:245–249
11. Devaraj S, Munichandraiah N (2008) Effect of crystallographic structure of MnO_2 on its electrochemical capacitance properties. J Phys Chem C 112:4406–4417
12. Chen H, Wang Y, Lv YK (2016) Catalytic oxidation of NO over MnO_2 with different crystal structures. RSC Adv 6:54032–54040
13. Ghavami RK, Rafiei Z, Tabatabaei SM (2007) Effects of cationic CTAB and anionic SDBS surfactants on the performance of Zn–MnO_2 alkaline batteries. J Power Sources 164:934–946
14. Cheng BF, Chen J, Gou X, Shen P (2005) High-power Alkaline Zn–MnO_2 batteries using γ-MnO_2 nanowires/nanotubes and electrolytic Zinc Powder. Adv Mater 17:2753–2756

Recovery of Manganese from Scrap Batteries of Mobile Phones

Deblina Dutta, Rekha Panda, Manis Kumar Jha and Sudha Goel

Abstract Present work is focused on the recovery of Mn as a value added product from the leach liquor of scrap lithium-ion batteries (LIBs) of mobile phones by the method of precipitation. The LIBs were crushed and beneficiated by wet scrubbing method to separate cathodic material, plastic and metallic parts. The cathodic material was found to contain 11.3% Mn, 10% Co, 2.4% Cu and 2.4% Li. The cathodic material was processed for leaching under the optimized condition developed by our group at CSIR-NML. Solvent extraction method was used to extract acid using organic extractant Tris(2-ethylhexyl)amine (TEHA) in order to reduce the consumption of alkali required during precipitation studies. The acid free leach liquor was subjected to purification for removal of Fe, Li, Cu and Co as precipitate at different pH. Systematic precipitation studies were carried in batch and continuous mode to recover Mn as $Mn(OH)_2$ at pH ~ 10 which was further roasted at 450 °C for 4 h to get pure Mn_3O_4.

Keywords Lithium-ion batteries · Mobile phones · Precipitation studies
Manganese precipitate

Introduction

Manganese (Mn) being an abundant metal in the earth's crust is present in rocks, soil and water. The largest producers of Mn are Australia, Brazil, Africa, China, India and Gabon. Mn is present in ores namely pyrolusite (MnO_2), rhodochrosite ($MnCO_3$) manganite ($Mn_2O_3.H_2O$) and haussmannite (Mn_3O_4) [1]. About 24% Mn

---

D. Dutta (✉) · S. Goel
School of Environmental Science & Engineering, Indian Institute of Technology Kharagpur, Kharagpur 721302, India
e-mail: deblina69envs@gmail.com

D. Dutta · R. Panda · M. K. Jha
Metal Extraction and Recycling Division, CSIR-National Metallurgical Laboratory, Jamshedpur 831007, India

H. Kim et al. (eds.), *Rare Metal Technology 2018*, The Minerals, Metals & Materials Series, https://doi.org/10.1007/978-3-319-72350-1_16

are found in the sea nodules present under the oceans for recovering of which various processes were followed and are reported [2–4]. According to mineral commodities survey, 2017 [5], 78% of the manganese resources are found in South Africa and 10% in Ukraine. In the USA, low grades of manganese resources are available and are irregularly distributed. India has a large reserve of Mn distributed in different states. Mn has applications mainly in the iron and steel making industries, in the manufacturing of mobile batteries, plant fertilizers and animal feed and as brick colorant. It is too brittle to be used as pure metal and so it is usually blended with other metals for commercial applications. In recent times, Mn plays a very important role in the production of the local mobile phone batteries for its easy availability and in terms of cost in comparison to cobalt. The life span of mobile phones decreases thereby increasing the quantity of batteries entering the municipal waste stream. Urban mining of metals can be considered as a boon to this situation. Recovering Mn from various electronic and electrical equipment before their disposal can meet the need of Mn in industries. Alkaline batteries, Lithium Ion Batteries (LIBs) comprise of Mn in its cathodic material blended with cobalt, lithium and aluminium which can be recovered after their end of life. This process can lead to a safe environment and society.

Various literatures are available for the processes used for the recovery of Mn from its ores. Li et al. [6], reported 96.47% Mn recovery from low grade ores by following leaching process optimizing the various process parameters. Leaching of high grade Mn ore with H_2SO_4 leads to reduction of the iron and sulphide content [7]. Baba et al. [1], reviewed all the processes involved in recovering Mn through hydrometallurgical process, its concentration, purification and extraction processes from ores and waste materials. Besides ores, Mn can also be recovered from the waste materials used in our daily life after their end of life. Very few studies are available to recover Mn from LIBs. He et al. [8] reported recovery of 99.31% Mn, 99.07% Li, 98.64% Co, and 99.31% Ni from LIBs in a leaching process in presence of L-tartaric acid and H_2O_2 99.7% Mn was recovered by using guar meal as a reducing agent with H_2SO_4 under optimized conditions [9]. Salgado et al. [10], reported the recovery of Zn and Mn from spent alkaline batteries by leaching and solvent extraction processes. Chemical precipitation studies have been considered a suitable process for the separation and purification of metals from its solution. Literatures are available for chemical precipitation of metals. About 60% cobalt, 99.99% iron and 10% copper were precipitated at pH 3 at 25 °C varying the pH with the addition of KOH [11]. Baba et al. [1], have reported four different precipitation processes which include hydroxide precipitation of Mn, oxidative precipitation of MnO_2, sulphide precipitation of metals from Mn^{2+} solution and ammonia/carbonate precipitation of $MnCO_3$. The present study represents the recovery of Mn from LIBs following the steps of leaching, solvent extraction of acid and selective precipitation study. The final product obtained after the roasting of dried precipitate at optimized condition is pure Mn_3O_4.

Materials and Methods

LIBs from different mobile phones were collected and crushed in a scutter crusher (Make: Hoysung, South Korea). LIBs comprised of metal, plastics, organics and cathodic material which contains blended mixture of Co, Mn, Li etc. Following scrubbing and air sparging processes, the cathodic material was separated out from metals and plastics and dried. For this experiment, the cathodic material is considered as the sample for characterization study and recovery of Mn. The cathodic powder was analyzed in X-ray Powder Diffraction (XRD) (PANalytical X Pert Powder, UK) for the phase identification of the metals present. Analysis of the samples was done in Atomic Absorption Spectrophotometer (AAS) (Perkin Elmer model, Analyst 200, USA).

Results and Discussion

Characterization Study

Figure 1 depicts the result obtained in XRD analysis of the cathodic material of LIBs, which indicates the majority of MnO_2 present in cathodic material.

Leaching Procedure

Leaching studies were carried out for dissolution of the metals present in the cathodic material of LIBs. The various process parameters such as acid

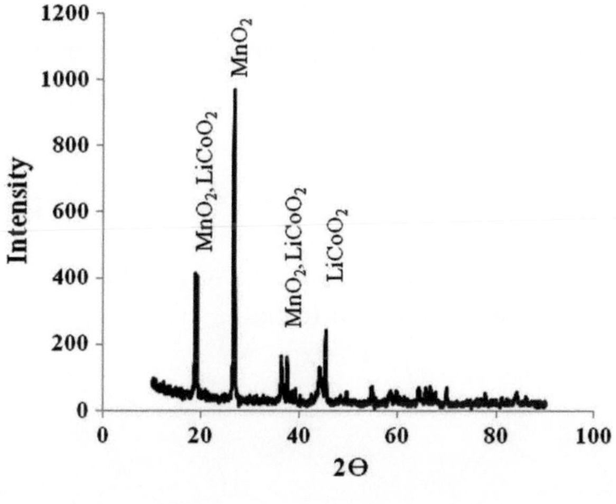

Fig. 1 XRD analysis of the cathodic material of LIBs

concentration, pulp density, temperature and time were optimized. 2 M H_2SO_4 at elevated temperature, mixing time 1 h and pulp density 100 g/L was found to be optimized condition for 99.99% recovery of Mn. After leaching, the solution was filtered and the residue obtained was processed for its Toxicity Characteristic Leaching Procedure (TCLP) analysis for safe disposal and the leach liquor generated containing the different metals was analyzed in Atomic Absorption Spectrophotometer (AAS) (Perkin Elmer model, Analyst 200, USA) for its metal concentration and further studied for separation and purification of the metals.

Recovery of Acid by Solvent Extraction

To control the pH for precipitation studies, experiments were carried out to recover acid from the leach liquor using 70% TEHA (Sigma-Aldrich) mixed with kerosene at O/A ratio 2/1 in three stages in 5 min. 99.99% of acid was extracted from the leach liquor [12]. Stripping of the loaded organic was carried out after each set of experiment. After the extraction of acid, the leach liquor was collected and precipitation study was done in a systematic and scientific way.

Systematic Precipitation Studies

After extraction of acid, the leach liquor was processed for the precipitation studies. pH plays the important role in this study. Different pH is adjusted to separate out the selective metals from the solution. Systematic studies were carried out to acquire the purified Mn concentrate. The general equation for the equilibrium of metal hydroxide can be written as:

$$M^{n+} + nOH \leftrightarrow M(OH)_n \quad (1)$$

For the hydroxide precipitation, solubility product (K_s) is a main function for the equilibrium for selective metal precipitation and the equilibrium constants can be represented as

$$K_1 = \frac{1}{M^{n+}[OH]^n} = \frac{1}{K_s} \quad (2)$$

where K_s represents the solubility product [1]. The present study has been done at room temperature in two studies.

Batch Precipitation Studies

Initially the experiments were carried out to see the precipitation behavior of the various metals viz. Mn, Li, Co, Cu and Fe present in the leach liquor at different pH (5, 7, 8, 9, 10, 11). 50 ml of each leach liquor were taken in the conical flask and pH was maintained by the addition of alkali/acid as required.

The results as photographs presented in Fig. 2 show the various salt in natural color, which supports the precipitation data presented in Fig. 3. At equilibrium pH 5 the analytical results indicates the 99% precipitation of Fe and Li, 28% Cu and 36% Co and no Mn precipitation. However, at equilibrium pH 7, 99% of Fe and Li, 62% Cu and 75% Co was obtained. Further, at pH 8, 99% of Fe, Li, Cu, Co was obtained. When increase in pH 8–9, 99% Fe, Li, Cu, Co and 28% Mn were observed. Therefore, to precipitate Mn, pH was increased to 10 and at last, pH 11

Fig. 2 Batch precipitation of metals at different pH with fresh leach liquor

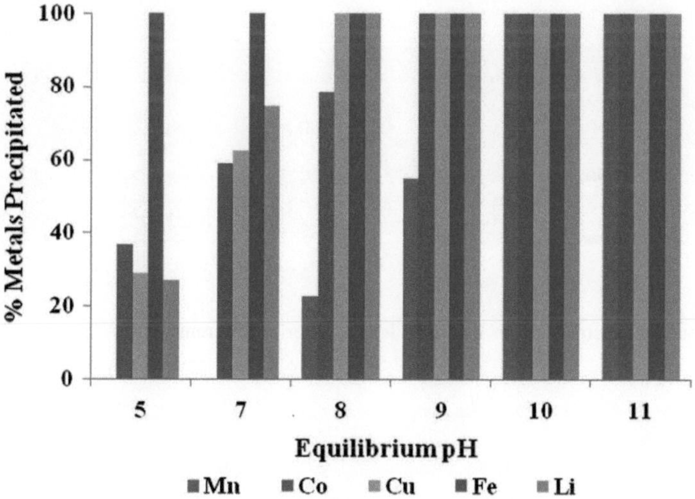

Fig. 3 Precipitation of metals at different pH

was maintained to acquire 99% Mn as precipitate. Finally the results indicate a higher pH of more than 10 is suitable for Mn precipitation salt.

Repeated Precipitation of Same Leach Liquor

Based on the previous results obtained in batch condition, the experiment was planned to simulate in continuous mode applicable to industry. Therefore, 200 ml of leach liquor was taken and pH was maintained to 5. The precipitate was filtered and analyzed. The result indicates the precipitation of all Fe and Li. Then the filtered solution, free of Fe and Li, was processed to achieve the pH 7 by addition of alkali where Co and Cu were precipitated mostly and the solution was filtered out. However, an increase in pH from 7 to 8 completely precipitated the Co and Cu. Increase in the pH from 8 to 9 resulted in partial precipitation of Mn obtained in the residue after filtration. Therefore, the pH was adjusted to 10 and most of the Mn present was precipitated out. Lastly, pH of the filtered solution was maintained at pH 11 and complete precipitation of the pure Mn was observed. Figure 4 shows the precipitate and solution obtained after each set of pH adjustment and filtration. Figure 5 represents the pH on which complete precipitation of selective metals was obtained from the leach liquor at different pH. Finally the precipitate of Mn salt was subjected to roasting to get Mn_3O_4 which is a marketable product (Fig. 6).

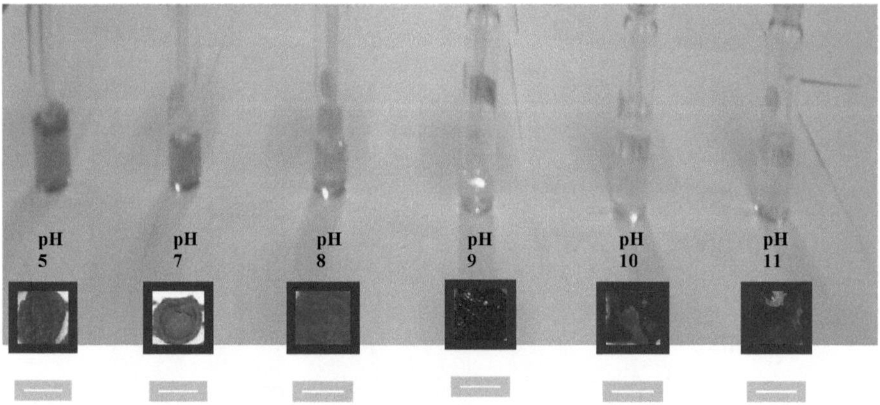

Fig. 4 Repeated precipitation of the same leach liquor at different pH

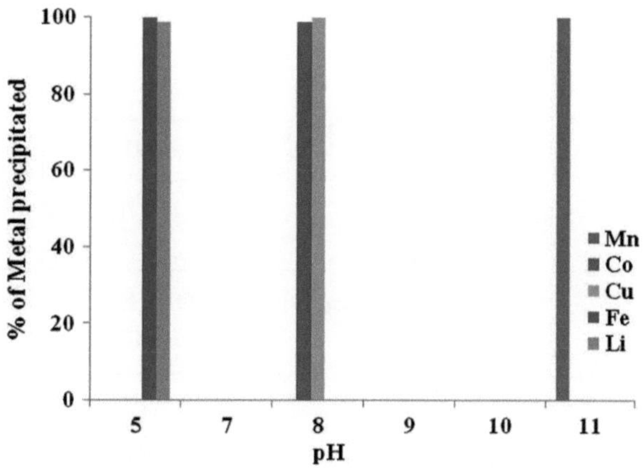

Fig. 5 Repeated precipitation of the same leach liquor at different pH

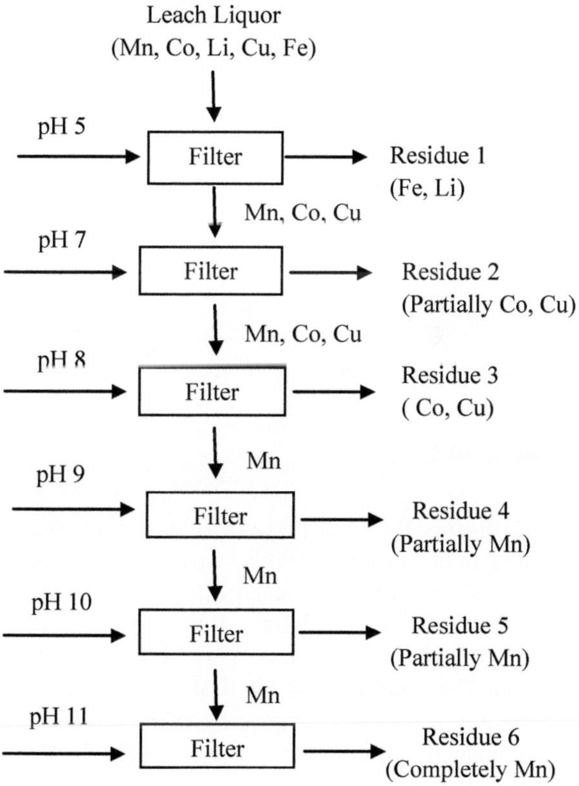

Fig. 6 Repeated precipitation of the same leach liquor at different pH to get various metal salts

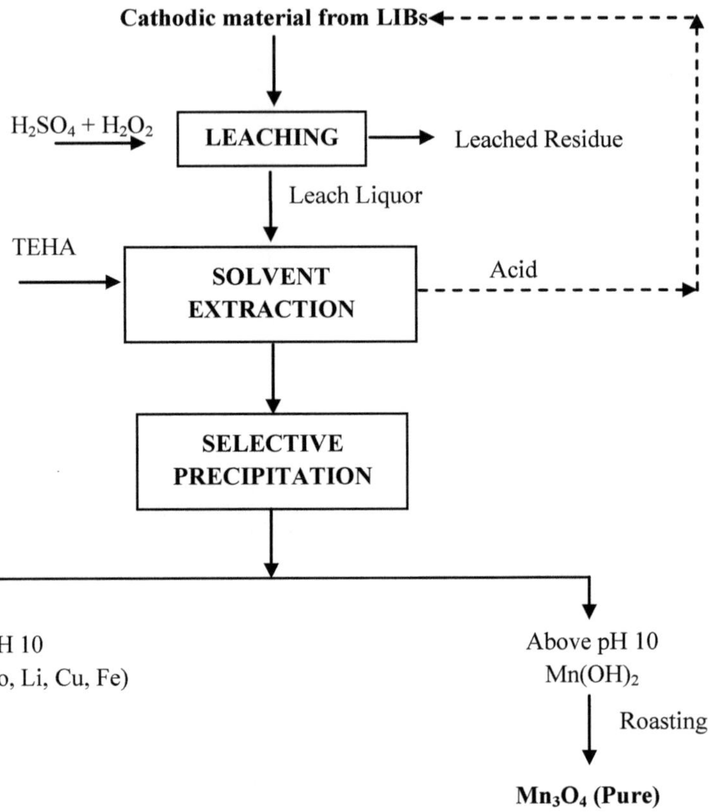

Fig. 7 Process flow-sheet for the recovery of Mn salt from waste LIBs

Conclusion

The hydrometallurgical process followed in this study for the separation and purification of Mn from LIBs is safe and environmentally friendly. Procuring all the metals and acid from the leach liquor, precipitation studies were carried out and resulted in pure Mn precipitate in hydroxide form. Further roasting of the residue obtained at pH 10 & 11 can provide Mn_3O_4 which is a marketable and value added product. The total process aims at a zero waste concept and development of a process for recovery of various metals from the e-wastes thus preventing any environmental harm. Figure 7 represents the process flow-sheet for the recovery of Mn from LIBs.

Acknowledgements Authors acknowledge the joint venture of IIT-Kharagpur and CSIR-NML, Jamshedpur for the e-waste recycling project and for giving the permission to publish this article. One of the authors Ms. Deblina Dutta, Senior Research Fellow is thankful to IIT-Kharagpur for financial support.

References

1. Baba AA, Ibrahim L, Adekola FA, Bale RB, Ghosh MK, Sheik AR, Pradhan SR, Ayanda OS, Folorunsho IO (2014) Hydrometallurgical processing of manganese ores: a review. JMMCE 2:230–247
2. DeCarlo EH, Zeltlln H (1982) Separation of copper, cobalt, nickel, and manganese from deep-sea ferromanganese nodules by adsorbing colloid flotation. Anal Chem 54:898–902
3. Kohga T, Imamura M, Takahashi J, Tanaka N, Nishizawa T (1995) Recovering iron, manganese, copper, cobalt, and high purity nickel from sea nodules. JOM 47(12):40–43
4. Premchand, Jana RK (1999) Processing of polymetallic sea nodules: an overview. In: Proceedings of the third ocean mining symposium. The International society of offshore and polar engineers. ISBN 1-880653-45-1
5. Mineral Commodities Summaries. https://minerals.usgs.gov/minerals/pubs/mcs/
6. Li CX, Zhong H, Wang S, Xue JR, Wu FF, Zhang ZY (2015) Manganese extraction by reduction-acid leaching from low-grade manganese oxide ores using CaS as reductant. Trans Nonferrous Met Soc China 25:1677–1684
7. Ibrahim IA, Abdel-Aal EA, El-Safty NA, Ismail AK (1995) Hydrometallurgical beneficiation of managenese ore from Sinai. Fizykochemiczne Problem y Mineralurgii. 29:73–79
8. He LP, Sun SY, Mu YY, Song XF, Yu JG (2016) Recovery of lithium, nickel, cobalt, and manganese fromspent lithium-ion batteries using L-tartaric acid as a leachant ACS Sustainable Chem. Eng. doi:https://doi.org/10.1021/acssuschemeng.6b02056
9. Kursunoglu S, Kaya M (2013) Recovery of manganese from spent batteries using guar meal as a reducing agent in a sulfuric acid medium. Ind Eng Chem Res 52:18076–18084
10. Salgado AL, Veloso AMO, Pereira DD, Gontijo GS, Salum A, Mansur MB (2003) Recovery of zinc and manganese from spent alkaline batteries by liquid-liquid extraction with Cyanex 272. J Power Sources 115:367–373
11. Kirchain RE, Blanpain B, Meskers C, Olivetti E, Apelian D, Howarter J, Kvithyld A, Mishra B, Neelameggham NR, Spangenberger (2016) Precipitation of metals from liquor obtained in nickel mining. Rewas 2016: towards Materials Resource Sustainability. doi: https://doi.org/10.1002/9781119275039.ch52
12. Kumari A, Jha MK, Lee JC and Singh RP (2016) Clean process for recovery of metals and recycling of acid from the leach liquor of PCBs. J. Clean Prod. 112:4826–4834

The Management of Lead Concentrate Acquisition in "Trepca"

Ahmet Haxhiaj and Bajram Haxhiaj

Abstract Based on the placement of lead and its consumption in industry branches, the paper deals with the composition of lead in the massive of Kopaonik. Removal, tenuity and grinding of galenit. The paper in particular treats the flotation process, and also treats physico-chemical bases of reagents in the process, as well as decanting, drying, filtering and storage of concentrates. Verification of the chemical composition of Pb concentrates with Pb, Zn, Bi, As, Sb, In, Au, Ag, etc. It is special that in Kopaonik massive the ratio of Pb: Zn composition is 1.4:1.0 and during the flotation lead is inclined to go with pirotines concentrate more than is allowed. We are strongly focused to maximally decrease the Pb amount that goes with pirotine concentrates and wastes. The paper as such gave the first effects in optimizing of these parameters with positive effects in flotation process in Trepca.

Keywords Mineral · Reagents · Grinding · Flotation · Pb concentrate
Pirotine · Sterile

Introduction

The management of lead concentrates quantity, lead's quantity in Pb and Zn concentrates, also Pb left quantity in flotation process in "Stan Terg", especially. The process is complex and necessary in gaining lead in technological processes. The article treats: equipment components, chemistry components dhe operation components. In flotation process is dosed sulfural mineral, which has this chemical composition

A. Haxhiaj (✉)
Faculty of Geosciences, University of Mitrovca "Isa Boletini",
Mitrovica, Republic of Kosovo
e-mail: ahaxhiaj52@yahoo.com

B. Haxhiaj
Faculty of Civil Engineering and Architecture, University of Prishtina
"Hasan Prishtina", Prishtina, Republic of Kosovo

H. Kim et al. (eds.), *Rare Metal Technology 2018*, The Minerals,
Metals & Materials Series, https://doi.org/10.1007/978-3-319-72350-1_17

$3.17\% \leq Pb \leq 3.85\%, 2.26\% \leq Zn \leq 2.54\%$ and 51.6 g/t $\leq Ag \leq 59.2$ g/t. As such for enrichment is preferred the use of selective flotation process. Lead composition residue and residue losses are complex and depends on many factors. Managing of Pb concentrate quantity and residues of lead is subject of study in this paper that refers to technological parameters of flotation process. The parameters that influence lead loses by residue are: load balance, load temperature, grinding and chemical composition of load for flotation. Industrial process of flotation analyses, chemical analyses of flotation products during benefit of lead concentrates made to achieve technico-technological and economical effects.

From Literature

Froth Flotation—Fundamental Principles

Froth flotation is a highly versatile method for physically separating particles based on differences in the ability of air bubbles to selectively adhere to specific mineral surfaces in a mineral/water slurry. The particles with attached air bubbles are then carried to the surface and removed, while the particles that remain completely wetted stay in the liquid phase. Froth flotation can be adapted to a broad range of mineral separations, as it is possible to use chemical treatments to selectively alter mineral surfaces so that they have the necessary properties for the separation. It is currently in use for many diverse applications, with a few examples being: separating sulfide minerals from silica gangue (and from other sulfide minerals); separating potassium chloride (sylvite) from sodium chloride (halite); separating coal from ash-forming minerals; removing silicate minerals from iron ores; separating phosphate minerals from silicates; and even non-mineral applications such as de-inking recycled newsprint. It is particularly useful for processing fine-grained ores that are not amenable to conventional gravity concentration.

Particle/Bubble Contact

Once the particles are rendered hydrophobic, they must be brought in contact with gas bubbles so that the bubbles can attach to the surface. If the bubbles and surfaces never come in contact, then no location can occur. Contact between particles and bubbles can be accomplished in a flotation cell such as the one shown schematically in Fig. 1.

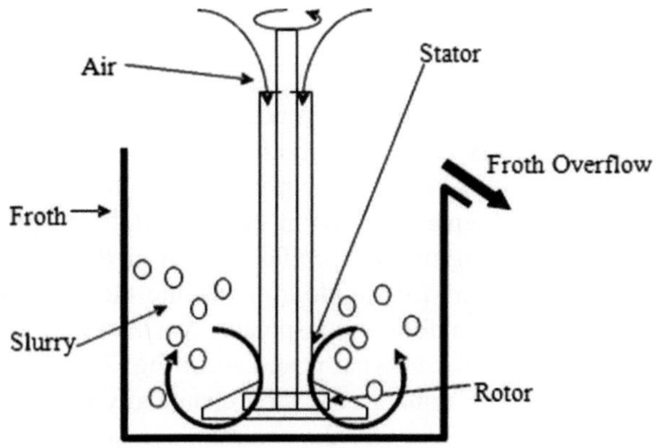

Fig. 1 Simplified schematic of a conventional flotation cell. The rotor draws slurry through the stator and expels it to the sides, creating a suction that draws air down the shaft of the stator. The air is then dispersed as bubbles through the slurry, and comes in contact with particles in the slurry that is drawn through the stator. Particle/bubble collision is affected by the relative sizes of the particles. If the bubbles are large relative to the particles, then fluid flowing around the bubbles can sweep the particles past without coming in contact. It is therefore best if the bubble diameter is comparable to the particle diameter in order to ensure good particle/bubble contact

Conventional Cells

Conventional flotation cells consist of a tank with an agitator designed to disperse air into the slurry, as was previously shown schematically in Fig. 5. These are relatively simple machines, with ample opportunity for particles to be carried into the froth along with the water making up the bubble films (entrainment), or for hydrophobic particles to break free from the froth and be removed along with the hydrophilic particles. It is therefore common for conventional flotation cells to be assembled in a multi-stage circuit, with “rougher”, “cleaner”, and “scavenger” cells, which can be arranged in configurations such as the one shown in Fig. 2.

Fig. 2 One possible configuration for a Rougher/Cleaner/Scavenger flotation circuit

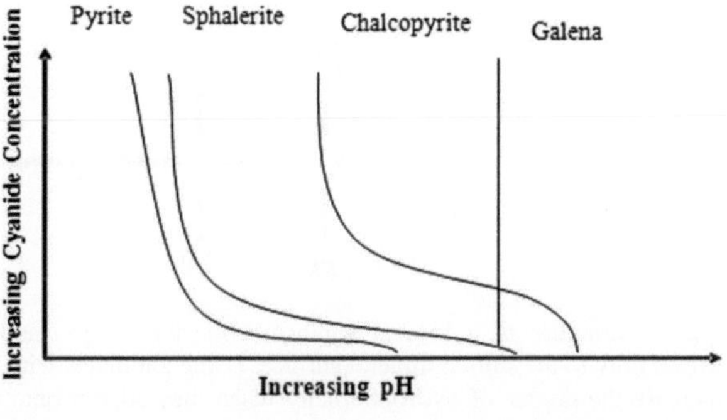

Reagents

The properties of raw mineral mixtures suspended in plain water are rarely suitable for froth flotation. Chemicals are needed both to control the relative hydrophobicities of the particles, and to maintain the proper froth characteristics. There are therefore many different reagents involved in the froth flotation process, with the selection of reagents depending on the specific mineral mixtures being treated.

The most common collectors for sulfide minerals are the sulfhydryl collectors, such as the various xanthates and dithiophosphates. Xanthates are most commonly used, and have structures similar to what is shown in Fig. 3. Xanthates are highly selective collectors for sulfide minerals, as they chemically react with the sulfide surfaces and do not have any affinity for the common non-sulfide gangue minerals. Other highly-selective collectors for use with sulfide minerals, such as ithiophosphates, have somewhat different adsorption behavior and so can be used for some separations that are difficult using xanthates.

The Management of Lead Mineral Resources

Place resources of Pb ores are composite place resources, that except Pb minerals also contain significant quantity of Zn minerals. So in most cases when it comes to Pb ores we mean lead and zinc ores. In smaller quantity are also present other metal minerals as Au and Ag minerals, Fe minerals, Cu minerals, Bi minerals, etc (Table 1).

Trepca's mine has modern infrastructure based on contemporary underground mining model. It has three wells developed in 12 horizons, while in the last three horizons are developed ramps that enable application of modern mechanisms for preparing the ore. Mine with its infrastructure as well as with modern methods of filling, has the capacity of ore use from 500,000 up to 650,000 tons a year.

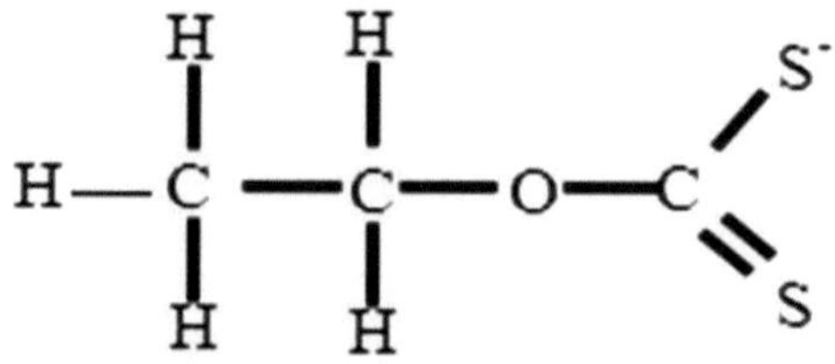

Fig. 3 Structure of a typical xanthate collector (ethyl xanthate). The OCSS- group attaches irreversibly to the sulfide mineral surface. Using xanthates with longer hydrocarbon chains tends to increase the degree of hydrophobicity when they adsorb onto the surface

Table 1 Presents the chemical composition of lead ores in Kopaunik massif

Type of mineral	% Pb	% Zn	g/t Ag	% Fe	% Cu	% S	% SiO_2
1	5.5	13.0	–	9.4	–	–	18.0
2	8.5	13.8	–	1.8	1.0	–	20.0
3	6.0	13.1	–	–	–	16.0	19.0
4	4.0	2.8	60	–	–	–	–
5	2.8	2.1	51	–	–	–	–
6	3.7	2.4	52	–	–	–	–

Galenit Quality

Lead with sulfur forms only a component which can be found in nature also in the form of clear mineral, as dark colored and metal shining Galenit. Temperature of its melting is 1135 °C. In absence of air the galena starts to sublimate in temperature about 7500 °C. Galena belongs to the group of components that evaporate easily and this attribute is the major cause of lead losses during its processing. Place resources of lead in Kosovo lie in the so-called metallurgical belt of Trepca, which is situated in the northeast of Kosovo starting from Leposavic to Gjilan. Length of the belt is over 80 km and average width of it is 30 km (Table 2).

Flotation Process

The ore from Stan Terg mine is carried by train to flotation, which has enough capacity for Stan Terg mine. Flotation process includes three important stages that are:

1. Grinding and teunity,
2. Classification of sizes, and
3. Conditioning.

Finalized grinding and teunity we consider it when the mineral derived from mine with size 400 mm is teunited in the size 0.074 μm. The broken mineral in 0.6 mm together with reagents that weight 74 t/h is dosed in mill for final teunity.

Table 2 Presents total ore reserves and the quantity of lead and ore in metallurgical belt of Trepca

Resource place	Ores (ton)	% Pb	Pb metal in (ton)
Stan Terg	35,081,000	3.85	1,349,579
Cernnac BBGom	7,544,227	6.85	516,645
ArtanesCB complex	16,837,227	4.67	749,354
Total	58,662,569	4.46	2,615,578

Measure/Pulp in particles with diameter 0.074 μm from mill with funnel Φ = 100 mm is carried to lead conditioner. Flotation process of minerals is based on Physical-chemical attributes of each mineral phase and their interaction laws. The basic goal of the flotation process is selective separation of useful compounds from the others. To achieve such a goal it is necessary that the mineral components of one ore, to remain in abeyance in aquatic environment. While being at the same time smooth/distributed together with air bubbles sent to this environment. Under conditions of such a mix where is putted into motion the phases dynamic, stiff, liquid, flotation environment gases, also happens selective separation of minerals where a part of them is attached to bubbles and the other part remains in water.

Bubbles/Aggregate (Froth)—Bubbles and mineral grindings have a smaller density than aqueous environment and get in surface of pulp by forming the product of flotation process (concentrate). The mineral particles interact with air bubbles, is possible in those cases where surface virtues of these particles have hydrophobic and natural abilities, or created by the activity and impact of reagents used in flotation process (Table 3).

In general reagents are separated in three groups: collecting reagents, fixing reagents, and foam making reagents. Their use is very important in industry.

Flotation process of lead mineral has three main processes: base flotation, controlling flotation and cleaning flotation.

Base and controlling flotation is done in pneumatic machines with "Denver" impeller type. In series are established 12 cells, each cell has a volume $V = 2.83\ m^3$. Scavenger flotation is done in mechanical machines with impeller and air. The machine has 14 cells, each cell has a volume $V = 1.1\ m^3$. The whole technological process of flotation is presented in [1].

Teunity mass 0.074 μm as digestion is dosed in base flotation in the machine with 12 cells, and surface product with the help of shovels is carried to machine number 8 of digestion flotation. While the lower leakage part is carried to entrance of controlling flotation for reflotation. The controlling flotation product is managed as follows. Quantity of first 6 cells is carried in cell number 4 of digestion flotation as final product (concentrate), while the product of 6 other cells of controlling flotation is returned in conditioner. Whereas the low mass of 6 cells of controlling flotation, goes as sterile.

Table 3 Presents the type and quantity of reagents used in flotation process of lead mineral

Reagents	Concentration (kg t^{-1})
Foams	0.012–0.12
Collectors	0.02–1.2
Carbohydrate oil	0.5–2.5
pH regulators	0.55–5.0
Activators	0.025–1.0
Depressors	0.025–0.5
Deactivators	0.025–0.25

Experimental Part

Mathematical Models for Calculating the Flotation Product

There is no one universal method for expressing the effectiveness of a separation, but there are several methods that are useful for examining froth flotation processes:

(a) Ratio of Concentration, the weight of the feed relative to the weight of the concentrate, The Ratio of Concentration is F/C, where F is the total weight of the feed and C is the total weight of concentrate. While this data is available in laboratory experiments, in the plant it is likely that the ore is not weighed and only assays will be available. However, it is possible to express the ratio of concentration in terms of ore assays. Starting with the mass balance equations, and the definition of the ratio of concentration:

$$F = C + T,\ Ff = Cc + Tt,\ \text{Ratio of Concentration} = F/C$$

where F, C, and T are the % weights of the feed, concentrate, mathematical formula, and tailings, respectively; and f, c, and t are the assays of the feed, concentrate, and tailings.
We now need to eliminate T.

$$C_{Pb} = C \times c\,t.h^{-1} \tag{2}$$

From these equations so that we can solve for F/C:
Ff = Cc + Tt, and multiplying (F = C + T) by t gives us:
Ft = Ct + Tt, so subtracting this equation from the previous eliminates T and gives:
F(f − t) = C(c − t), and rearranging produces the equation for the ratio of concentration:

$$F/C = (c - t)/(f - t) \tag{1}$$

(b) % Metal Recovery, or percentage of the metal in the original feed that is recovered in the concentrate. This can be calculated using weights and assays, as (Cc)/(Ff) · 100. Or, since C/F = (f − t)/(c − t), the % Metal Recovery can be calculated from assays alone using 100(c/f)(f − t)/(c − t).
(c) % Metal Loss is the opposite of the % Metal Recovery, and represents the material lost to the tailings. It can be calculated simply by subtracting the % Metal Recovery from 100%.
(d) % Weight Recovery is essentially the inverse of the ratio of concentration, and equals.

$$100 \cdot C/F = 100(f - t)/(c - t)$$

(e) Enrichment Ratio is calculated directly from assays as c/f, weights are not involved in the calculation.
The main parameters of technological process of lead flotation in "Trepca" company.
$F = 150\ t\ h^{-1}$ load dosed for flotation.
C—concentrate weight that depends on quality of technological flotation process.
T—osess weight, sterility that depends on the quality of mineral and in technological flotation process.
f—percentage of steel in ore.
c—percentage of steel in concentrate that depends on the quality of technological flotation process.
t—percentage of steel in sterile that depends on the quality of technological flotation process.
Alternative I, the calculation of lead concentrate quantity, Pb and Zn's percentage in concentrate and Pb losess with residu when mineral contains 3.17% Pb.

(a) The quantity of Pb concentrate is calculated with mathematical formula.

$$F/C = (c - t)/(f - t) \tag{1}$$

F = 150 t.
Laboratory analysis of mineral, concentrate and sterile made in the Laboratory of Trepca's company dated 18.06.2013, results:
f = 3.17% Pb c = 75.3% Pb t = 0.29% Pb.
$C = F(f - t)/(c - t) = 150\ (3.17 - 0.29)/(75.3 - 0.29) = 5.7993\ t\ h^{-1}$.
$C = 5.7993\ t\ h^{-1}$

(b) The quantity of Pb in Pb concentrate is calculated with

$$C_{Pb} = 5.7993 \times 0.753\ t\ h^{-1}\ C_{Pb} = 4.366\ t\ h^{-1}$$

(c) Pb quantity in losess that goes with Zn concentrate with pirot and sterile, is calculated with mathematical formula.

$$C_{losess} = C_{Pb}\,\text{mineral} - C_{Pb} t\, h^{-1} \tag{3}$$

$C_{losess} = 4.755 - 4.366\ t\ h^{-1}$.
$C_{losess} = 0.389\ t\ h^{-1}$.

(d) % Weight Recovery is essentially the inverse of the ratio of concentration, and equals mathematical formula.

$$100 \cdot C/F = 100\,(f - t)/(c - t). \tag{4}$$

$100 \cdot C/F - 100 \cdot (f - t)/(c - t) = C/F - (f - t)/(c - t) = 0.00023\%$
0.00023% is Pb concentration from load.

(e) Mineral enrichment, load is calculated with mathematical formula.

$$c/f\,\% \tag{5}$$

75.3/3.17 = 23.753% is solid enrichment of mineral.

(f) Losess weight is calculated with mathematical formula.

$$F = C + T\,t\,h^{-1} \tag{6}$$

$T = F - C\ t\ h^{-1}$
$T = 150-5.7993 = 144.200\ t\ h^{-1}$
$144.200\ t\ h^{-1}$ is losess weight which in itself contains Zn concentrate, pirot concentrate and the sterile.

(g) The lead enrichment report is calculated with mathematical formula.

$$R_{Enrichment\ Pb} = \%Pb_{koncentrat} / \%Pb_{mineral} \tag{7}$$

$R_{Enrichment\ Pb} = 75.3/3.17 = 23.753\%$
$R_{Enrichment\ Pb} = 23.752\%$ is lead enrichment during flotation process
*Alternative II, t*he calculation of lead concentrate quantity, percentage of Pb in Pb and Zn's concentrate and Pb losess with residu when mineral contains 3.17% Pb.

(a) The quantity of Pb concentrate is calculated with mathematical formula (1).
F = 150 t
Concentrate laboratory analysis of mineral and sterile made in the Laboratory of Trepca's company dated 19.06.2013, results:
f = 3.17% Pb
c = 73.60% Pb t = 0.14% Pb
$C = F\ (f - t)/(c - t) = 150\ (3.17 - 0.14)/(73.60 - 0.14) = 6.207\ t\ h^{-1}$
$C = 6.207\ t\ h^{-1}$

(b) The quantity of Pb in Pb concentrate is calculated with mathematical formula (2).
$C_{Pb} = 6.207 \times 0.736\ t\ h^{-1}$ $C_{Pb} = 4.568\ t\ h^{-1}$

(c) Pb quantity in losess that goes with Zn concentrate with pirot and sterile, is calculated with mathematical formula (3).
$C_{losses} = 4.755\text{-}4.568\ t\ h^{-1}$ $C_{losses} = 0.167\ t\ h^{-1}$

(d) % Weight Recovery is essentially the inverse of the ratio of concentration, and equals mathematical formula (4).
$100\ C/F-100\cdot(f-t)/(c-t) = C/F-(f-t)/(c-t) = 6.207/$
$150-(3.17-0.14)/(73.60-0.17) = 0.00018\%$

0.00 018% Pb concentration from load.

(e) Mineral enrichment, load is calculated with mathematical formula (5).
73.60/3.17 = 23.217% solid enrichment of mineral

(f) Losess weight is calculated with mathematical formula (6).
$T = F - C\ t\ h^{-1}$
$T = 150 - 6.207 = 143.793\ t\ h^{-1}$
$143.793\ t\ h^{-1}$ is losess weight which in itself contains Zn concentrate, pirot concentrate and the sterile.

(g) The lead enrichment report is calculated with mathematical formula (7).
$R_{Enrichment\ Pb} = 73.60/3.17 = 23.217\%$
$R_{Enrichment\ Pb} = 23.217\%$ is lead enrichment during flotation process
Alternative III, the calculation of lead concentrate quantity, percentage of Pb in Pb and Zn's concentrate and Pb loses with residu when mineral contains 4.67% Pb and it has been taken from "Artanes" ore.

(a) Quantity of Pb concentrate is calculated with mathematical formula (1).
F = 150 t
Laboratory analysis of mineral, concentrate and sterile made in the Laboratory of Trepca's company dated: 20.06.2013, results: f = 4.67% Pb
c = 74.4% Pb t = 0.29% Pb
$C = F(f - t)/(c - t) = 150\ (4.67 - 0.29)/(74.4 - 0.29) = 8.865\ t\ h^{-1}$ $C = 8.865\ t\ h^{-1}$

(b) The quantity of Pb in Pb concentrate is calculated with mathematical formula (2).
$C_{Pb} = 8.865 \times 0.744\ t\ h^{-1}$ $C_{Pb} = 6.595\ t\ h^{-1}$

(c) Pb quantity in losess that goes with Zn concentrate with pirot and sterile, is calculated with mathematical formula (3).
$C_{losess} = 7.005–6.595\ t\ h^{-1}$ $C_{losess} = 0.47\ t\ h^{-1}$

(d) % Weight Recovery is essentially the inverse of the ratio of concentration, and equals mathematical formula (4).
$100 \cdot C/F - 100 \cdot (f - t)/(c - t) = C/F - (f - t)/(c - t) = 8.865/150 - (4.67 - 0.29)/(74.4 - 0.29) = 0.0060\%$
0.0060% Pb enrichment from load.

(e) Mineral enrichment, load is calculated with mathematical formula (5).
74.4/4.67 = 15.931% is a not good enrichment of mineral f Loses weight is calculated with mathematical formula (6).
$T = F - C\ t\ h^{-1}$
$T = 150 - 8.865 = 141.135\ t\ h^{-1}$
$141.139\ t\ h^{-1}$ is losess weight which in it self contains Zn concentrate, pirot concentrate and the sterile.
g The lead enrichment report is calculated with mathematical formula (7).
$R_{Enrishment\ Pb} = 74.4/4.67 = 15.931\%$
$R_{Enrishment\ Pb} = 15.931\%$ is a not good enrichment of Pb during flotation process (Tables 4 and 5; Figs. 4 and 5).

Table 4 Presents quantity of Pb in mineral, quantity of Pb concentrate, quantity of Pb in concentrate and Pb quantity in losess for Alt. I, II, III

Alternatives	Quantity of Pb in mineral (t h^{-1})	Quantity of Pb concentrate (t h^{-1})	Quantity of Pb, in Pb concentrate (t h^{-1})	Quantity of Pb in losess (t h^{-1})
Alt. I	4.753	5.7993	4.366	0.389
Alt. II	4.753	6.207	4.568	0.167
Alt. III	7.005	8.865	6.595	0.47

Table 5 Presents Pb and mineral enrichment in Alt. I, II, III

Alternative	Mineral enrichment (%)	Pb enrichment (%)
Alt. I	23.753	23.752
Alt. II	23.217	23.217
Alt. III	15.931	15.931

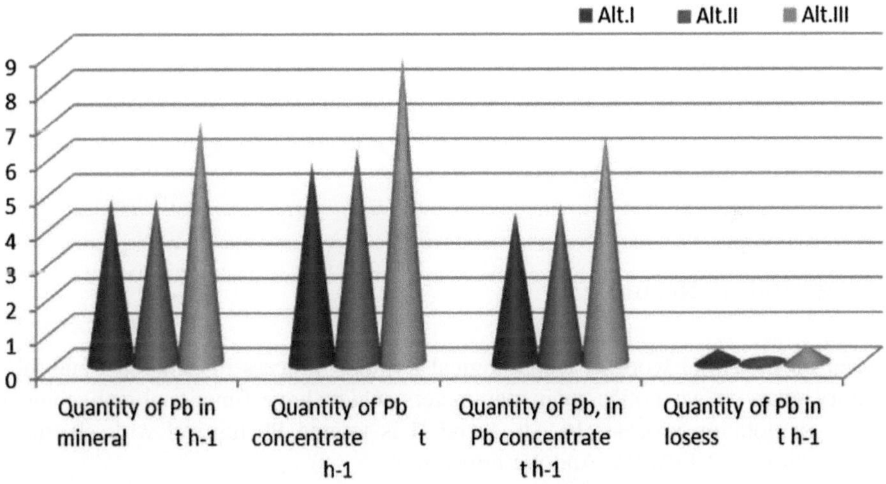

Fig. 4 Presents quantity of Pb in mineral, quantity of Pb concentrate, quantity of Pb in concentrate and Pb quantity in losess for Alt. I, II, III

Ag Quantity in Lead Minerals

Sulfur minerals of Pb–Zn are rich with precious metals. Chemical analysis in the laboratory of Trepca's company dated 19.06.2014, result as below (Table 6).

Ag quantity in mineral is calculated with mathematical expression.

$C_{losese} = 0.8455\ t\ h^{-1}$

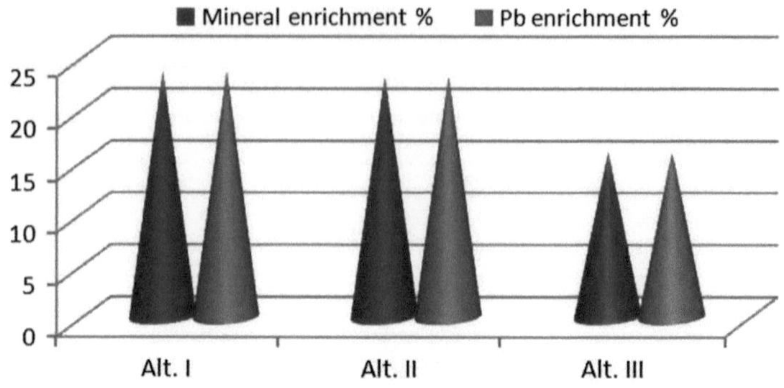

Fig. 5 Presents Pb and mineral enrichment in Alt. I II, III

Table 6 Presents the composition of some samples

Sample	Shift	Quantity (Ag t^{-1})
1	1	11.38
2	2	38.50
3	3	16.67

Enrichment of Pb during flotation process for Alt. III has a decrease with average value Enrichment Pb = 7.5315%. These parameters of technological flotation process identify a good management and a not good management of flotation process of Pb mineral in "Trepca".

Discussion of Results

Based on industrial work of technological flotation process in "Trepca", and laboratory analysis we notice main parameters which have functional/management power in flotation process. In Alt. I and II is treated Pb mineral with chemical composition of 3.17% Pb. And we notice that:

Quantity of Pb in mineral is 4.755 t h^{-1}.

Quantity of Pb concentrate and quantity of Pb in Pb concentrate for Alt. II has an Increase with value C = 0.4053 t h^{-1} while Pb quantity in Pb concentrate han an increase with value CPb = 0.202 t h^{-1}.

Quantity of Pb losess for Alt. II has a decrease with value 0.222 t h^{-1}

Mineral enrichment for Alt. II has a decrease with value 0.536 t h^{-1}

Losses weight for Alt. II has a decrease with value 0.407 t h^{-1}

Pb enrichment during the flotation process in Alt. II has a decrease with value 0.535%

In Alt. III the flotation process has used mineral of Artana which has 4.67% Pb and based in production of the flotation process in "Trepca" and laboratory analysis we notice:

Quantity of Pb concentrate and quantity of Pb in Pb concentrate increases with average value C = 2.961 t h^{-1} and CPb = 2.192 t h^{-1}.

Quantity of Pb in losses increases with average value

In Alt. III, in flotation process we have used the mineral with composition 4.67% Pb. Quantity of mineral and main parameters of flotation process as grinding, conditioning, Ph and drying are the same as in Alt. I, II.

Based on the product of flotation process and laboratory analysis we can conclude:

Pb quantity dosed in flotation process with mineral is

7.005 t h^{-1} which is bigger than in Alt. I, II for 2.25 t h^{-1}. Quantity of Pb concentrate has an increase with average value 2.851 t h^{-1} toward Alt. I, II. Quantity of Pb in losses as an increase with average value 0.192 t h^{-1}, a quantity that has a bad impact on economical sustainability in flotation procesess in Trepca.

Chemical percentage of Pb in mineral must be in borders 3.17% $\leq$ % Pb $\leq$ 4.67%.

Conclusion is made by laboratory and graphic analysis, and also by losess of lead with residu with values from 0.167 t h^{-1} until 0.471 t h^{-1}, which is a crucial parameter for managerial quality of flotation process in Trepca.

Conclusions

Based in the industrial process of Pb concentrate production in Trepça and mathematical calculations made, can be concluded that all lead concentrate balance positions are variable at 4.568 t h^{-1} $\leq$ C $\leq$ 8865 t h^{-1}. Which depend on the parameters of the technological process such as the percentage of lead in the mineral and the temperature of the process. Sterile deposits and the amount of Pb in sterile waste has a value of 0.167 t h^{-1} $\leq C_{Losess} \leq$ 0.470 t h^{-1} and depends on the pH of the flotation process, the percentage of Fe and Zn in the mineral. As such, this value can be reduced by observing the pH and reactors of the flocculatory process. The article as such treated the Ag quantity in the Pb concentrate which has a value of 1707 kg h^{-1} $\leq$ Ag $\geq$57,750 kg h^{-1} significant amount as well as reducing the amount of Pb in unused sterile waste in order to increase the economic viability of the process of winning the lead concentrate.

References

1. Kawatra SK, Eisele TC (1992) Recovery of Pyrite in coal flotation: entrainment or hydrophobicity? Miner Metall Process 9(2):57–61
2. Froth Flotation—Fundamental Principles
3. Kawatra SK, Eisele TC (2001) Coal desulfurization: high-efficiency preparation methods. Taylor and Francis, New York
4. Agolli F (1985) Metalurgjia e metaleve me ngjyrë. UP Prishtinë

5. Bittrich JH (1976) Leitdaten der chemischen thermodynamic. Berlin
6. Haxhiaj A, Elezi D, Sh SHKOLOLLI (2006) Mjedisi dhe menaxhimi i gazrave termike në zonën e parangrohjes të furrave Vatergjakete në Trepçë. Simpozium VI, Materials and their use, Tiranë
7. Haxhiaj A (1989) Materialna i toplotna bilansa PORTPIRI pec′i, magistarski rad. Sveučilište u Zagrebu, MF u Sisku
8. Haxhiaj A, Rizaj M, Elezi D (2007) The management of coke and the optimum percentage of lead in agglomerate in Port-Piri furnace. In: 10th national conference of metallurgy, Bullgari
9. Haxhiaj A (2003) Intensifikimi dhe optimizimi I procesit të shkrirjes reduktuese të aglomeratit të plumbit në furrat shahte, punim i doktoratës. UP FXM, Mitrovicë
10. Jovanovic M (1978) *Nova tehnologija i pos trojenja za proizvodnja olova*, Zbornik radova II Jugoslovenskog simpozijuma o metalurgij, TMF (8, 31, 51) Beograd
11. Rizaj M, Terziqi A, Beqiri E, Kamberaj N (2008) Concentration and Distribution Characteristics of Trepça Shaft Furnace Slag, World of Metallurgy-Erzmetall 61 N0. 2, 109–114, Clausthal-Zellerfeld, Germany
12. Haxhiaj A, Haxhiaj E (2010) The optimization of the coke and agglomerate quantity in lead production in "water-jacket" furnace. TMS Annual Meeting & Exhibition, Seattle Washington, pp 249–257

The Recovery of Cesium Salts from the Taron Deposit

Mohammad Mokmeli, David Dreisinger, William (Bill) McWilliam, Colin Oloman, David Trueman, Bruce Downing and Rod McElroy

Abstract The Taron deposit in Argentina contains significant concentrates of cesium in a complex oxidized mineral assemblage associated with iron-arsenic-aluminum-manganese-silica mineralization. The recovery of cesium has been studied using sulfuric acid leaching with recovery of cesium as alum. The waste solution from cesium recovery is oxidized and neutralized to form stable iron–arsenate residues. The cesium alum is further treated to form high purity cesium hydroxide, which may be further processed to a range of cesium salts including cesium formate. The initial process developments for the Taron deposit are presented.

Keywords Taron · Cesium · Acid · Alum

M. Mokmeli · D. Dreisinger (✉)
Department of Materials Engineering, University of British Columbia, 6350 Stores Road, Vancouver, BC V6T1Z4, Canada
e-mail: david_dreisinger@yahoo.com

W. (Bill) McWilliam
Cascadero Copper Corporation, 554 East Kings Road, North Vancouver, BC V7N1J3, Canada

C. Oloman
Department of Chemical and Biological Engineering, University of British Columbia, 2360, East Mall, Vancouver, BC V6T1Z3, Canada

D. Trueman
5360 Bunting Ave, Richmond, BC V7E5W1, Canada

B. Downing
20200 Grade Crescent, Langley, BC V3A4J6, Canada

R. McElroy
4178 Vine Street, Vancouver, BC V6C3B9, Canada

H. Kim et al. (eds.), *Rare Metal Technology 2018*, The Minerals, Metals & Materials Series, https://doi.org/10.1007/978-3-319-72350-1_18

Introduction

The Taron project is located 160 km northwest of the city of Salta, Argentina. The property consists of five (5) Contiguous Mineral Tenures, approximating 8179 ha (83 Units) in area. The Tenures are registered to Cascadero Minerals S.A. (CMSA), which is 100% owned by Cascadero Minerals Corporation (CMC), a Canadian Company, which is 70% owned by Cascadero Copper Corporation (CCD) and 30% owned by Regberg Ltd. (RB). CMC operates as a 70% CCD and 30% RB joint venture [1].

The project is at approximately 4250 m above sea level. The annual rainfall (in the summers) ranges from 200–400 mm. Mid-summer high temperatures range from 14–21 °C with overnight lows of 6 °C. Mid-winter temperatures range to −8 °C with extremes of −15 °C. The project location is shown in Fig. 1.

Fig. 1 Taron project location

The Taron deposit has been subjected to exploration through trenching and drilling (2017). The mineralization has been described as follows [1].

The main zone of +200 ppm cesium mineralization lies in the Core and North Zones which cover a 700 × 1500 m area intermittently exposed over a vertical range of 80–100 m. Within this zone, the average Cs grade based on the 2017 drill results is about 1400 ppm Cs. Rb is closely correlated with Cs and averages about 220 ppm.

A sample from trench 109 WW was submitted to SGS Mineral Services for petrographic, XRD, and SEM study. The initial sample analyses ran 7% Mn, 5% As, and 2% Cs. The study concluded the following:

- Arsenate cements account for about 25% of the sample and Mn-oxyhydroxides another 5%.
- Mn reports as both Mn-oxyhydroxides and as a suite of Ca–Fe- and Ca–Mn-Arsenate hydrate minerals, probably wallkilldellite. Only about 15% of Mn reports as Ca–Mn-Arsenate in the sample.
- Remaining Mn reports mostly as cryptomelane in massive form as well as radiating, concentrically banded Mn-oxyhydroxides. Among these minerals, coronadite, hollandite, and romanechite have been confirmed by XRD.
- There is a complex suite of arsenate mineral present, ranging from ludlockite, wallkilldellite and its Fe-analogue as well as pharmacosiderite, and yukonite.
- Cs reports predominantly as a Cs-substituted pharmacosiderite at levels of up to 12% Cs in this phase. Mixtures of Cs-pharmacosiderite and other phases are also present and as a result, producing a pure Cs-pharmacosiderite concentrate is unlikely.

Cesium Extraction and Purification Process

Overview

The goal of this study was to develop a hydrometallurgical process for the production of marketable cesium formate solution, on the laboratory scale, starting from a cesium mineral sample obtained from the Taron deposit.

Figure 2 shows a conceptual flow chart of the tested process. Each unit operation was tested (conditions shown) sequentially without recycle. Atmospheric sulfuric acid leaching at 90 °C was applied to dissolve the cesium and other valuable metals into solution. After leaching, aluminum sulfate was added to the PLS (Pregnant Leach Solution) which was then stored in a refrigerator for 7 days. Cesium precipitated out of solution in the form of impure aluminum cesium sulfate crystals (alum) during this period. Afterwards, the liquid phase was removed, and the cesium containing crystals were left to air dry. These crystals were re-dissolved in boiling water. The resultant solution was filtered and returned to the refrigerator for re-crystallization. This process purified the initially collected crystals. After another

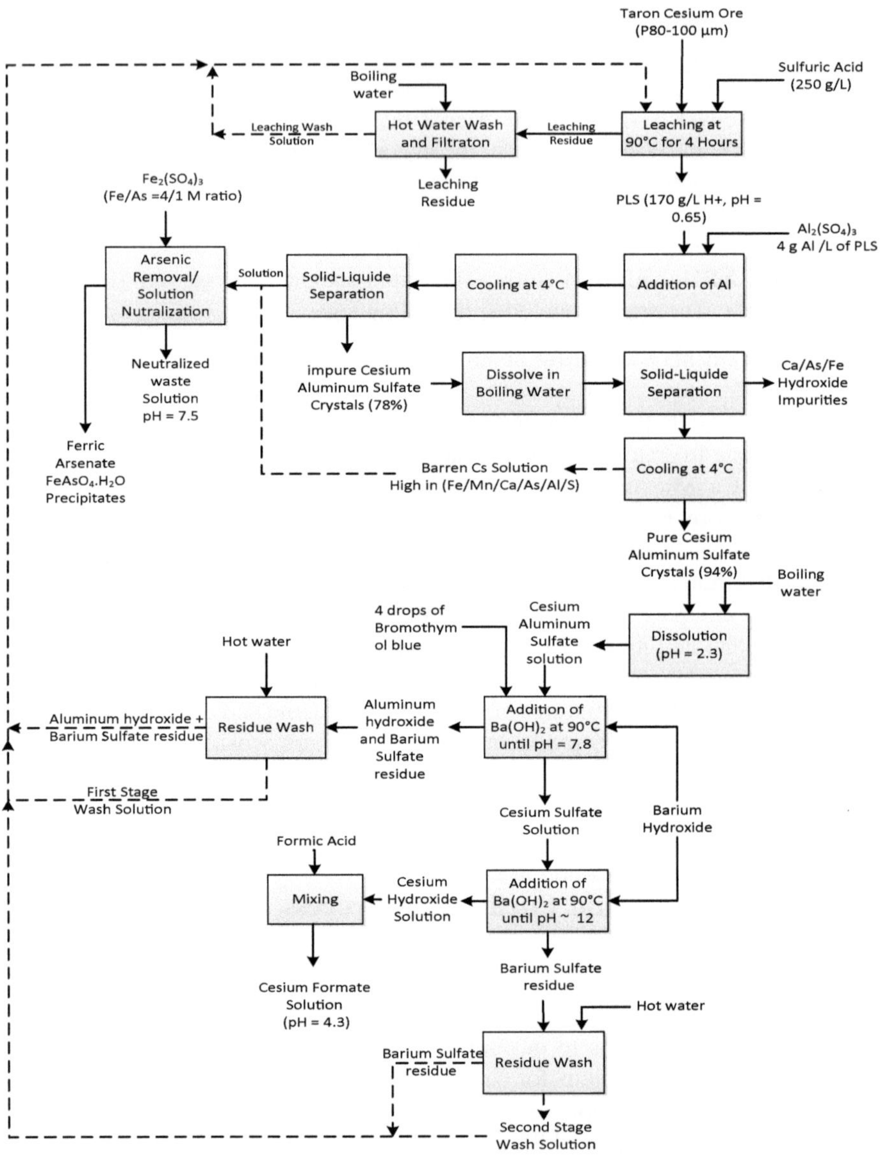

Fig. 2 Cesium extraction process. Dashed lines are potential recycle streams

10 days in the fridge, the solution was decanted, and the crystals were filtered from the solution. These purified aluminum cesium sulfate crystals were dissolved in water, and barium hydroxide was added in two stages with two-step filtrations. The addition of barium hydroxide to the dissolved cesium aluminum sulfate first precipitated the aluminum as hydroxide and second precipitated sulfate as barium sulfate. The remaining solution contained cesium hydroxide. Formic acid was then

added to the collected solution producing cesium formate. The overall procedure had three stages: leaching and crystal formation, crystal purification, and cesium sulfate purification, and the corresponding recoveries of cesium were 94.7, 99.9, and 96.1%, respectively. This gives a total of 90.9% cesium recovery over the entire process assuming all wash solutions are recycled without cesium losses.

As depicted in Fig. 2, all wash solutions and two of the residues (the barium sulfate and aluminum hydroxide residues) would be recycled in a commercial process. All wash solutions can be concentrated by heating and evaporation. All other solid residues generated during the process have low cesium content and therefore are not economically feasible to be returned to the leaching stage. It is envisioned that these solids would report to waste disposal after neutralization, however further experimentation would be required to confirm the makeup of the combined waste stream.

Sample Preparation

The Taron mineral sample from a 2009 drill hole (No 5) was received in four bags (part A, B, C and D) of different moisture content and weight. Samples were crushed and ground to a P80 of 100 μm. Table 1 provides a typical sample analysis. The cesium content of the sample was 4761 ppm or 0.476%. There is a significant concentration of As, Fe, Mn and SiO_2 in the material.

Leaching

Leaching was conducted in 2-liter jacketed and baffled glass reactors maintained at 95 °C. The reactors had condensers to minimize evaporative loss. In a typical experiment, 1700 mL of 250 g/L H_2SO_4 solution was charged to the reactor and heated. A 450 g Taron mineral sample was then added and leaching continued for 4 h. Intermediate samples were sometimes taken from the reactor for kinetic analysis. The reactors were agitated at 1000 rpm with overhead impellers. The final slurry was vacuum filtered via Büchner funnels. The solids were washed and dried before analysis and wash solutions kept separate from primary filtrate for analysis and metallurgical accounting. The primary filtrate was filtered again with finer mesh filter paper to ensure a clear solution for downstream processing.

Aluminum Cesium Sulfate Primary Precipitation

The leachate from each test was collected and treated with aluminum sulfate solution to promote formation of aluminum cesium sulfate precipitates (alum).

Table 1 Typical Taron sample analysis

Elements. All values are ppm (wt).												
Ag	Al	As	Ba	Be	Bi	Ca		Cd	Ce	Co	Cr	Cs
1	33,500	40,495	2506	5	<0.5	28,600		7	34	111	177	4761
Cu	Dy	Er	Eu	Fe	Ga	Gd		Hf	Ho	K	La	Li
1483	3	2	1	67,300	13	2		1	1	17,800	17	52
Lu	Mg	Mn	Mo	Na	Nb	Nd		Ni	P	Pb	Pr	Rb
<0.1	4100	>10,000	11	14,300	8	15		126	3400	22	4	407
S	Sb	Sc	Se	Sm	Sn	Sr		Ta	Tb	Te	Th	Ti
<500	4	3	3	2	2	542		1	<0.1	2	5	2060
Tl	Tm	U	V	W	Y	Yb		Zn	Zr	F		
2533	0	11	82	1	16	1		2173	32	635		
Oxides and Loss on Ignition. All values are wt%.												
SiO_2	Al_2O_3	Fe_2O_3	CaO	MgO	Na_2O	K_2O	MnO	TiO_2	P_2O_5	Cr_2O_3	Ba	LOI
48.86	8.7	9.7	4.07	0.86	1.74	2.14	8.13	0.42	0.82	0.03	0.24	5.76

A typical addition was 4 grams of aluminum per liter added as aluminum sulfate powder (and well mixed with the solution). The solutions were placed in a fridge (4 °C) and left to crystallize alum for one week. The supernatant was sampled for analysis and the solids removed from solution and air-dried.

The purification of the alum crystals was necessary. Approximately 25–30 g of alum were leached in 750 mL of water at 90 °C for one hour. Any remaining solids (containing Ca, Fe and As) were filtered. The filtrate and solids were collected for assay. The remaining filtrate was again placed in a fridge for 7–10 days. The purified alum crystals were then recovered by filtration and drying. Washing was not applied in order to avoid alum re-dissolution.

Cesium Hydroxide and Cesium Formate Production

A 2 g sample of the synthesized aluminum cesium sulfate crystals was fully dissolved in 100 mL of water held at 90 °C throughout the duration of the experiment. Barium hydroxide was added sequentially to a pH setpoint. All aluminum hydroxide and the majority of barium sulfate will precipitate at around pH 7.6 (bromothymol blue used as a pH indicator). Considering the overall precipitation reaction, 2 mol of barium hydroxide required per mole of aluminum cesium sulfate crystals assuming the following reaction goes to completion:

$$AlCs(SO_4)_2 \cdot 12H_2O + 2\ Ba(OH)_2 \cdot 8\ H_2O \rightarrow Al(OH)_3\downarrow + 2\ BaSO_4 \downarrow + CsOH + 20\ H_2O$$

After each addition the pH, and the mass of remaining barium hydroxide was measured and recorded. When a pH of ~7.5 to 8 was reached, the addition of barium hydroxide was stopped. The first stage of the experiment stopped at this point to ensure the precipitated aluminum hydroxide didn't dissolve back into the solution after a further rise in pH. At this stage all aluminum hydroxide and 1.5 mol of barium sulfate should have precipitated as seen in the following equation:

$$0.5\ Al_2(SO_4)_3 + 1.5\ Ba(OH)_2 \rightarrow Al(OH)_3\downarrow + 1.5\ BaSO_4\downarrow \ (\text{pH} \sim 7.5\text{–}8)$$

The slurry was then filtered of aluminum hydroxide and barium sulfate solids that had precipitated.

Barium hydroxide was then continually added to the solution and the pH of the solution was monitored after each addition. Barium hydroxide addition stopped when a pH of 11.83 was reached. It is expected that all barium sulfate precipitated at this pH according to following reaction:

$$0.5\ Cs_2SO_4 + 0.5\ Ba(OH)_2 \rightarrow 0.5\ BaSO_4\downarrow + CsOH$$

This final pH of 11.8 was chosen as the end point for production of cesium hydroxide. Once the final pH was reached, the solution was filtered once again to remove the barium sulfate.

A further conversion of cesium hydroxide to cesium formate was performed. Formic acid was added in one to one stoichiometric ratio as shown in the equation below:

$$CsOH + HCOOH \rightarrow CsHCO_2 + H_2O$$

Arsenic Precipitation and Waste Treatment

Arsenic was removed from the PLS by first adding ferric sulfate at a 4:1 molar ratio of Fe^{3+} to As at 50 °C, then adding calcium carbonate to precipitate ferric arsenate and raise the pH of solution.

Experimental Results and Discussion

Leaching

The results of a typical leaching test are summarized in Table 2. The cesium extraction was excellent at 94.7% with high extractions of iron, arsenic and rubidium. The reported extraction was based on calculated head assays and the mass balance error relates to the comparison of the calculated and assay head.

Arsenic Precipitation Before Cesium Recovery

A sample of leach solution was treated with iron (III) sulfate and limestone ($CaCO_3$) to precipitate arsenic as a ferric hydroxide/ferric arsenate precipitate. This was done with a 4:1 Fe/As ratio (mol/mol) for 4 h at 50 °C. The results showed excellent precipitation of arsenic but the poor washing characteristics caused a loss of ~25% of the Cs with the precipitate. For this reason, the removal of arsenic by co-precipitation with iron is best performed after cesium recovery (as shown in Fig. 2).

Table 2 Results for 4 h of Leaching at 90 °C with 250 g/L H_2SO_4 and 20% solids

Element	Al	As	Cs	Fe	Rb	Tl
Extraction (%)	17.5	95.9	94.7	78.1	76.6	23.5
Mass balance error (%)	−4.9	9.9	8.5	7.0	1.4	4.1

Aluminum Cesium Sulfate Precipitation and Purification

A series of experiments were performed to determine the requirement for aluminum addition to form aluminum cesium sulfate from the leachate. Additions of 2–8 g/L Al were tested with a further 7 days at 4 °C in a refrigerator to favour precipitation. These results are shown in Table 3.

From this test it was confirmed that refrigeration of the samples after aluminum addition significantly increases the cesium and rubidium recoveries. The addition of 4 g/L aluminum was chosen for further testing because it produced 98.9% recovery of cesium as well as an adequate rubidium recovery. An SEM-EDX analysis of the precipitates formed after aluminum addition (cesium aluminum sulfate crystals and gypsum) is shown in Fig. 3.

Analysis showed that the impure aluminum cesium sulfate crystal had a purity of 77.9% on a weight metal basis. After the purification process the purity on a weight metal basis increased to 94.2%. The purification results are summarized in Table 4.

Table 3 Precipitation of Cs and other elements with increasing aluminum sulfate addition

Addition/precipitation (%)	Al	As	Cs	Fe	Rb	Tl
2 g/L Al	5.2	4.1	97.8	4.2	62.2	8.0
4 g/L Al	12.6	9.0	98.9	11.3	70.2	8.2
6 g/L Al	5.8	6.1	99.2	6.2	77.2	11.4
8 g/L Al	14.2	12.3	99.4	14.5	82.1	15.3

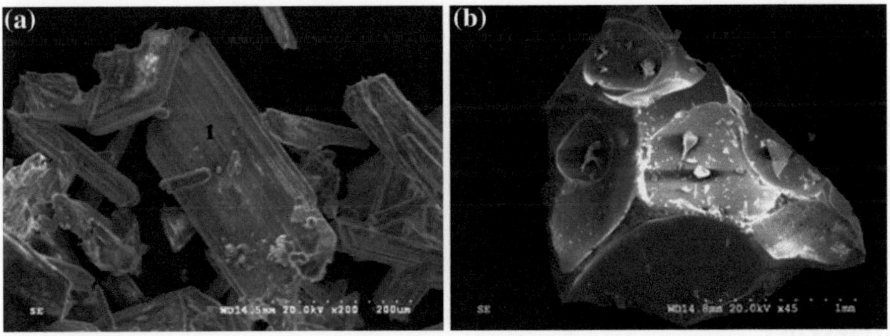

Fig. 3 SEM pictures of **a** gypsum and **b** cesium aluminum sulfate crystals

Table 4 Purification of cesium aluminum sulfate crystals

Crystal/ element (ppm)	As	Ca	Cu	Fe	Mg	Mn	P	Rb	S	Sr	Tl
Impure	9227	48,100	241	23,900	80	2170	80	6522	181,000	589	133
Purified	24	<100	2.3	13,400	<10	6	<10	1106	106,000	7	7

Table 5 Cesium hydroxide and cesium formate solution production

Stream/element (ppm)	Cs	Al	S	Ca	Fe	K	Rb	Si
Alum dissolution solution	4885	902.3	2944	48.4	65	4	32.7	7.4
Cs Sulfate solution (pH ~ 8)	6265	1	933	15	<10	7	35	4.6
Cs Hydroxide solution (pH ~ 12)	7431	1	218	2	<10	11	43	22
Cs Formate solution	7425	2	515	1	<10	9	42	20

Cesium Formate Production

The cesium aluminum sulfate was dissolved in water and purified in two steps as per the process description. The final cesium hydroxide solution was then treated with formic acid to make cesium formate solution.

The results of the purification are summarized in Table 5.

Iron dropped from 65 ppm in the initial solution to undetectable amounts (<10 ppm) after the first filtration. Similarly, calcium levels dropped from 48.4 ppm in the initial solution to 0.9 ppm in the final cesium formate solution commercial brines. Further testing is required to develop methods to concentrate and purify the final product solution.

Conclusions and Recommendations

This study has covered the process development for the production of cesium formate from the Taron deposit using batch tests in the laboratory scale. Continuous tests on the laboratory scale and mini pilot scale (~1 kg/h) are suggested before moving onto an industrial scale.

The investigation of alternative processing schemes for Taron mineralization will be presented in future publications.

Reference

1. GeoSim Services Inc (2017) Taron Project NI 43–101 Technical Report, Salta Province, Argentina, September 14, 2017

Recovery of Lithium from Brine with MnO_2 Nanowire Ion Sieve Composite

Rajashekhar Marti and York R. Smith

Abstract Lithium is anticipated to be a key, strategic material in the clean technology economy. The majority of lithium reserves are contained in continental brines and sea water (>60%). Currently the most economical approach to recover lithium from such resources is through selective adsorption. Spinel type lithium manganese oxide and lithium titanium oxide have demonstrated to be excellent candidate materials (after de-lithiation). However, due to the small size of the powders, they have limited industrial application and require immobilization. In this work, we examine the application of novel manganese oxide nanowire composites for the extraction of lithium from dilute solutions. Manganese oxide nanowires were synthesized and simultaneously immobilized on diatomaceous earth via hydrothermal method. Lithium manganese oxide were then formed by solid-state reaction, de-lithiated and tested for lithium ion adsorption.

Keywords Lithium · Adsorption · Manganese oxide · Nanowires

Introduction

Lithium is a critical and technologically important metal that has widespread use, particularly in energy storage applications (e.g., lithium-ion battery or LIB). With Electric Vehicles (EV) and hybrid-EV's anticipated to become a widely adopted cleaner alternative to combustion engines, the demand is project to increase drastically over the next 100 years. For example, the projected demand for lithium to power EV's is estimated to be between 184,000 t and 989,000 t per year by 2050 [1]. In comparison, about 31,500 t were produced globally in 2016 [2]. In order to meet the increasing demand, it is necessary to extract lithium from low-grade lithium resources effectively. Continental brines and sea water consists of around 61.8% of lithium resources and are the most inexpensive and available source for

R. Marti · Y. R. Smith (✉)
Metallurgical Engineering Department, University of Utah, Salt Lake City, UT 84115, USA
e-mail: york.smith@utah.edu

H. Kim et al. (eds.), *Rare Metal Technology 2018*, The Minerals,
Metals & Materials Series, https://doi.org/10.1007/978-3-319-72350-1_19

lithium extraction [3]. Due to the low concentrations of continental brines and sea water (0.1–100 ppm), the use of selective adsorbents have gained popularity. Among all the adsorbents explored till date, spinel-type lithium manganese and lithium titanium oxide ion sieves have been most successful due to their higher selectivity and memory for Li ion specific sites [4]. However, these adsorbents have limited industrial applications due to the loss of powder because of their small particle size thus making powder recovery difficult and costly [5]. In order to improve lithium ion sieve (LIS) applications for industrial use, LIS composites materials have been synthesized by various techniques such as granulation, foaming, and membrane formation [4].

In this work, a novel lithium adsorbent material was synthesized and investigated. Manganese oxide (MnO_2) nanowires were immobilized on diatomaceous earth (DE) via hydrothermal treatment (MnO_2@DE). The obtained MnO_2@DE was then subject to solid-state reaction with lithium salt to obtain spinel structured lithium manganese oxide (LMO) immobilized on DE (LMO@DE) composite. The LMO@DE composite was then treated with HCl to remove lithium and form a spinel structured delithiated lithium manganese oxide on DE composite (DLMO@DE) adsorbent. This DLMO@DE adsorbent was then characterized and tested for lithium adsorption.

Experimental Methods

Preparation of Adsorbent

Diatomaceous earth was acquired from EP Minerals and used without any prior treatment. First, 2.0 g of DE was vigorously stirred in 30 ml of deionized water (DI) for 30 min. To this, equal moles of $KMnO_4$ and $(NH_4)_2S_2O_8$ were added to the slurry and stirred for 10 min. In order to study the optimum MnO_2 loading, the ratio of MnO_4^- to 1.0 g DE was varied between 2.5 and 10 mmol MnO_4^-/g DE. The slurry was transferred to a Teflon lined stainless steel autoclave and hydrothermally treated in an oven for 12 h at 150 °C. The obtained solids were then washed with DI water, filtered, and dried in a vacuum oven at 60 °C overnight. MnO_2 loading on DE was calculated by measuring the mass difference between the precipitate and initial amount of DE in the slurry.

Li_2CO_3 was then added to MnO_2@DE with a molar ratio of Li/Mn = 0.55 and ground with mortar and pestle for 15 min. The mixture was then calcined at 650 °C for 6 h in air to obtain spinel structured LMO immobilized on DE (LMO@DE). The LMO@DE was then added to 400 ml of 0.5 M HCl and stirred for 24 h to form delithiated lithium manganese oxide immobilized on DE (DLMO@DE). The DLMO@DE was then subject to vacuum filtration and washed with copious amounts of DI water. DLMO loaded on DE was measured by measuring mass difference between adsorbent and initial DE. In order to compare Li uptake of the composite with DLMO powder the above process was repeated without Diatomaceous Earth.

Characterization

The crystal structure of the adsorbents was analyzed by X-ray diffraction (XRD). XRD analysis was carried out using a Rigaku Miniflex XRD (CuKα = 1.54059 Å) from 2θ = 10°–80° with a step size of 0.01° and dwell time of 0.5 degrees/min. The XRD data was analyzed using Rigaku PDXL2 analysis software. Scanning electron micrographs and electron dispersion spectroscopy mapping were collected using a Hitachi S-4800 SEM with an Oxford EDS detector.

Adsorption

Approximately 1.0 g of DLMO@DE adsorbent was added to 100 ml of 100 ppm LiCl solution (pH = 10.20, buffered by ammoniacal buffer) for 24 h. Li ion adsorption was determined using ICP—OES and calculated using the following expression (Eq. 1)

$$\text{Li uptake } \frac{\text{mg}}{\text{g}(DLMO)} = (C_0 - \text{Ce}) * \left(\frac{\text{V}}{\text{m*fraction of DLMO loaded}}\right) \quad (1)$$

where, C_0 is the initial Li concentration, Ce the equilibrium Li concentration, V the volume of LiCl buffer solution, and m the mass of adsorbent.

Results and Discussion

Table 1 shows the weight % of MnO_2. The MnO_2 loading increased with increased concentration of MnO_4^- ions during hydrothermal treatment.

Figures 1 and 2 show the XRD pattern for MnO_2@DE and LMO@DE for various ratios of mmol MnO_4^-/g DE. In Fig. 1, all the peaks correspond to α-MnO_2. A broad peak around 23.7° can also be observed, which is attributed to

Table 1 Weight % loading of MnO2 loaded for different values of MnO_4^-/DE

mmol MnO_4^-/g DE	% MnO_2 loaded
10	43.97
5	28.54
2.5	14.83

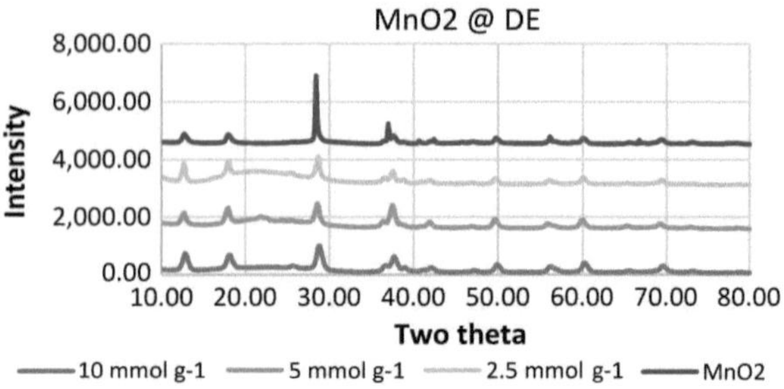

Fig. 1 XRD pattern of MnO_2@DE for various MnO_2 loadings

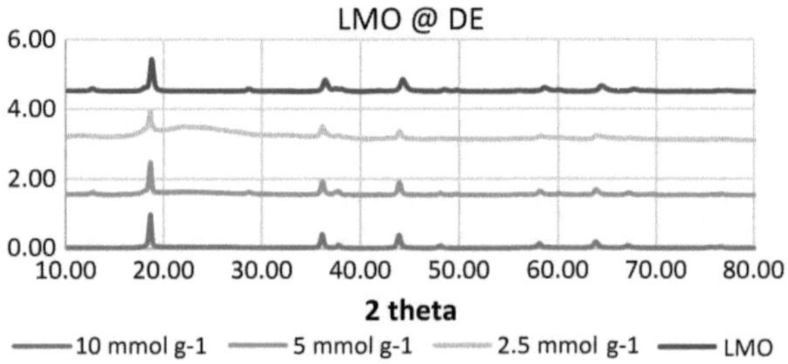

Fig. 2 XRD pattern of LMO@DE for various MnO_2 loadings

DE. Regardless of the concentration of MnO_4^-, predominantly α-MnO_2 is observed. Figure 2 shows the XRD patterns LMO@DE composites with various MnO_2 loadings. The diffraction peaks indicates the formation of spinel type $LiMn_2O_4$. After delithiation of $LiMn_2O_4$, spinel structured λ-MnO_2 (DLMO) is obtained. The λ-MnO_2 is obtained by topotactical extraction of lithium from spinel structured LMO in 0.5 M HCl retaining the spinel structured LMO framework [5].

SEM images (Fig. 3a–c) show that MnO_2 was immobilized on DE and the coverage of MnO_2 increased with increasing MnO_4^- concentration. It is interesting to note that at lower concentration (i.e., 2.5 mmol of MnO_4^- ions) a nanowire structure is not formed. Increasing the concentration above 2.5 mmol of MnO_4^- results in the formation of nanowires. At higher concentration (i.e., 10 mmol of MnO_4^- ions) the MnO_2 nanowires not only cover the entire DE surface, but also form aggregates of MnO_2 nanowires, which are not adhered to DE (Fig. 4).

Energy dispersive X-ray spectroscopy (EDS) was used to determine the composition of the composite adsorbent materials. A summary of average composition for various MnO_2 loadings is given in Table 2. Elements Si, Al, and Fe are due to the DE support.

After delithiation, the adsorbents were subject to lithium adsorption in buffered solution to maintain a constant pH condition since lithium adsorption is highly dependent on pH [6]. Table 3 shows the lithium adsorption for various MnO_2 loadings. It is observed that as the MnO_4^- concentration increases from 2.5 mmol MnO_4^-/g DE to 5 mmol MnO_4^-/g DE, the lithium uptake increases. This is due to an increase in the loading of spinel structured DLMO. However, for 10 mmol

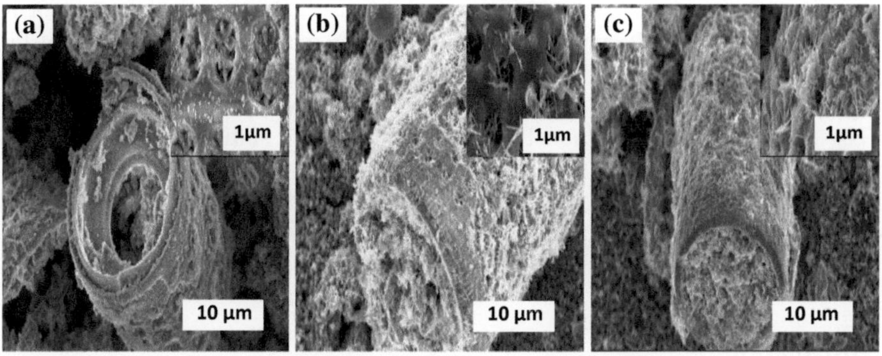

Fig. 3 SEM micrographs of MnO_2@DE for **a** 2.5 mmol MnO_4^-/g DE **b** 5 mmol MnO_4^-/g DE **c** 10 mmol MnO_4^-/g DE

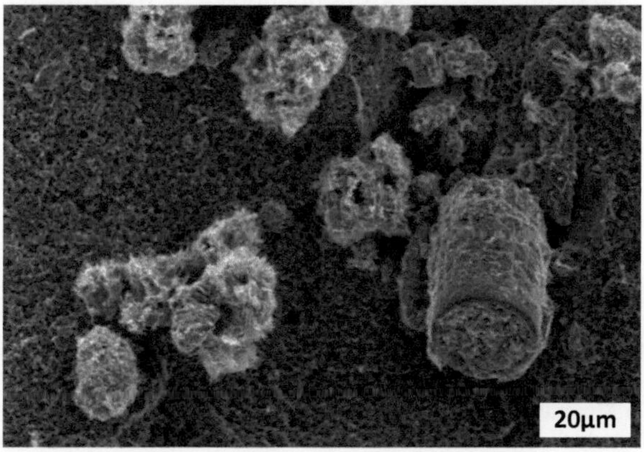

Fig. 4 Aggregates on MnO_2 formed which did not get immobilized on DE at 10 mmol MnO_4^-/g DE

Table 2 Average composition of MnO_2@ DE at various concentration of MnO_4^-

Element	2.5 mmol MnO_4^-/g DE	5 mmol MnO_4^-/g DE	10 mmol MnO_4^-/g DE
Mn	5.19 ± 0.12	15.65 ± 0.20	22.93 ± 0.17
Si	35.85 ± 0.19	29.31 ± 0.19	25.06 ± 0.18
O	56.24 ± 0.22	48.24 ± 0.28	49.31 ± 0.23
Al	2.08 ± 0.06	5.37 ± 0.10	2.08 ± 0.07
Fe	0.61 ± 0.02	1.37 ± 0.04	0.60 ± 0.06

Table 3 Lithium uptake (mg Li/g DLMO) for different adsorbents

mmol MnO_4^-/g DE	C_0 (ppm)	C_e (ppm)	Li uptake (mg/g DLMO)
10	96.39	28.335	15.633
5	96.39	50.167	16.434
2.5	96.39	83.205	9.382

MnO_4^-/g DE, the normalized lithium uptake decreases. This may be because not all of the DLMO is immobilized on DE whereas for 5 mmol MnO_4^-/g DE all of the DLMO is immobilized. Although α MnO_2 has better intercalation/deintercalation properties than λ-MnO_2 [7], the latter is better for selective extraction of lithium because of its 3D (1 × 3) tunnel structure which has a sufficient gap for lithium intercalation/deintercalation but too narrow for other metallic ions to enter the tunnel [8]. For α-MnO_2, the tunnel gaps are sufficient enough to accommodate Na^+, K^+, and Mg^{2+}, thus lacking selective extraction of lithium [7].

Conclusion

In this work, a novel lithium adsorbent is examined. Spinel structured LMO@DE composite material was successfully synthesized. The loading of MnO_2 on DE has a significant effect on lithium uptake. A lithium uptake of 16.43 mg/g demonstrates to be a promising lithium adsorbent composite. Future studies will focus on the effects of various operating conditions on lithium uptake and the nature of adsorption. Further, simulated and real salt brine solutions and adsorbent stability/recyclability will be examined.

References

1. Speirs J, Contestabile M, Houari Y, Gross R (2014) The future of lithium availability for electric vehicle batteries. Renew Sustain Energy Rev 35:183–193
2. Ober JA (2017) Mineral commodity summaries 2017. US Geological Survey
3. Grosjean C, Miranda PH, Perrin M, Poggi P (2012) Assessment of world lithium resources and consequences of their geographic distribution on the expected development of the electric vehicle industry. Renew Sustain Energy Rev 16:1735–1744
4. Xu X, Chen Y, Wan P, Gasem K, Wang K, He T, Adidharma H, Fan M (2016) Extraction of lithium with functionalized lithium ion-sieves. Prog Mater Sci 84:276–313
5. Hong H-J, Park I-S, Ryu J, Ryu T, Kim B-G, Chung K-S (2015) Immobilization of hydrogen manganese oxide (HMO) on alpha-alumina bead (AAB) to effective recovery of Li+ from seawater. Chem Eng J 271:71–78
6. Shi X, Zhou D, Zhang Z, Yu L, Xu H, Chen B, Yang X (2011) Synthesis and properties of $Li_{-1.6}Mn_{1.6}O_4$ and its adsorption application. Hydrometallurgy 110:99–106
7. Devraj S, Munichandraiah N (2008) Effect of crystal structure of MnO_2 on its electrochemical capacitance properties. J Phy Chem C 112:4406–4417
8. Ma LW, Chen B-Z, Chen Y, Shi X-C (2011) Preparation, characterization and adsorptive properties of foam-type lithium adsorbent, microporous and mesoporous materials 142:147–153

FEM Simulation of Nodulation in Copper Electro-refining

Ken Adachi, Yuya Nakai, Atsushi Kitada, Kazuhiro Fukami and Kuniaki Murase

Abstract In the copper electro-refining process, short circuiting between the cathode and anode caused by nodulation has the largest impact on the loss of current efficiency. In order to improve current efficiency, it is critically important to study the mechanism of the growth of the nodule. In this study, the nodulation was modeled using the finite element method to simulate the growth of copper bumps attached to the cathode. By considering the scale of the electrodes and their pitch, the relationship between the height of the nodule and its growth rate is investigated. In particular, a threshold height of the nodule that determines whether the nodule will rapidly grow to come in contact with the adjacent anode is identified. By comparing the result of the simulation and the experiments, the effect of leveling additives and the generation of dendritic growth are discussed.

Keywords Copper · Electro-refining · Nodulation · FEM · Simulation

Introduction

In copper refineries, recycling of metals is conducted using copper as a "collector metal", especially aimed at recovering precious metals from recycled raw materials such as scrap electronic devices. Recently, however, the quality of valuable metals in recycling raw materials is decreasing due to the intense competition for acquisition and the increasing concentration of impurities that come into the copper refinery process [1]. Improved technology is required to increase productivity efficiency since impurities remain a problem in the refining process.

In the copper tank house, electro-refining is conducted at the current efficiency of 93–98% in general. The largest amount of loss is due to the short circuiting of

K. Adachi (✉) · Y. Nakai · A. Kitada · K. Fukami · K. Murase
Department of Materials Science and Engineering, Kyoto University, Yoshida-Honmachi 36-1, Kyoto 606-8501, Japan
e-mail: adachi.ken.25e@st.kyoto-u.ac.jp

H. Kim et al. (eds.), *Rare Metal Technology 2018*, The Minerals, Metals & Materials Series, https://doi.org/10.1007/978-3-319-72350-1_20

electrodes caused by contact between copper anodes and pure copper cathodes [2] (see Table 1). The short circuiting is caused by the nodules formed on the cathodes that come in contact with the adjacent anodes (Fig 1).

The cause of the nodule formation was reported to be determined by several factors [4],

(1) Conductive particles derived from impurities stuck onto the surface of cathodes.
(2) The local shortage of leveling additives such as Cl^-, thiourea and gelatin.
(3) The deformation or scratching of SUS cathode electrodes.

Studies on nodulation have so far mainly been focused on the effect of the various parameters on the formation of the tiny nodule of the scale of 1–2 mm. However, few reports have focused on the process by which nodules grow large enough to come in contact with the adjacent anode with a scale over 15 mm. It is important to reveal the entire process in order to fully understand the driving forces or mechanisms, and to best maintain high current efficiency by preventing short circuiting.

In this study, finite element method (FEM) modeling of nodule growth in the electro-refining process was conducted. The effects of leveling additives and

Table 1 Loss of current in electro-refining

Cause of loss of current	Contribution to current efficiency (%)
Anode-to-cathode short circuiting	<3
Stray current ground	<1
Reoxidation of cathode by O_2 or Fe^{3+}	<1

Fig. 1 Photograph of a nodule in an industrial tankhouse [3]

dendritic growth, which are difficult to simulate, were discussed by comparing FEM modeling results with experimental results under similar conditions. We especially focused on two points: (i) the range of nodule size in which leveling additives suppress the growth of a nodule and (ii) the threshold nodule size in which the growth rate increases rapidly.

Modelling Description

(1) Geometry and mesh setting

The electro-refining processes were simulated by considering the secondary current density in a 2D model. Figure 2 shows the geometry of the model with a vertical wall on each side as are the electrodes, the bottom and the top are the electrically insulated boundaries, and the rest is the electrolyte, which is an aqueous solution of copper sulfate and sulfuric acid. The right wall is the anode copper and the left wall is cathode copper. The cathode has a bump at the center that acts as a model of a nodule. The growth of the nodule was studied by changing the initial height of the bump from 0.1 to 15 mm and analyzing the relationship between the initial height and the growth rate of the bump.

In this model, the domains were discretized into triangle mesh elements as shown in Fig. 3. The maximum element size was 1 mm and the minimum element size was 0.2 mm. Finer structured layers of the elements along the surface of the bump on the cathode were used to investigate the growth in more detail.

(2) Current density distribution

In this model, it was assumed that the electrolyte is uniform and that the local current density is governed by the secondary Butler-Volmer equation (see Fig. 4). The overall electric current I is held constant to keep the average local current density at about 350 A/m^2. The parameters of i_0 and αn are obtained by conducting

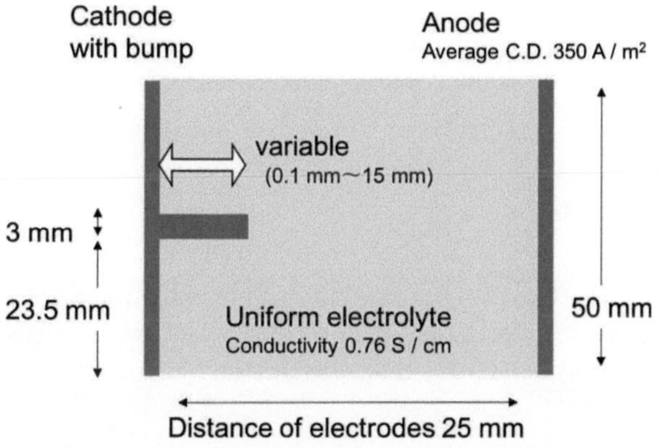

Fig. 2 Illustration showing geometry of simulated model

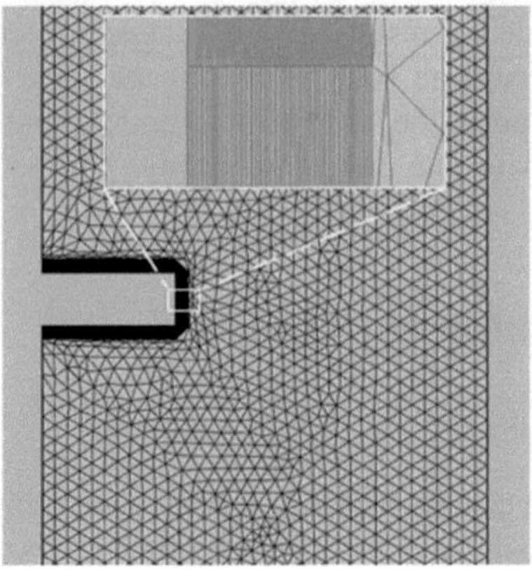

Fig. 3 Mesh setting for FEM model composed of a triangle mesh of 0.2–1 mm scale and a finer rectangle mesh along the bump

Fig. 4 Equation of current density distribution and polarization parameters for FEM model

$$i_{loc} = i_0 \left[exp\frac{(1-\alpha)nF\eta}{RT} - exp\frac{-\alpha nF\eta}{RT} \right]$$

Variable

i_{loc}: local C.D. $[A/m^2]$

η: overpotential $[V]$

Measured value

i_0: exchange C.D. $200[A/m^2]$

αn: symmetry factor 0.14

Constant

F: Faraday constant $96485[C/mol]$

R: gas constant 8.31

T: temperature $338[K]$

the corresponding electrolysis experiment under the same scale and the same conditions as the common copper electro-refining process (see Table 2 and Fig. 5).

Results

Figure 6 shows an example of the simulation result, demonstrating change in the shape of cathode and the distributions of current density (a) at the beginning of the electrolysis and (b) after the electrolysis for 40 h.

Table 2 Experimental condition to obtain polarization curve

Composition of electrolyte	$CuSO_4$	48 g/L	(0.76 M)
	H_2SO_4	190 g/L	(1.94 M)
	Cl^-	60 mg/L	(1.94 mM)
	Thiourea	5 mg/L	
	Glue	0.5 mg/L	
Temperature	65 °C		
Distance between electrodes	25 mm		
Electrodes	Copper plate 50 mm • 50 mm • 2 mm		

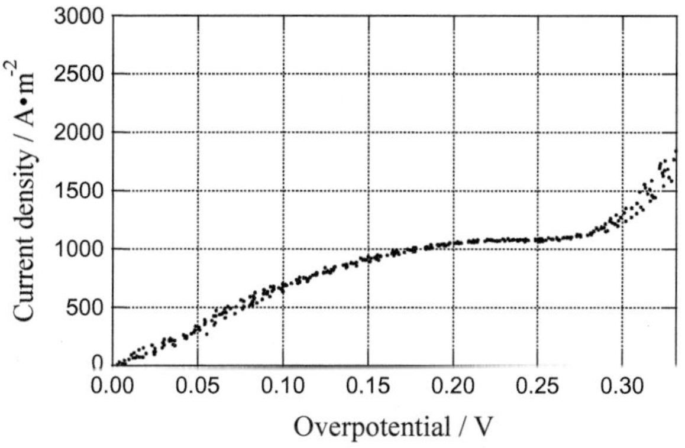

Fig. 5 Cathodic polarization curve under the conditions described in Table 2, scanning rate is 0.1 mV/sec

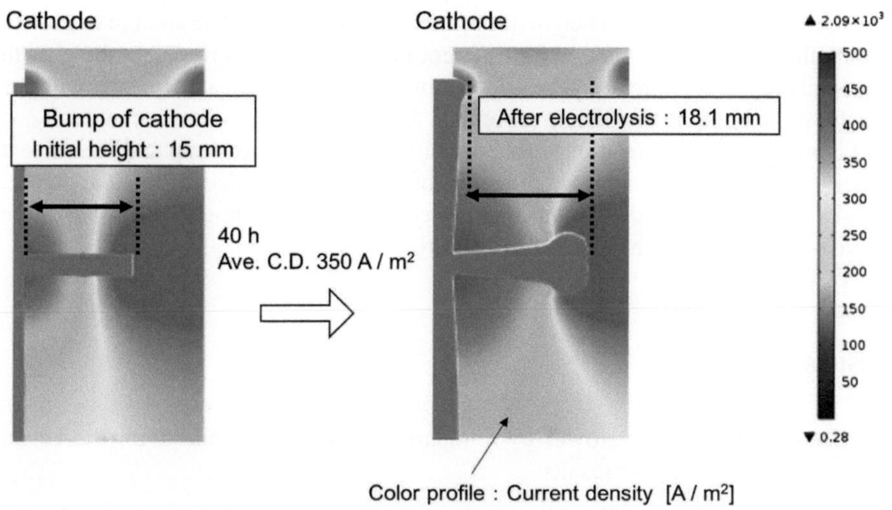

Fig. 6 Simulated growth of bump on cathode

Fig. 7 Relationship between initial height and growth of bump

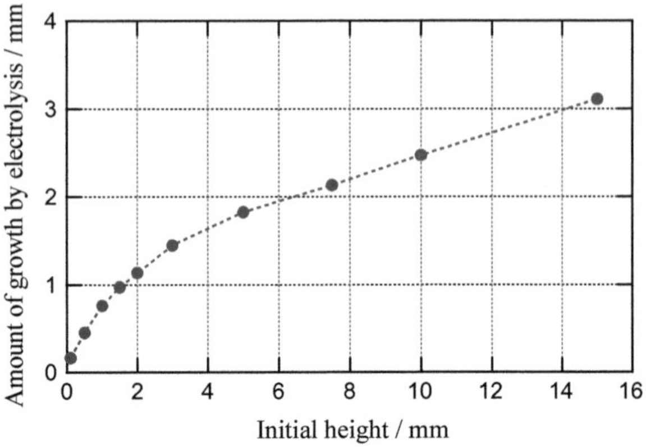

Fig. 8 Simulated relationship between bump height and current density at the tip

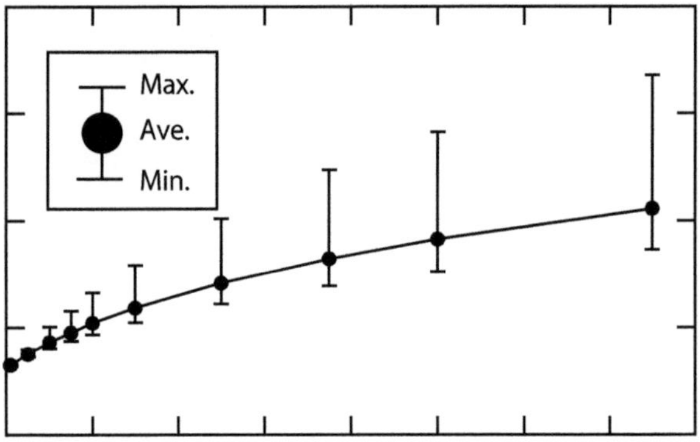

Figure 7 shows the relationship between the initial height of a bump on the cathode and the amount of growth in height during electrolysis for 40 h. The amount of growth during the specific time interval for different initial heights can represent the growth rate indirectly. Therefore, it can be said that the growth rate increases linearly with the height of the bump after a height more than 5 mm.

Figure 8 shows the relationship between the height of the bump and the current density at the tip of the bump. In the range of heights greater than 5 mm, the current density exceeds 1000 A/m^2, a limiting current density according to the polarization curve obtained during the electrolysis experiment (see Fig. 5).

Discussion

(1) Considering nodule growth by comparing simulation results with the corresponding experiment results

We investigated the growth of bumps on the cathode by modeling the nodule in a series of corresponding experiments (see the presentation of "Experimental

Fig. 9 Comparison of simulation and experiment for the growth of a bump with respect to the initial bump height after electrolysis for 40 h

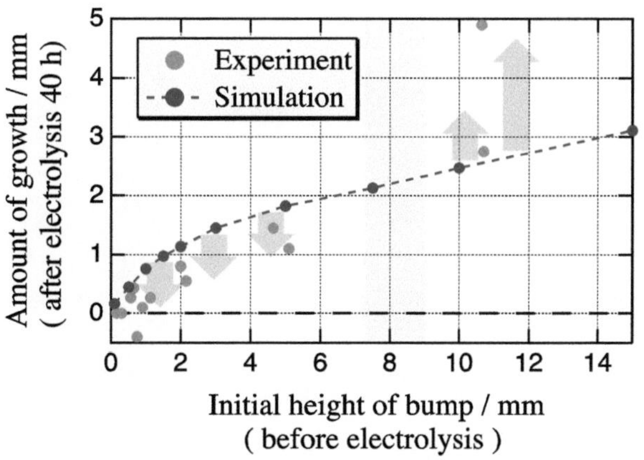

Modeling of Nodulation in Copper Electro-refining" in this symposium). Figure 9 compares the results of the simulation and the experiment showing the relationship between the initial height of the bump and the amount of growth during electrolysis for 40 h. For an initial height less than 5 mm, the growth obtained by the simulation exceeded that in the experiment. This is considered to be due to the effect of leveling additives which is reflected just in the experiment. On the other hand, for initial heights more than 10 mm, the growth found in the experiment exceeded that in the simulation. This is due to the rapid growth of dendrites, which is caused by the large current density at the tip of the bump growing closer to the anode. Therefore, the threshold height for the growth of nodules is considered to exist in the range between 6 and 10 mm.

(2) Considering nodule growth by analyzing limiting current density

The threshold height for growth of nodules can be discussed by considering the current density as well. According to the polarization curve (see Fig. 5), the limiting current density in this system is 1000 A/m^2. Over the range of current density, copper ions are depleted in the vicinity of the cathode and the shape of deposition becomes less anisotropic and more dendritic. According to the simulation of the current density distribution, the current density at the tip of the bump exceeds the limiting current density for heights of more than 5 mm. The corresponding experiments confirm that dendritic deposition can be observed after similar heights (Fig. 10).

Conclusions

We conducted FEM modelling of copper bumps on a copper cathode during electrolysis as a simulation for the growth of nodules in the copper electro-refining process that result in short circuiting at the anode. The results of FEM simulations

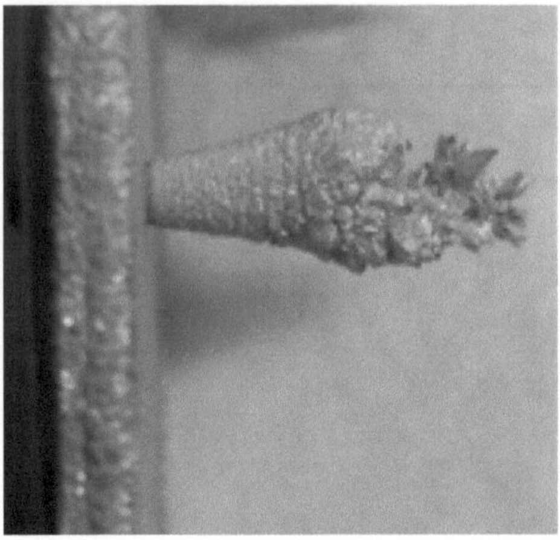

Fig. 10 Image showing onset of dendritic deposition in an experiment

were compared with corresponding experiments, and the process of nodule growth as a function of nodule height was discussed.

In the case of a small bump, the simulated growth was larger than that observed in experiment; this is considered to be due to the leveling effect of additives which cannot be easily simulated. In the case of a large bump, the growth in the experiment is larger than that in the simulation; this is considered to be due to the dendritic deposition caused by the high current density at the tip, above the experimentally determined limiting current density, which must be accounted for. The threshold height of the growth is suggested to be in the range from 6 to 10 mm, which is when nodules become large enough to overcome growth suppression by leveling effects and transition to more rapid dendritic growth.

References

1. Ikeda M, Unei A (2015) JOGMEC Reports, 2015, vol 45
2. Davenport W (2011) Extractive metallurgy of copper. Elsevier, New York
3. Shimokawa K (2013) Studies on the improvement of industrially operated copper electrorefining. Ph.D. thesis, Kyoto University
4. Andersen TN, Pitt CH, Livingston LS (1983) J Appl Electrochem 13(4):429–438

Part IV
Ti, V, Mo and W

Review of TiO_2-Rich Materials Preparation for the Chlorination Process

Shiju Zhang, Songli Liu, Wenhui Ma and Yongnian Dai

Abstract The resources of vanadium–titanium magnetite in China are very abundant, most of which are used to prepare titanium-rich materials. Titanium pigment, electric welding bar and titanium sponge are made from titanium-rich materials such as synthetic rutile and titanium slag etc. To fully utilize titanium resources, many production methods of titanium-rich materials were proposed. In this article, authors review on methods of preparing titanium-rich materials which are Electric Smelting Process, Reduction-rust Process, Selective Chlorination Process, Acid Leaching Process, and sum the advantages and disadvantages of them. Existing problems and possible tendency of production methods of titanium-rich materials are also discussed.

Keywords Titanium-rich materials · Preparation technology · Status and development

---

S. Zhang · W. Ma · Y. Dai
Faculty of Metallurgical and Energy Engineering, Kunming University of Science and Technology, Kunming 650093, Yunnan, People's Republic of China

S. Zhang
Resources and Environmental Engineering College, Panzhihua University, Panzhihua 617000, Sichuan, People's Republic of China

S. Liu
Machinery and Electrical Engineering, Yangtze Normal University, Fuling 408100, Chongqing, People's Republic of China

W. Ma (✉) · Y. Dai
State Key Laboratory of Complex Nonferrous Metal Resources Cleaning Utilization in Yunnan Province, Kunming University of Science and Technology, Kunming 650093, Yunnan, People's Republic of China
e-mail: mwhsilicon@126.com

W. Ma · Y. Dai
Engineering Research Center for Silicon Metallurgy and Silicon Materials of Yunnan Provincial Universities, Kunming University of Science and Technology, Kunming 650093, Yunnan, People's Republic of China

H. Kim et al. (eds.), *Rare Metal Technology 2018*, The Minerals, Metals & Materials Series, https://doi.org/10.1007/978-3-319-72350-1_21

Introduction

Titanium-rich materials have a high grade of TiO_2 which is more than 85% (mass fraction), and include natural rutile, synthetic rutile and titanium slag. Natural rutile is the titanium-rich material obtained by simple beneficiation from raw ore with the content of TiO_2 not less than 90%. Most of titanium-rich materials, such as synthetic rutile and titanium slag are prepared by Electric Smelting Process, Reduction Rust Process, Choice Chlorination Process, Acid Leaching Process, or other methods from ilmenite concentrate.

Titanium-rich materials are the raw materials for producing high-grade titanium dioxide and sponge titanium. The processes of producing high quality titanium-rich materials are essential for the development of titanium industry, especially for fully utilize titanium resources in Panzhihua China where contains more than 8.7×10^8 tons of TiO_2 accounting for 90.6% of the national reserves [1–5].

So far many production methods of titanium-rich materials have been proposed, but every process has its merits and weaknesses. In this article, authors briefly introduced the processes of titanium-rich materials production methods and their existing problems, and put forward some proposals for development also.

Natural Rutile

Natural rutile, a high quality raw material for preparing titanium tetrachloride, is one of the important titanium bearing minerals with a TiO_2 content of 90%–99wt%. However, natural rutile holds little proportion in the proved titanium resources all over the world. There is only 195 million tons of natural rutile with only 9.5% of total titanium resources. About 80% natural rutile resources are mainly distributed in Brazil, India and Australia, of which Australia is the world's natural rutile reserves than any other country and with the largest production capacity as well. For easy mining, simple beneficiation and low production costs, the stock of available resources is decreasing sharply, and Australia government has begun limiting the exploitation.

With the development of world industry, the demand of rutile is increasing, while their reserves are decreasing year by year. The annual demand for raw materials only chlorination titanium dioxide chloride is more than 3.5 million tons, while the output of natural rutile is only about 0.4 million tons each year around the world. This causes the price of rutile to keep going up and the contradiction between supply and demand of the resources to be more acute [6, 7].

Titanium-Rich Materials Production Methods

With the gradual exhaustion, price rise of natural rutile resource and the development of titanium industry, titanic iron ore and titanium concentrate have become the main raw materials of titanium production industry. Titanic iron ore and titanium concentrate need to enrich into synthetic rutile and titanium slag as the Ti content in them are not high enough for production.

The titanium-rich material produced by smelting with electric furnace is called titanium slag in general. High titanium slag is the product with TiO_2 content more than 90%. The titanium-rich materials made from ilmenite in other ways are called artificial rutile. Titanium slag is mainly used in the production of titanium dioxide by sulfuric acid process; High titanium slag and artificial rutile are mainly used as raw materials for Chlorinated titanium dioxide and titanium sponge production [8, 9]. Using titanium-rich materials as raw materials of titanium production can significantly reduce emission of the three wastes, environmental pollution, cost of production and simplified the process meanwhile.

Many methods of preparing titanium-rich materials have been studied and proposed which can be divided into pyrometallurgical processes and hydrometallurgical processes. Pyrometallurgical processes include Electric Smelting Process, carbonization Process, Choice Chlorination Process, Plasmamelt Process, Microwave Heating plasma Process, and etc. Hydrometallurgical processes include Acid Leaching Process, Reduction Rust Process, Reduction magnetic Separation Process, and etc. In all of these methods, Electric Smelting Process, Acid Leaching Process, Reduction Rust Process and Choice Chlorination Process have been put into industrialized production [7].

Electric Smelting Process

Electric Smelting Process is the main method used in the current production of titanium-rich materials. Ferriferous oxides are reduced to liquid iron, and TiO_2 is enriched in the slag at temperature above 1650 °C in electric furnace. The slag is broken to obtain titanium-rich material. The advantages of this method are: less environmental pollution, simple process, recovery and utilization of the electric furnace gas and high efficiency of utilization. However, impurities especially non-ferrous impurities cannot be removed perfectly with this method, so the content of impurities in ilmenite concentrate has an important influence on titanium slag grade. Large electricity consumption is a problem of this method [10–12].

It consumes 65% of the global total output of titanium concentrates annually to prepare titanium-rich materials by Electric Smelting Process, and most of capacity is distributed in South Africa, Canada, the Commonwealth of Independent States and Norway. The slag with TiO_2 grade of about 80–90% produced by the factories in these areas is the main raw material of production of sulfate process titanium dioxide. Most electric furnaces abroad are hermetic or Semi-hermetic type with the advantages of advanced technologies, high efficient utilization of semi steel and large scale of production, especially for hermetic electric furnaces in Canada and South Africa.

About 65% of titanium-rich materials in china are prepared by Electric Smelting Process, and the main equipment is open-top electric furnace [13]. Compared with foreign countries, there still exits a big gap both in techniques and equipment. There are many questions existing in this process in China, such as small production scale, low processing and equipping level, unstable furnace condition, difficulties to recycle electric furnace gas, high energy consumption, and etc. The good news is that Yunnan Xinli Non-Ferrous Metals Co., Ltd has introduced the South Africa's technology and constructed 80,000 t/a TiO_2 slag project and the matched mining project in Wuding County. In addition, Panzhihua Iron and Steel Co. has introduced the Ukrainian technology of the Semi-hermetic electric furnace.

Hydrochloric Acid Leaching Process

The domestic and foreign scholars have delved into the preparation of artificial rutile via hydrochloric acid leaching, with research findings mainly including BCA dilute hydrochloric acid recycled leaching, Wahchang method, ERMS method and TSR method, etc. [14–22]. Wherein, the BCA dilute hydrochloric acid recycled leaching method is representative and has wide industrial application; the Wahchang method, also called concentrated Hydrochloric Acid Leaching Process, has no essential distinction from the dilute Hydrochloric Acid Leaching Process, just is different in that it uses concentrated hydrochloric acid as leaching agent; while ERMS method and TSR method are improvements on BCA method by emphatically improving the pretreatment technique of raw materials and technology of hydrochloric acid regeneration via leach liquor.

The BCA dilute hydrochloric acid recycled leaching method (also called Benilite method) invented by American Benilite Company in early 1970s was recognized as an advanced method to prepare artificial rutile and was adopted by multiple factories. By the end of 1970s, the total productive capacity has reached more than 400,000 t/a. BCA hydrochloric acid recycled leaching method has quick leaching speed and strong impurities removal capacity (such as ferrum, calcium, magnesium, aluminum and manganese, etc.), and product obtained has high porosity and large reactive activity, thus becoming quality raw materials for preparation of TiO_2 via chlorination; the recycling of hydrochloric acid realizes closed circulation and depollution. Hence this process is suitable for treating a most complete line of ilmenite concentrates. But the process has a drawback of severe erosion to equipment by hydrochloric acid as a highly corrosive acid, leading to need for dedicated equipment manufactured from antiseptic materials, hence large investment; besides, the process of hydrochloric acid regeneration via spray air roasting dissipates much energy [21].

Sulfuric Acid Leaching Process

Also called ISK method and invented by Japanese Ishihara Sangyo Kaisha Ltd. [21, 23], the sulfuric acid leaching leaches artificial rutile from ilmenite concentrate immersed in the waste sulfuric acid (concentration is about 22%) produced in production of titanium dioxide using sulfuric acid method The ISK method includes such processes of reduction, pressure leaching, filtration and washing, incineration, etc. The artificial rutile factory using this method built by ISK Company has an annual productivity of up to 100,000 tons.

The ISK method adopts high grade Indian ilmenite placer (the TiO_2 content is 59.62%, the raw materials of the lowest grade ever used contains 53% of TiO_2), where the iron mainly occurs in the form of Fe^{3+}. The method first makes a weak reduction for ilmenite placer with refinery coke as reducer to reduce the Fe^{3+} in the mine into Fe^{2+} at 900–1000 °C, with percent reduction up to over 90%. The reduction charge is cooled to below 80 °C in the rotary cooler with air being isolated, then is discharged to remove the residual refinery coke via magnetic separation. The leaching process is conducted at 120–130 °C under pressure of 0.2 MPa using the waste dilute sulphuric acid discharged in production of titanium dioxide using sulfuric acid method, with TiO_2 hydrated colloidal solution added as seed crystal to accelerate deferrization, improve deferrization rate and conduce to control the granularity of end products and prevent substantive product of fine particles from being produced. The solid and liquid of leached product are separated using band filter. The obtained liquid phase solution mainly contains ferrous sulfate, which can be used to prepare ammonium sulfate and iron oxide red; while the solid phase separated is washed and incinerated to get artificial rutile with TiO_2 content of 96% and recovery rate of about 90%.

When removing the ferrum in ilmenite, the ISK method can also partially remove the soluble impurities such as calcium, magnesium, aluminum and manganese, hence the product obtained is of high grade [23–26]. Besides, the method effectively utilizes the waste dilute sulphuric acid discharged in production of titanium dioxide using sulfuric acid method, and makes acid leaching under low concentration and at low temperature, hence the production cost is low and the problem of "three wastes" in production of titanium dioxide is also solved. However, the method is only applicable to high-grade ilmenite with low content of calcium and magnesium, and low-grade ilmenite complicates the process engineering and lowers product quality. In addition, the three wasters are largely produced and subsidiary flows are complicated.

Reduction-Rust Process

Also called Becher method [27–32], the reduction-rust process was developed by Australian CSIRO and is a method with low pollution and cost, which uses coal as fuel and reducer in reduction process and only consumes a handful of hydrochloric acid or ammonium chloride in rusting process, with nearly neutral red mud and waste water produced in the rusting process. The method reduces all the ferriferous oxides in ilmenite into metallic iron in rotary kiln at 1000–1200 °C, recovers surplus carbon from the product, and places the magnetic materials into solution containing NH_4C1 of about 2.0% (pH = 6–7), then blows in air to make the zero-valent iron electro-chemical corroded into Fe^{2+} ions to enter solution and combines with OH^+ ions produced by electrochemical action to form fine $Fe(OH)_2$ to be further oxidized into deposit of $Fe(OH)_3$. The ferric oxides are washed away by dilute solution in multilevel hydroclone to concentrate titanium dioxide, then dilute acid washes away the remnant ferrum and manganese in titanium-rich materials, which is filtered and dried to get titanium-rich materials containing TiO_2 of 93–96% (in the case of placer as raw material). Hydrated ferric oxide (rust) can be made into iron red product after dried and incinerated. The titanium-rich materials can be directly used as chlorination stock to produce $TiCl_4$, or artificial rutile after air roasting for sale.

The titanium-rich material produced by reduction-rust process has even granularity, stable color and high grade. In addition, compared with Electric Smelting Process, its production process consumes low power, while compared with Acid Leaching Process its environmental pollution is small, and hence this process is highly favored. However, the method has drawbacks of severe ringing in reduction kiln caused by high requirements on reduction temperature (1000 °C), short period of continuous production, large technology difficulty in strong reduction, long rusting time, etc., plus its high requirements on ilmenite, it is generally regarded as applicable to treatment of high grade placer. While the ilmenites in China are mostly granite mine and domestic production of titanium-rich materials using this method fails to be widely applied due to small scale, large dispersity, imperfect process flow, and absence of solutions for critical technologies and equipment, etc.

Australia utilizes the local inexpensive coal and ilmenite (containing TiO_2 of 54%) by this method to produce artificial rutile containing TiO_2 of 92–94%, which serves as quality raw materials to produce titanium dioxide using chlorination process. The constructed factories in Australia have annual productivity of up to 790,000 tons in total. Chinese Zhuzhou Dongfeng Smeltery and Beihai Dressing Plant have adopted reduction-rusting method to produce artificial rutile with quality ilmenite as raw material. The annual productivity was about 4000t artificial rutile containing TiO_2 of 88%. But the product has halted due to severe kiln ringing and short continuous production period [33, 34].

Selective Chlorination Process

The selective chlorination process was developed by Japanese Mitubish Metal Company, which built a pilot plant with monthly productivity of 1000t artificial rutile [21, 26, 35]. The method takes advantage of different capacities of the components in ilmenite concentrate when reacting with chlorine. The sequence for the components in the concentrate to react with Cl_2 at 850–950 °C in the presence of reducer carbon is as follows: $CaO > MnO > FeO > V_2O_5 > MgO > Fe_2O_3 > TiO_2 > Al_2O_3 > SiO_2$. Thus, the components located before TiO_2 can be chlorinated with priority in chlorination through controlling the carbon addition, and ferrum can be volatilized and discharged in the form of $FeCl_3$ to be congealed and recovered in dust precipitator; The TiO_2 is not chlorinated and is discharged along overflowing on material bed in the oven, then is dressed to remove un-chlorinated mineral and residual reducer, and then is washed to separate out the residual chloride impurities such as $CaCl_2$, $MnCl_2$, $MgCl_2$, etc. in TiO_2, to get the product of artificial rutile. To improve the selective chlorination rate of ferrum and inhibit the generation of $FeCl_2$ hindering good running of fluidized bed furnace, the ilmenite concentrate is usually preoxidated in oxidative atmosphere of 900–950 °C before chlorination.

Despite advantages of short flow, simple equipment, high productivity and low cost, the chlorination process is hard to treat ilmenite concentrate with high content of calcium and magnesium, and the melted $CaCl_2$, $MgCl_2$ and $FeCl_2$ produced in chlorination process is prone to damage the boiling state, making boiling chlorination hard to be performed. In addition, the large use of corrosive gas chlorine, severe erosion of equipment, and troublesome tail gas treatment make this process fail to find wide use [31, 36].

Except above mentioned Electric Smelting Process, acid leaching process, reduction-rusting process and selective chlorination process which are applied to industrial production, other methods of preparing titanium-rich materials remain under research in labs.

The Problems in Preparation of Titanium-Rich Materials in Panzhihua China

Titanium sponge and titanium dioxide are two prime products in titanium industry. The industrial methods of producing titanium dioxide include Sulfuric Acid Process and Chlorination Process. Coupled with enhanced environmental considerations of people, the Chlorination Process which has advantages of short process flow, large productive capacity by single line, low production cost, high-grade product, continuous automation of operation, and environmental friendliness, etc. is gradually superseding sulfuric acid method as the future trend of development for titanium industry. Currently with output accounting for about 60% of total yield of titanium oxide in the world, the chlorination process is the first choice for production of high-grade titanium oxide [17, 31].

$TiCl_4$ is a requisite intermediate in preparing titanium oxide and titanium sponge via chlorination process, and can be prepared by three methods of shaft chlorination, fused salt chlorination and fluid bed chlorination [17]. The shaft furnace chlorination process fine-grinds the chlorinated titanium slag (or rutile) and refinery coke, which then are added with agglomerant to be averaged, agglomerated and coked, and the fabricated agglomerate is piled in shaft type chlorinator and takes on fixed layer to react with chlorine. Hence the method is also called fixed layer chlorination or agglomerate chlorination. The complex profile and low productive capacity of equipment have made it washed out as early as 1980s. The fused salt chlorination suspends the levigated titanium slag or rutile and refinery coke in fused salt medium (mainly composed of KCl, NaCl, $MgCl_2$ and $CaCl_2$), then lets in chlorine for chlorination. In this method, the chlorination leftover is mixed in fused salt and exhausted in the form of waste salt. As when raw materials used are the same, the fused salt chlorination produces more offal than boiling chlorination, and the offal is more intractable, thus hindering its development. As a relatively advanced method, the boiling chlorination adopts fine titanium-rich materials and solid carbon as reducers, which take on fluid state at high temperature under action of chlorine flow, with chlorination reaction occurring in the meanwhile. The method can accelerate the mass and heat transfer between gas phase and solid phase to intensify production, and has become the mainstream method of preparing $TiCl_4$, yet it requires high on raw materials, generally requiring the TiO_2 grade of raw materials to be greater than 90% and the total content of CaO and MgO to be smaller than 1.5%, plus requirements of appropriate particle size distribution.

Most of titanium mines, especially the vanadium-titanium magnetite in Panzhihua, are intractable granite mines, which is of compact structure and solidifies and melts much MgO and CaO, so the grade of TiO_2 in the dressed titanium concentrate is low while the content of MgO and CaO is high, thereby posing great challenge to preparation of titanium-rich materials [33]. Although many efforts have been done at research on the preparation methods for titanium-rich materials via various ilmenites such as Electric Smelting Process, different acid leaching methods, reduction-rusting method and selective chlorination process, etc., people still fail to develop an efficient path of large-scale industrial production of high-quality titanium-rich materials suitable for the process of boiling chlorination till now for the vanadium-titanium magnetite in Panzhihua.

Currently, the ilmenite concentrate in Panzhihua is mainly used to produce high titanium slag via Electric Smelting Process, yet the quality of high titanium slag prepared has always been low due to problems of calcium and magnesium content, etc., which gravely hinders the development of titanium oxide preparation using chlorination process. Thus, to study the efficient method of preparing Titanium-rich materials from ilmenite concentrate is of extremely important significance for development of titanium industry in Panzhihua.

Conclusion

Based on the introduction and analysis of these technologies on preparation of titanium-rich materials, all of these technologies are restricted by the problems of technology advantages, industrialization, energy saving, environmental protection, etc. Although a great deal of researches has been done, production methods of titanium-rich materials have not been able to make a breakthrough progress. It is great significant to develop a highly efficient, environmentally friendly green process technology, especially for the resources in Panzhihua.

References

1. Mowei TD (2009) Selection of titanium. Metallurgical Industry Press, Beijing, pp 10–72
2. US Geological Survey (2015) Mineral commodity summaries 2015. Washington
3. Liu J, He Z, Liu Y et al (2015) Geological and geochemical characteristics of vanadium titanium magnetite deposit in Hongge Sichuan. J Jilin Univ (Earth Sci Ed) 45(1): 62
4. Zhang Y (2009) Research on the tiny particle ilmenite experiment by slon magnetic separator. Ph.D. thesis, Jiangxi University of Science and Technology
5. Bai C (2003) Study on some physical chemistry problems of blast furnace slag-bearing titanium. Ph.D. thesis, Chongqing University
6. Li X, Wen S (2011) Actuality of titanium dioxide and titanium sponge feedstock industry in the word and key points of domestic developing. Titanium Ind Prog 28(3):9–13
7. Chen L (2012) Investiation on selective carbon thermal reduction of ilmenite for preparation of rich titanium material. Ph.D. thesis, Central South University
8. Samal S (2016) Synthesis and characterization of titanium slag from ilmenite by thermal plasma processing. JOM 68(9):1–10
9. Dong F, Liu X (2015) Discussion on production technology of Ti-rich material. China Chem Trade 18:101
10. Luran J (2014) Chinese high titanium slag production technology status and development of. China Metal Bull 10:43–45
11. Hu K, Xi G, Yao J, Xi X (2007) Domestic research and production status of titanium slag. Rare Metals Lett 26(3):7–8
12. Tang Z (2006) Production situation of titanium slag and application trend in titanium dioxide. China Coat 21(10):53–56
13. Yang F (2014) Study on preparation of rich titanium material form titanium slag. Ph.D. thesis, Northeastern University
14. Sun Y (2006) Research on new technology for producing high quality titanium rich materials with high calcium magnesium ilmenite. Ph.D. thesis, Kunming University of Science and Technology
15. Chen JH (1976) Pre-leaching or reduction treatment in the beneficiation of titaniferous iron ores: US, US3967954
16. Walpole EA (1997) Process for separating ilmenite: US, US5595347
17. Balderson GF, Macdonald CA (1999) Method for the production of synthetic rutile: US, US5885324A
18. Akhgar BN, Pazouki M, Ranjbar M et al (2010) Preparation of nanosized synthetic rutile from ilmenite concentrate. Miner Eng 23(7):587–589
19. Ogasawara T, Araújo RVVD (2000) Hydrochloric acid leaching of a pre-reduced Brazilian ilmenite concentrate in an autoclave. Hydrometallurgy 56(2):203–216
20. Rado TA (1980) Process for the production of synthetic rutile: US, US4199552[P]
21. Zhang L (2011) A new process for preparing high quality titanium rich materials from Panzhihua ilmenite. Ph.D. thesis, Central South University
22. Hu W, Qi B, Jiang X et al (2012) Semi-industrial expansion test of artificial rutile from ilmenite with hydrochloric acid leaching process. Nonferrous Metals (smelting parts) (2):27–29
23. Li D, Liu H (2009) Titanium smelting process. Chemical Industry Press, Beijing, pp 22–40

24. Sun H, Zhou G, Peng T et al (2015) Recovery of titanium from titanium-rich product prepared from high Ti-bearing blast furnace slag by sulfuric acid leaching. Min Metall 24(3):54–58
25. Li C, Liang B, Wang HY (2008) Preparation of synthetic rutile by hydrochloric acid leaching of mechanically activated Panzhihua ilmenite. Hydrometallurgy 91(1):121–129
26. Dengke T, Sun Y et al (2016) Production technology and development prospects of artificial rutile. Chlor-Alkali Ind 52(8):31–36
27. Sui L (2015) Theory and technology of green utilization of titanium slag. Ph.D. thesis, Northeastern University
28. Filippou D, Hudon G (2009) Iron removal and recovery in the titanium dioxide feedstock and pigment industries. JOM 61(10):36–42
29. Hu K, Xi G, Yao J, Xi X (2006) The status of artificial rutile produced by reduction-rusting process and the feasibility of its use in pan steel. Adv Titanium Ind 23(4):17–22
30. Benjamin SE, Sykes JM (1990) Chloride-induced pitting corrosion of swedish iron in ordinary portland cement mortarsand alkaline solutions: the effect of temperature. Corrosion of reinforcement in concrete. The third international symposium on corrosion of reinfor
31. Ensley KL, Moles OW, Perkins HA (1985) Process for the treatment of metal-bearing ores: US, US 4562048A
32. Gueguin M (1986) Process of producing synthetic rutile from titaniferous product having a high reduced titanium oxide content: US, US4629607
33. Wang Y (2003) Study on preparing Ti-rich feedstock by selective leaching the modified titanium slag with hydrochloric acid solution. Ph.D. thesis, Kunming University of Science and Technology
34. Cao H (2007) Study on preparing Ti-rich feedstock by leaching the modified Ti-bearing blast furnace slag with hydrochloric acid. Ph.D. thesis, Northeastern University, 3–4
35. Wen W (2003) Thermodynamics and kinetics of producing synthetic rutile by selective chlorination of ilmenite. Iron Steel Vanadium Titanium 24(1):8–15
36. Peng X, Shi Z (2015) Waste treatment and reuse of titanium dioxide production. Resour Conserv Environ Prot (7):60

Effect of CaO Additive on the Interfacial Reaction Between the $BaZrO_3$ Refractory and Titanium Enrichment Melt

Guangyao Chen, Juyun Kang, Pengyue Gao, Wajid Ali, Ziwei Qin, Xionggang Lu and Chonghe Li

Abstract The $BaZrO_3$ and CaO-doped $BaZrO_3$ ceramics were fabricated at 1750 °C for melting the Ti_2Ni alloys, respectively. Employing XRD, OM and SEM, the influence of CaO on the composition of $BaZrO_3$ and the interaction between the $BaZrO_3$ ceramic and titanium alloy was investigated. The results showed that the CaO doped $BaZrO_3$ was consisted of CaO and $Ba_{1-x}Ca_xZrO_3$. The grain on the surface of $BaZrO_3$ ceramic was loose after melting, and the thermodynamic analysis indicated that the dissolution-erosion was responsible for the interaction mechanism. An amount of $BaZrO_3$ refractory was attached to the Ti_2Ni alloy bottom after the alloy was cooled. However, no refractory was found on the bottom of $T_{i2}Ni$ alloy, which was cooled down on the CaO doped $BaZrO_3$ ceramic. CaO additive can effectively reduce the interaction between the $BaZrO_3$ refractory and the titanium melt, indicating that it was a very promising refractory for preparing the titanium alloy.

Keywords Barium zirconate · Calcia · Doping · Titanium alloy
Interfacial reaction

G. Chen
Materials Genome Institute of Shanghai University, 201900 Shanghai, China

G. Chen · J. Kang · P. Gao · W. Ali · Z. Qin · X. Lu · C. Li (✉)
State Key Laboratory of Advanced Special Steel, Shanghai Key Laboratory of Advance Ferrometallurgy, School of Materials Science and Engineering, Shanghai University, 200072 Shanghai, China
e-mail: 675934321@qq.com

X. Lu · C. Li
Shanghai Special Casting Engineering Technology Research Center, 201605 Shanghai, China

H. Kim et al. (eds.), *Rare Metal Technology 2018*, The Minerals, Metals & Materials Series, https://doi.org/10.1007/978-3-319-72350-1_22

Introduction

The titanium melt possesses the high chemical activity [1, 2], and it can easily react with the normal refractory (such as SiO_2 [3], MgO [4]) to contaminate the melt. Kostov [5] presented that ZrO_2, CaO and Y_2O_3 exhibited the good thermodynamic stability to titanium melts based on thermodynamic calculation, and these refractories may be suitable for melting titanium alloys. However, the reactant α-Ti(Zr, O) was found after melting Ti6Al4 V alloy in ZrO_2 mold [6]. By comparing with ZrO_2, CaO and Y_2O_3 are suitable for melting titanium alloys attributing to their lower Gibbs free energy of formation than that of ZrO_2 [7, 8], but there remain many significant drawbacks that are hard to be solved before manufacturing them on an industrial scale, i.e., the inherently poor thermal shock of Y_2O_3 [9] and the hygroscopic nature of CaO [10]. Based on the above considerations, it is essential to seek a new high-stable refractory for induction melting of titanium alloys.

$BaZrO_3$, is a high stable compound, which was often used as the electron conductor material as well as a crucible refractory to grow high quality of $YBa_2Cu_3O_{7-x}$ (YBCO) single crystals [11, 12] Recently, it was introduced in the refractory field for melting the titanium alloys. Zhang tested the $BaZrO_3$ crucible with TiO_2 as a sintering aid in contact with TiNi melt [13], and He studied the interfacial reaction between the $BaZrO_3$ crucible and TiAl melt [14]. Li prepared the TiFe-base hydrogen storage alloy by means of the $BaZrO_3$ crucible [15]. All results mentioned-above indicated that $BaZrO_3$ refractory showed good corrosion resistance to TiNi, TiFe and TiAl melts, whereas our previous study revealed that it still exhibited insufficient stability against the alloys rich in titanium [16]. Hence, to improve the stability of $BaZrO_3$ is necessary for broadening its scope in preparing the titanium alloys.

Currently, doping is a simple and potentially effective method to improve the stability of $BaZrO_3$, the oxide, i.e., In_2O_3 [17], Bi_2O_3 [18] and CaO [19], are the common additives. Due to the high activity of titanium melt, some additives, such as the In_2O_3 and Bi_2O_3, can react with and contaminate the melts, obviously, they should be excluded as the additives for melting titanium alloys. CaO exhibits sufficient stability to titanium melts, and implies that it may be an appropriate additive [20, 21]. Gui [22] revealed that BaO–CaO–ZrO_2 compound may be a new composite refractory material for melting titanium alloys based on the assessment of BaO–CaO and $BaZrO_3$–$CaZrO_3$ binary system by using the CALPHAD (calculation of phase diagram) technology. However, the use of Ca-doped $BaZrO_3$ for melting titanium alloys were never reported.

In the present study, firstly, the $BaZrO_3$ and CaO-doped $BaZrO_3$ powders were synthesized by a solid phase method using the industrial grade $BaCO_3$, ZrO_2 and CaO powders, respectively, then the $BaZrO_3$ and Ca-doped $BaZrO_3$ crucibles were prepared by the cold isostatic pressing as well as the solid phase sintering process, the effect of CaO on the phase composition and microstructure of $BaZrO_3$ was studied. Next, Ti_2Ni alloys were melted in both crucibles, the interaction between the crucibles and the alloys were investigated to evaluated the stability of

CaO-doped $BaZrO_3$ by comparing with the $BaZrO_3$, and the interaction mechanism was also proposed.

Experimental

The Preparation of Samples

Firstly, the industrial-grade $BaCO_3$ (purity >98%), ZrO_2 (purity >99%) and Y_2O_3 (purity >99.9%) powders were used for preparing the $BaZrO_3$ (N1) and Ca-doped $BaZrO_3$ (N2) powders per the stoichiometric proportion in Table 1. These powders were mixed, then ball-milled in ethanol with yttria stabilized zirconia (YSZ) balls in the concrete mixer at a speed of 300 rotation per minute (rpm). The weight ratio of powder, YSZ balls and ethanol were 3:4:0.8. The slurries were then dried in an oven at 150 °C for 3 h. Secondly, the mixed N1 and N2 powders were calcined at 1400 °C for 12 h. Thirdly, the final N1 and N2 powders were fabricated into the crucibles biscuit in a U-shape steel mandrel of 3.5 cm width and 4.5 cm height by using the cold isostatic pressing with a pressure of 120 Mpa for 3 min, and fired with dwell times of 4 and 6 h at 1450 and 1750 °C, respectively. The crucible after sintering had a width of 2.9 cm and a height of 3.7 cm. A slow heating rate was controlled at 2 °C/min to avoid the generation of cracks. For the melting experiment, Ti_2Ni master alloys were prepared in the water-cooled cooper induction furnace using the titanium sponge (purity >99.9%) and nickel plate (purity >99.9%), and the oxygen concentration of the master alloy was about 0.045 wt%. The two alloys were melted in the N1 and N2 crucibles, respectively. Before the melting, the crucible was placed in the VIM furnace and then backfilled with Al_2O_3 ramming mass to prevent damage of the induction coil in case of crucible failure. The alloy about 70 g was placed in the crucible and the furnace was evacuated to 10^{-3} mbar and backfilled with high purity argon gas for at least three times. Then the melting was performed under the high vacuum. Once the melt was visible, backfilled the argon gas quickly and raised gradually the temperature up to 1500 °C, and kept at this temperature for 5 min. Then the melt could solidify and cooled down in the crucible. The crucibles after melting were sectioned longitudinally using a water-cooled diamond wheel and then cut to a suitable size to be mounted in the epoxy resin. The mounted specimens were ground using silicon carbide papers and the polished to 10 mm. The SEM measurement was used for investigating the interaction between the crucible and the alloy.

Table 1 Powder ratio of $BaCO_3$, ZrO_2 and CaO

Sample No.	Composition, *x*/% (mole fraction)		
	$BaCO_3$	ZrO_2	CaO
N1	49.0	51.0	0
N2	42.5	42.5	15.0

Results and Discussion

XRD patterns of the N1 and N2 powders after sintering at 1400 °C were shown in Fig. 1. Only the $BaZrO_3$ phase was observed in the N1 powders, and it agrees with the data of standard $BaZrO_3$ (JCPDS 06-0399). XRD pattern of the N2 powder was also similar to that of $BaZrO_3$, and no CaO phase was appeared. After the XRD data fitting, the cellular parameter of the N2 powder was about 0.4193 nm. No significant change was appeared by comparing with that of the standard $BaZrO_3$ (0.4191 nm) [23], and it indicates that the CaO was solid soluted into the $BaZrO_3$.

Figure 2 shows the SEM pictures of the surface of N1 and N2 crucibles after sintering at 1750 °C for 6 h. From Fig. 2a, the structure of the surface of N1 crucible was uniform and compact. However, the structure of the surface of N2 crucible was uneven, and the distribution of the grains and the grain boundaries was irregular, as shown in Fig. 2b. XRD pattern of the surface of N2 crucible shows that it was consist of CaO and $Ba_{1-x}Ca_xZrO_3$. Apparently, the dopant amount (15 mol %) of CaO was significantly excess. It was because that $r(Ca^{2+}) = 0.1$ nm, $r(Ba^{2+}) = 0.135$ nm, $\frac{(r_{Ba^{2+}} - r_{Ca^{2+}})}{r_{Ba^{2+}}} \times 100\% = 25.9\% > 15\%$, and only the finite solid solution can be formed between the $BaZrO_3$ and CaO. However, no CaO in the N2 powder was detected by using the XRD measurement, and it was because that concentration of CaO might have been too low in order to be indentified with XRD, and the lower detection limit for XRD is generally considered to be 1% [24], although lower detection limits could be achieved. By combining Fig. 2b with Fig. 3, it can be concluded that the precipitation of CaO resulted in the disordered structure of the surface of N2 crucible.

The SEM pictures of the cross-section of N1 and N2 crucibles were shown in Fig. 4. From Fig. 4a, the inside structure of N1 crucible was compact, and many pores of small sizes was uniformly distributed. However, after doping the CaO,

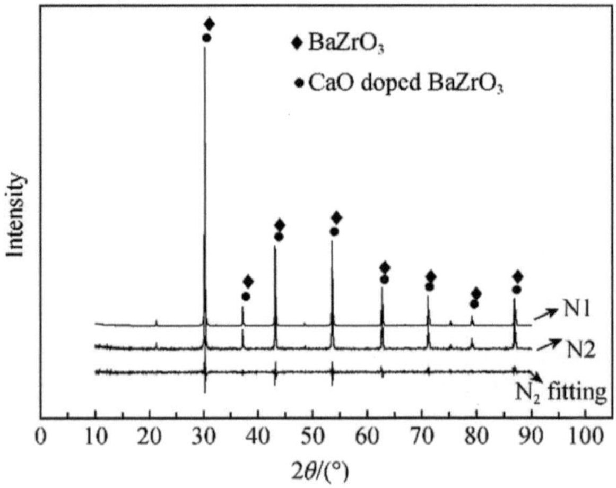

Fig. 1 XRD patterns of samples N1 and N2 after sintering at 1400 °C for 12 h

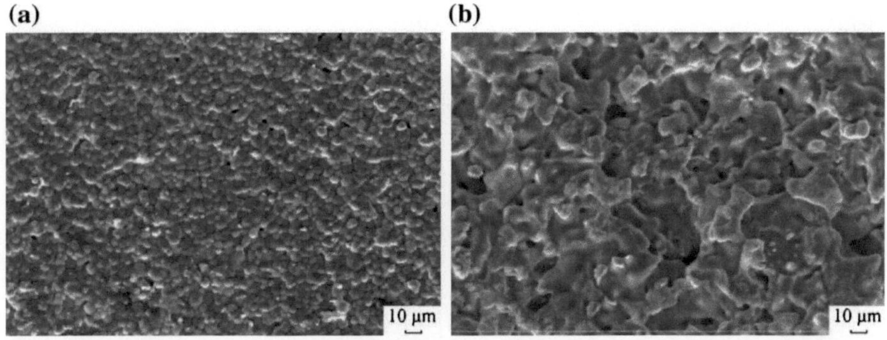

Fig. 2 SEM pictures of the surface of N1 and N2 crucibles after sintering at 1750 °C for 6 h; **a** N1: the $BaZrO_3$ crucible, **b** N2: the CaO-doped $BaZrO_3$ crucible

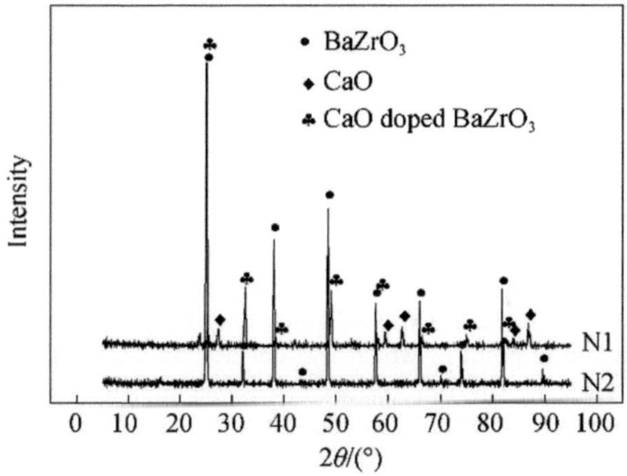

Fig. 3 XRD pattern of the surface of $BaZrO_3$ and CaO-doped $BaZrO_3$ crucibles N1: the $BaZrO_3$ crucible, N2: the CaO-doped $BaZrO_3$ crucible

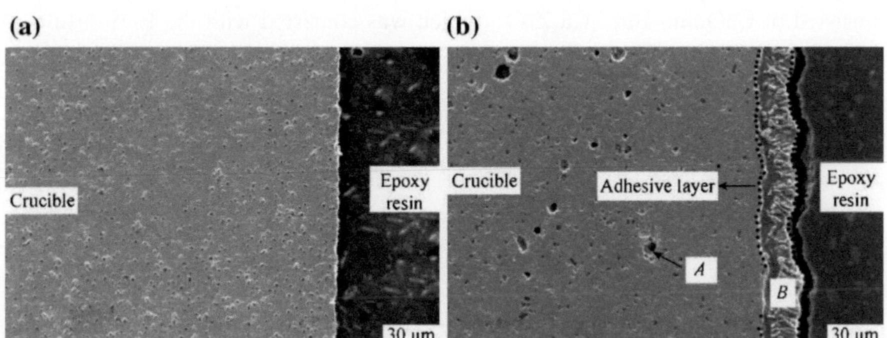

Fig. 4 SEM pictures of cross-section of the $BaZrO_3$ and CaO-doped $BaZrO_3$ crucibles; **a** the $BaZrO_3$ crucible, **b** the CaO-doped $BaZrO_3$ crucible

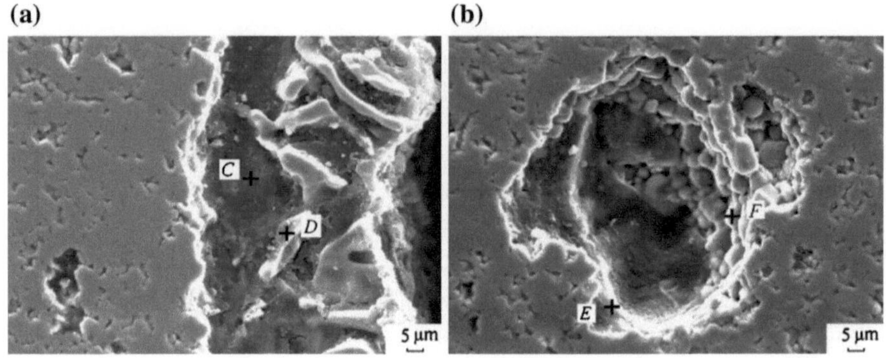

Fig. 5 SEM pictures of points *A* and *B* in Fig. 4b; **a** Point *A*, **b** Point *B*

the quantity of pores was reduced, and a small number of pore of big sizes was appeared, and the sidewall of the crucible was uneven, a precipitation of 70 μm thickness was observed, its structure was different from that of the insider crucible.

For investigating the formation mechanism of the pores of big size and the precipitation layer, the magnifying microstructures of the area A and B in Fig. 4b were shown in Fig. 5. From Fig. 5a, it can be seen that the precipitation layer was consisted of two phases with different intensity, EDS results in Table 2 revealed that the low-light substance in point C was CaO, and the high-light substance in point D was $Ba_{1-x}Ca_xZrO_3$. Meanwhile, the pore of big size inside the CaO-doped $BaZrO_3$ crucible also consisted of two kinds of substances, as shown in Fig. 5b. Some thick material in point E, which was covered on the small grains, was CaO. The grains in point F was $Ba_{1-x}Ca_xZrO_3$. Apparently, the excess CaO can be precipitated during the sintering, and the thickness of the precipitated layer was about 80 μm.

One part of CaO enriched inside the CaO-doped $BaZrO_3$ crucible exhibited the growing of $Ba_{1-x}Ca_xZrO_3$ leading to the formation of the pores of big size, and another part enrich in the sidewall of the crucible leading to the disordered structure in Fig. 2b. XRD pattern in Fig. 6 exhibited that the inside of the crucible was also consisted of CaO and $Ba_{1-x}Ca_xZrO_3$, which was consisted with the EDS results in Table 2.

Table 2 EDS results of points *C*, *D*, *E* and *F* in Fig. 5

Point	Composition, x/% (wt)			
	Ba	Zr	Ca	O
C		79.86	0.31	19.83
D	59.27	09.81	02.70	28.22
E	05.48	01.44	57.74	35.34
F	62.53	02.50	13.70	21.27

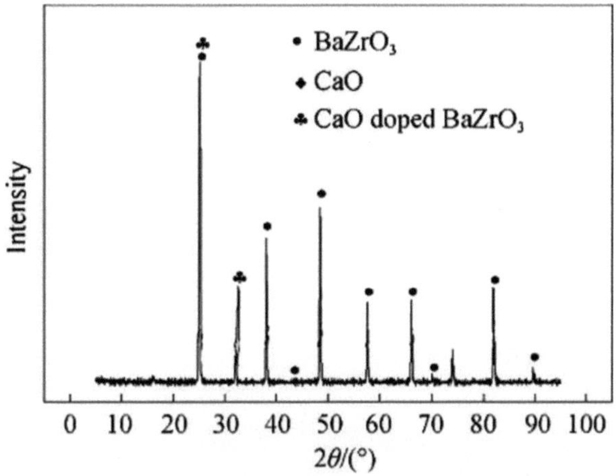

Fig. 6 XRD pattern of the inside of CaO-doped $BaZrO_3$ crucible

To further figure out the effect of CaO on the sintering behavior of $BaZrO_3$ crucible, the dense degree of $BaZrO_3$ and CaO-doped $BaZrO_3$ are calculated by using the measured apparent density and the theoretical density. Due to the complex phase composition of CaO-doped $BaZrO_3$ crucible, and the uncertainty ratio of each phase results that the theoretical density of CaO-doped $BaZrO_3$ crucible is hard to confirm, we tried to assume that the CaO cannot be solute into the $BaZrO_3$, and the CaO-doped $BaZrO_3$ crucible was consisted of CaO and $BaZrO_3$. So, the theoretical density of the crucible can be expressed as:

$$\rho_{tb} = 0.85 \times \rho_{BaZrO_3} + 0.15 \times \rho_{CaO} \quad (1)$$

where, ρ_{BaZrO_3} and ρ_{CaO} were the theoretical density of $BaZrO_3$ (6.23 g/cm^3) and CaO (3.35 g/cm^3).

The theoretical density (ρ_{tb}) of Ca-doped $BaZrO_3$ could be obtained by the Eq. (1), and the relative density (ρ_{rb}) of the crucible was calculated by combing with the measured apparent density (ρ_a), and the equation can be expressed as:

$$\rho_{rb} = \frac{\rho_a}{\rho_{tb}} \times 100\% \quad (2)$$

The resulted is shown in Table 3. It indicated the density of $BaZrO_3$ crucible was improved by doping the CaO. By combing with the Fig. 4, the decreasing of pores of small size can be improve the density of Ca-doped $BaZrO_3$ crucible, although the pores of big size were appeared.

Now, the Ti metal-crucible interaction is considered as a common method to evaluate the stability of the crucible refractory for induction melting of the titanium alloys. Figure 7 shows the microstructure of the interaction between the CaO-doped $BaZrO_3$ crucible and Ti_2Ni alloy, for comparison, the microstructure of interaction between the $BaZrO_3$ crucible and Ti_2Ni alloy is also provided. From Fig. 7a, it can

Table 3 Density of $BaZrO_3$ and CaO doped $BaZrO_3$ crucibles

Crucible	Theoretical density (ρ_{tb})/(g cm^{-3})	Bulk density (ρ_a)/(g cm^{-3})	Relative density (ρ_{rb})/%
$BaZrO_3$	6.23	6.05	97.1
Ca-doped $BaZrO_3$	5.798	5.71	98.5

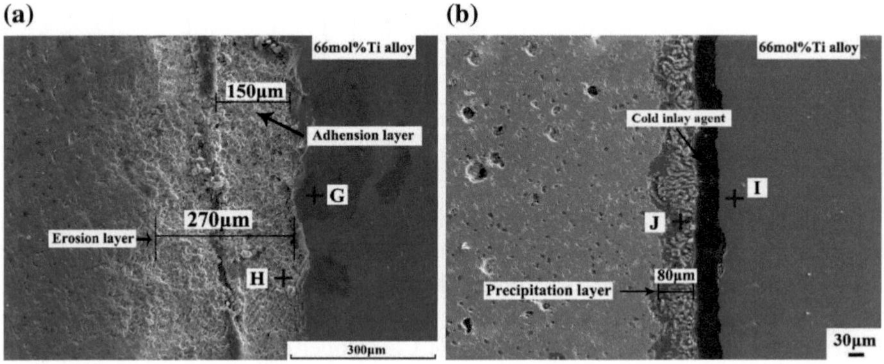

Fig. 7 SEM pictures of the interaction between the crucibles and Ti_2Ni alloy; **a** the $BaZrO_3$ crucible, **b** the CaO-doped $BaZrO_3$ crucible

be seen the erosion layer of the $BaZrO_3$ crucible after melting Ti_2Ni alloy was about 270 μm thickness, and there exists a loose structure and large amount of naked grains in it, meanwhile, the alloy can easily separate from the crucible leading to an amount of cold inlay agent penetrated into the erosion layer. The cold inlay agent separates this layer into two parts: one is remained in the crucible, and another is attached to the metal surface, the thickness of adhesive layer increases to 150 μm. Whereas, no erosion layer is appeared after melting Ti_2Ni alloy in the CaO-doped $BaZrO_3$ crucible. The structure of the precipitated layer, which was contacted with Ti_2Ni melt, was still unchanged, and the thickness of the precipitated layer was still about 80 μm, and no refractory is attached to the metal surface, as shown in Fig. 7b. EDS results in Table 4 indicated that the precipitated layer was not eroded by Ti_2Ni melt. Apparently, the erosion extent of CaO-doped $BaZrO_3$ crucible by Ti_2Ni melt is lower than that of the $BaZrO_3$ crucible, which exhibits the higher thermodynamic stability of CaO-doped $BaZrO_3$ by comparing with the $BaZrO_3$. It indicated that the CaO-doped $BaZrO_3$ refractory may be a potential refractory for induction melting of the titanium alloys.

Table 4 EDS results of points *G*, *H*, *I* and *J* in Fig. 7

Point	Composition, x/% (wt)					
	Ba	Zr	Ca	O	Ti	Ni
G	/	/	/	/	62.56	37.44
H	55.96	35.23	/	08.81	/	/
I	/	/	/	/	63.57	36.43
I	/	79.86	0.31	19.83	/	/

Conclusion

(1) The 15 mol% CaO was not completely soluted into the $BaZrO_3$, and one part of overdose of CaO reunited inside the CaO-doped $BaZrO_3$ crucible, and hindered the growth of $Ba_{1-x}Ca_xZrO_3$ grains leading to the generation of the holes of big size in the crucible. Another part of the CaO was precipitated in the surface of the crucible, and the thickness of precipitated layer was about 70 μm.

(2) The calculated relative density of $BaZrO_3$ crucible was about 97.1%, moreover, a 98.5% relative density for the CaO-doped $BaZrO_3$ crucible was achieved. It indicated that the CaO was helpful to the sintering of $BaZrO_3$ crucible.

(3) After melting Ti_2Ni alloy in the $BaZrO_3$ crucible, a 270 μm thickness erosion layer was observed. However, no significant erosion phenomenon was appeared for the CaO-doped $BaZrO_3$ crucible after melting the Ti_2Ni alloy. The CaO-doped $BaZrO_3$ refractory exhibited a better thermodynamic stability to the titanium melt than that of $BaZrO_3$ refractory, which indicated that it was a potential candidate material for induction melting of titanium alloys.

References

1. Lutjering G, Williams JC (2007) Titanium. Springer, Berlin, pp 1–2
2. Kroll W (1940) The production of ductile titanium. Trans Electrochem Soc 78(1):35–47
3. Gopienko VG (1972) Contact reaction of metallic titanium with oxide refractory materials. Refractory 78(1):405–407
4. Kuang JP, Haring A, Campbell J (2000) Investigate into refractories as crucible and mould materials for melting and casting γ-TiAl alloys. Mater Sci Technol 16:1007–1016
5. Kostov A, Friedrich B (2006) Predicting thermodynamic stability of crucible oxides in molten titanium and titanium alloys. Comp Mater Sci 38(2):374–385
6. Lin KF, Lin CC (1999) Interfacial reactions between Ti6Al4 V alloy and zirconia mold during casting. J Mater Sci 34(23):5899–5906
7. Gomes F, Barbosa J, Ribeiro CS (2008) Induction melting of γ-TiAl in CaO crucibles. Intermetallics 16:1292–1297
8. Cui RJ, Gao M, Zhang H (2010) Interaction between TiAl alloys and yttria refractory material in casting process. Mater Process Technol 210(9):1190–1196
9. Zhang XY, Zhao YQ, Bai CG (2005) Titanium alloys and application. Beijing: Chemical Industry Press, pp 21–24

10. Tetsui T, Kobayashi T, Mori T et al (2010) Evaluation of yttria applicability as a crucible for induction melting of TiAl alloy. Mater Trans 51(9):1656–1662
11. Dongyun GAO (2010) Electrical conductivity of the high-temperature proton conductor (in Chinese, dissertation). Tianjin: Tianjin University
12. Liang RX, Bonn DA, Hardy WN et al (1998) Growth of high quality YBCO single crystals using $BaZrO_3$ crucibles. Phys C 304:105–111
13. Zhang Z, Zhu KL, Liu LJ et al (2013). J Chin Cream Soc 41(9):1272–1283
14. He J, Wei C, Li MY et al (2015). Chin J Nonferr Met (in Chinese) 25(6):1505–1511
15. Li CH, Zhou H, Chen GY et al (2016). J Chongqing Univ: Nat Sci Ed (in Chinese) 39(2): 107–113
16. Chen GY, Cheng ZW, Wang SS et al (2016). J Chin Ceram Soc 44(6):890–895
17. Lei B, Emiliana F, Ziqi S et al (2011) Sinteractivity proton conductivity and chemical stability of $BaZr_{0.7}In_{0.3}O_{3-\delta}$ for solid oxide fuel cells (SOFCs). Solid State Ion 196(1):59–64
18. Yang ZN, Li J, Qiu JB et al (2013) Synthesis and photoluminescence properties of Eu^{3+}, Bi^{3+} Co-doped $BaZrO_3$ phosphors. Spectrosc Spectr Anal 33(1):19–22
19. Levin I, Amos TG, Bell SM et al (2003) Phase equilibria, crystal structures, and dielectric anomaly in the $BaZrO_3$-$CaZrO_3$ system. J Solid State Chem 175:170–180
20. Fu BG, Wei HW, Zhou CM et al (2014) Interfacial reactions between Ti-1100 alloy and CaO crucible during casting process. Trans Nonferr Met Soc China 24(10):3118–3125
21. Gomes F, Barbos J, Ribeiro CS (2008) Induction melting of γ-TiAl in CaO crucibles. Intermetallics 16:1292–1297
22. Gui N, Meng DZ, Lu XG et al (2015). Hot Working Technol (in Chinese) 44(7):98–107
23. Song ZT, Zhao H, Chen TC et al (2011). J Tianjin Norm Univ: Nat Sci Ed (in Chinese) 31(1): 35–41
24. Zevin L, Kimmel G (1995) Quantitative X-ray diffractometry. Springer, New York, pp 10–20

Extracting Uranium and Molybdenum from Refractory U–Mo Associated Ore

Kang Liu, Zhiping Yang, Fengqi Zhao, Liuyin Shi, Yan Song and Xing Fan

Abstract A refractory U–Mo associated ore containing jordisite which enclose U, Mo component, and impede penetration, diffusion process of leaching, oxidation agent was investigated through three approaches of focusing on U, Mo extraction. The results indicate flotation beneficiation without followed metallurgical process is not appropriate because of high target metals loss rate via water leaching. H_2SO_4, H_2O_2 dosage recommended to oxidation leaching are 10 and 6–8% of ore mass, and U, Mo leaching rate are 92.71 and 52.16% respectively. For the acid pressure leaching, U, Mo leaching rate 93.75 and 76.07% are achieved under optimal temperature 130–140 °C and H_2O_2 dosage 8% combined with pressure air 0.6 MPa. Fluidized bed furnace roasting parameters include air flow rate 0.64 m/s, air ore ratio 0.3 m^3/kg at roasting temperature 550 °C with U, Mo leaching rate 93.2 and 89.5%, so enhanced metallurgical process effectively destroy wrapper and increase metal recovery rate.

Keywords U–Mo ore · Common leaching · Acid pressure leaching
Oxidation roasting

K. Liu (✉) · Z. Yang · F. Zhao · L. Shi · Y. Song · X. Fan
Beijing Research Institute of Chemical Engineering and Metallurgy, CNNC, Jiukeshu 145, Tongzhou 101149, Beijing, China
e-mail: langzity1106@163.com

Z. Yang
e-mail: zhpyangf@aliyun.com

F. Zhao
e-mail: zhaofengqi@bricem.com.cn

L. Shi
e-mail: shiliuyin825@163.com

Y. Song
e-mail: 294269618@qq.com

X. Fan
e-mail: friking@sina.com

H. Kim et al. (eds.), *Rare Metal Technology 2018*, The Minerals, Metals & Materials Series, https://doi.org/10.1007/978-3-319-72350-1_23

Introduction

Uranium (U) and Molybdenum (Mo) are typical critical metals resource [1–4], which have significant applications as food for nuclear power station and in the fields of iron & steel, petrochemical industry respectively. Because of the special similar chemical property, it is usually presented via the mineral type of U–Mo associate ore in nature. Therefore, comprehensive recovering U and Mo from U–Mo associate ore is very important to high efficient utilization of natural resource and economic benefit of manufacturing enterprise.

Several countries have experiences for resource utilization containing U and Mo through different technical processes [5–9]. American Grants plant extract U & Mo from lignite using direct H_2SO_4 leaching with $NaClO_3$ as oxidation agent, and the annual yield of U_3O_8 on account of U leaching rate >90% and raw Mo reach 4200 and 34 t. Annual 1000 t U with recovery rate 94% and amounts of by-product raw Molybdenum (Mo 15%) is achieved by alkaline pressure leaching system for treating U–Mo associate ore conducted at Lodeve company of France. U ore involving 0.39% Mo content with U, Mo total recovery rate 68.7 and 74.8% respectively is carried out by common alkaline leaching in Mexico. Niger COMINAK company uniformly mix the concentrated sulfuric acid and sandstone U ore for the heating duration 3 h on the belt conveyer. Meanwhile, the annual yield of U based on U leaching rate 94% and Mo reach 2000 and 122 t.

A refractory U–Mo associated ore has special Mo content existing form jordisite except molybdite, ferrimolybdite, molybdenite, powellite and wulfenite, which is presented as homogeneous amorphous colloidal MoS_2. Black fine powder colloidal MoS_2 as hypergenic mineral form compact wrapper which enclose U, Mo mineral component, and impede penetration and diffusion process of leaching agent and oxidation agent from mineral surface to internal site. Up to now, the process of extracting U & Mo from this refractory U–Mo associated ore has not been studied sufficiently yet, and the technical flow and essentials parameters of process procedures is still absent. Therefore, this work is aimed to indicate the occurrence modes of U, Mo composition using process mineralogy, and compare three technological routes, for instance, conventional acid leaching [10–12], acid pressure leaching [13–16] and oxidation roasting [17, 18] followed with acid leaching.

Experiment

Experimental Materials

The ore used in this research was gathered from HeBei province, China with the density 2.726 g/cm^3. Particle size for experiments is 100% minus 100 meshes. The chemical composition of raw ore is listed in Table 1 where the content about 82.39% is presented as SiO_2 and Al_2O_3. Target metals U, Mo are 0.192 and 1.250%

Table 1 Chemical analysis of U–Mo associated ore

Chemical	Content(wt%)	Chemical	Content(wt%)
SiO_2	73.810	U	0.192
Al_2O_3	8.580	U^{6+}	0.097
Fe_2O_3	3.590	Mo	1.250
MgO	0.820	ΣFe	2.810
CaO	0.389	Fe^{3+}	2.470
P_2O_5	0.044	Fe^{2+}	0.340
K_2O	1.157	S	4.000

respectively, furthermore the ratio of U^{6+} to U^{4+} is approximate 1:1. Sulfur content 4% indicate that the metals of ore deposit in the form of sulfide. Rhyolite porphyry, rhyolites and cryptoexplosive breccia are 3 kinds main petrographic composition in the ore, and the polarizing microscope micrograph of rhyolite porphyry is shown by Fig. 1. U, Mo minerals in the ore mainly include jordisite, umohoite, moluranite and molybdenite. Figure 2 is the ore EPMA micrograph which present the umohoite (Um) and moluranite (Mo) with the diameter as 5 and 30–40 μm.

Analytical Method

The raw ore were dried by oven (DHG-9075A) and ground in a vibrating grinder (KNM-1). Energy dispersive X ray fluorescence analyzer (XRF-1800 X) was used to analyze chemical composition of raw and processed samples. Concentration of U and Mo in the leachate was conducted with an ICP-AES analyzer (OPTIMA

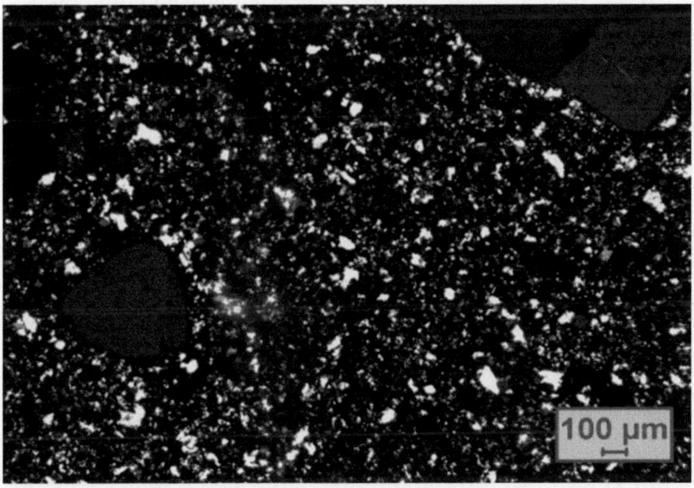

Fig. 1 Polarizing microscope micrograph of rhyolite porphyry

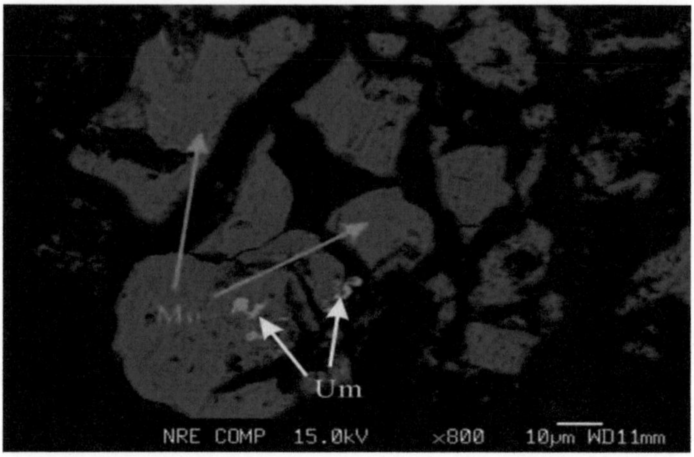

Fig. 2 Electronic prode micrograph of ore

7000DV). U^{6+}, Fe^{3+}/Fe^{2+} content in the ore were determined through H_2SO_4–$TiCl_3$–NH_4VO_3 volumetry and $K_2Cr_2O_7$ titrimetry. Petrographic composition of ore was observed via polarizing microscope (CX31P-OC-1). The morphology of ore particles was presented by a electronic prode(EMPA, JXA-8100).

Leaching Test

A 500 mL three neck, flat bottom flask as a leaching vessel put on the underbed in a thermostatic water bath was used in the water leaching and common acid leaching process. Taking account of leaching temperature, H_2O_2 dosage as main factors, the pressure acid leaching was performed in the autoclave. The oxidation roasting process was conducted in the fluidized bed furnace, and the roasted sample was used to the followed acid leaching procedure. U, Mo extraction rate was calculated by U, Mo content in the leached residue and concentration of the leachate. Every data of all experiments was determined 3 times, and the average value was taken to obtain the final results.

Results and Discussion

Water Leaching

For the too low grade of U, Mo composition in raw ore, it wants to involve flotation procedure in metal beneficiation, which can effectively decrease ore treatment amount and reagent consumption. Based on the aim above, the water leaching process of ore was conducted under 24 h, ambient temperature and mass ratio of liquid to solid 5:1. Table 2 shows the leachate and residue results of water leaching process, which indicate flotation beneficiation without followed metallurgical process is not appropriate because of high target metals U, Mo loss rate 8.33 and 8.80% via water leaching.

Conventional Acid Leaching

In the practical chemical process, U^{4+} and Mo^{4+} in the raw ore almost do not dissolve in the acid solution, but U^{6+} and Mo^{6+} can dissolve. Furthermore, oxidation agent don't directly oxidize U^{4+} to U^{6+}, which oxidize Fe^{2+} to Fe^{3+}, then Fe^{3+} oxidize U^{4+} to U^{6+}. In other words, Fe^{2+} and Fe^{3+} in the ore make a role of electron transfer in the process of uranium oxidation acid leaching. Chemical reactions (1)–(6) describe the U, Mo content dissolution process in the acid solution with oxidation agent.

$$Fe + 2H^+ = Fe^{2+} + H_2 \tag{1}$$

$$2Fe^{2+} + 2H^+ + H_2O_2 = 2Fe^{3+} + 2H_2O \tag{2}$$

$$UO_2 + 2Fe^{3+} = UO_2{}^{2+} + 2Fe^{2+} \tag{3}$$

$$UO_3 + 2H^+ = UO_2{}^{2+} + H_2O \tag{4}$$

$$MoS_2 + 9H_2O_2 = H_2MoO_4 + 2H_2SO_4 + 6H_2O \tag{5}$$

$$H_2MoO_4 + H_2SO_4 = MoO_4{}^{2-} + SO_4{}^{2-} + 4H^+ \tag{6}$$

Based on the constant parameters as H_2SO_4 (92%) dosage 10% of the ore mass, leaching time and temperature 6 h and 60 °C, mass ratio of liquid to solid 1:1, the effect of H_2O_2 (35%) dosage of ore mass on U, Mo leaching rate is described via Fig. 3. H_2O_2 was added in the second and third hour of leaching time by constant flow pump. It can be seen that the H_2O_2 dosage has a gentle influence on U leaching rate. The U leaching rate is 88.02 and 93.23% after adding H_2O_2 0 and 10%, while the rate reaches peak 96.88% with H_2O_2 6%. Mo leaching rate increases to 56.32% from 42.80% following H_2O_2 dosage enhancement between 0 and 10%. Increasing

Table 2 Results of water leaching process

Leachate				Residue		Leaching rate	
pH	Eh(mV)	U(g/L)	Mo(g/L)	U(wt%)	Mo(wt%)	U(%)	Mo(%)
3.51	242	0.040	0.279	0.176	1.14	8.33	8.80

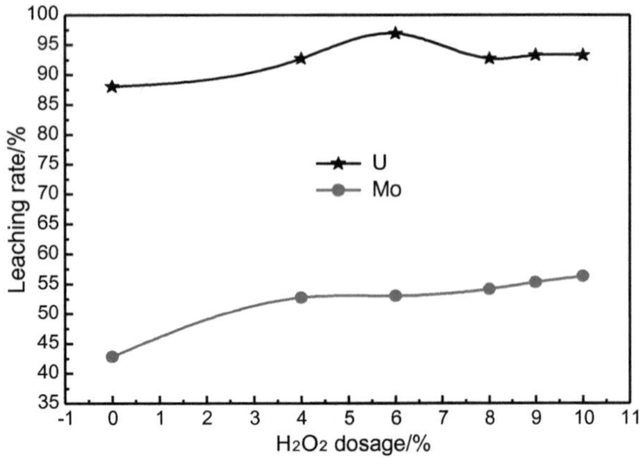

Fig. 3 Effect of H_2O_2 dosage on U and Mo leaching rate

H_2O_2 dosage can enhance oxidation degree of leaching system, so it can improve leaching rate of U and Mo. Furthermore, the minerals occurrence mode present almost 1:1 U^{6+}/U^{4+} and more molybdite (MoO_3), ferrimolybdite ($Fe_2(MoO_4)_3{\cdot}8H_2O$), which can easily react with sulfuric acid according with chemical reactions (4), (7), (8). Considering the leaching efficiency, reagent cost and dosage, H_2O_2 (35%) dosage of ore mass 6–8% is recommended as optimal parameter of common acid leaching process.

$$MoO_3 + H_2SO_4 = MoO_2SO_4 + H_2O \tag{7}$$

$$Fe_2(MoO_4)_3{\cdot}8H_2O + 3H_2SO_4 = 3H_2MoO_4 + Fe_2(SO_4)_3 + 8H_2O \tag{8}$$

Acid Pressure Leaching

Under mass ratio of liquid to solid 5:1, H_2SO_4 dosage of ore mass 10%, H_2O_2 dosage of ore mass 8% and leaching time 3 h, Fig. 4 shows the relationship between U, Mo leaching rate and acid pressure leaching temperature. It is observed that temperature has very unobvious variation influence on U, Mo leaching rate which are enhanced from 91.88 to 92.34% and from 64.04 to 68.38% respectively between 130 and 150 °C. Considering the energy consumption and U, Mo leaching rate, temperature of acid pressure leaching 130–140 °C is chosen as optimal condition in this study.

Based on mass ratio of liquid to solid 5:1, H_2SO_4 dosage of ore mass 10%, leaching temperature 140 °C and leaching time 3 h, Fig. 5 shows the effect of H_2O_2 dosage on U, Mo leaching rate. In the range of H_2O_2 dosage from 5 to 8%, the U, Mo leaching rate gradually increases with increasing of H_2O_2 dosage, and it reached the maximum 92.34 and 68.38% from 87.35 and 64.80% at H_2O_2 dosage 8% respectively.

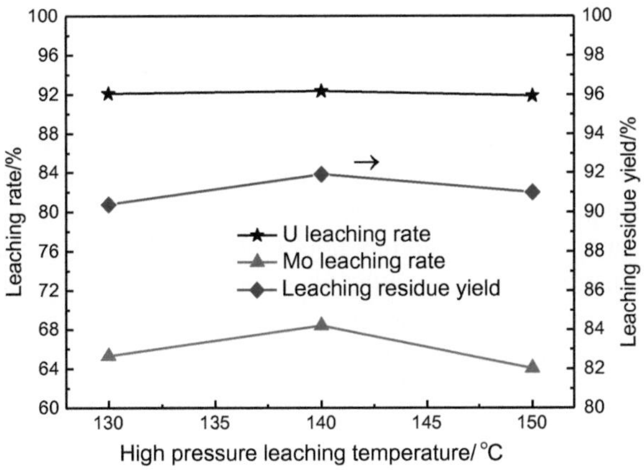

Fig. 4 Effect of pressure leaching temperature on U, Mo leaching rate

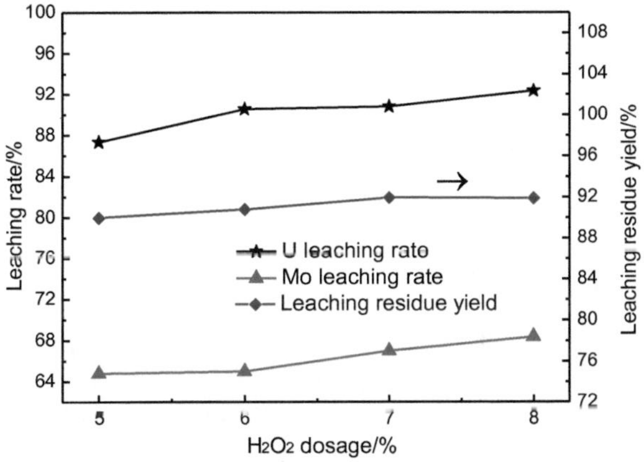

Fig. 5 Effect of H_2O_2 dosage on U, Mo leaching rate

From the results of acid pressure leaching above mentioned, U leaching rate keep approximate 92–93%, but Mo leaching rate only increase from 56.32 to 68.38% compared with common acid leaching. The experimental results are dissatisfied and obvious lower than expected. H_2O_2 as oxidation agent of the experimental process in the seated autoclave can be divided to two parts. The first part in the liquid phase participate in the oxidation reactions, and the second part decompose to generate O_2 according with chemical reaction (9) because of itself easy decomposition property. Oxygen solubility in the reaction system of liquid phase is constant, so its oxidation capacity is limited seriously. When the H_2O_2 dosage is lower than 7% based on result of Fig. 5, high pressure and temperature situation intensely decompose H_2O_2 and lower the oxidation capacity of liquid phase, so the U, Mo leaching rate both decrease in a certain level.

$$2H_2O_2 = O_2 + 2H_2O \tag{9}$$

Pressure air of 0.6 MPa involved to the acid pressure leaching process on account of H_2O_2 decomposition, the U, Mo leaching rate is achieved as 93.75 and 76.07% which have more growth of Mo leaching rate. Firstly, oxygen in pressure air of sealed situation effectively impede H_2O_2 decomposition and improve oxidation capacity in the liquid phase. Secondly, pressure air enhance dissolved quantity of oxygen in the liquid phase. Therefore, taking account of U, Mo leaching rate, energy and reagent consumption in the acid pressure leaching process, the temperature 130–140 °C, H_2O_2 dosage 8% and air pressure 0.6 MPa are recommended as technical factors.

Fluidized Bed Furnace Roasting with Acid Leaching

Oxidation roasting, a common procedure for molybdenite(MoS_2) treatment, can utilize high temperature air to oxidize U^{4+} to U^{6+}, and MoS_2 to MoO_3 or molybdate which can dissolve in H_2SO_4. Furthermore, oxidation roasting also dispose carbon-containing organic matters, sulfide and ferrous oxide, which influence oxidation, dissolution reaction of U, Mo oxide and increase reagent consumption. The main chemical process of oxidation roasting can be described via reactions (10)–(15).

$$2UO_2 + O_2 = 2UO_3 \tag{10}$$

$$2U_3O_8 + O_2 = 6UO_3 \tag{11}$$

$$2MoS_2 + 7O_2 = 2MoO_3 + 4SO_2 \tag{12}$$

$$2MoO_2 + O_2 = 2MoO_3 \tag{13}$$

$$2MeS + 3O_2 = 2MeO + 2SO_2 \tag{14}$$

$$MeO + MoO_3 = MeMoO_4 \tag{15}$$

The oxidation roasting can effectively destroy wrapper jordisite of U–Mo ore and enhance Mo leaching rate which is higher than 80% without oxidizing agent in the leaching step of roasted ore. Meanwhile, the U leaching rate is as good as enhanced oxidation stirring leaching process, and the filtration performance of ore pulp is greatly improved. Fluidized bed furnace is choose for oxidation roasting because of high temperature air direct heating ore, high heat transfer between liquid and solid phase, quick reaction rate, high thermal efficiency and large treatment capacity.

Operating air flow rate for fluidized bed furnace is a significant technical factor, which is more than critical boiling rate of all normal ore particles. Based on the roasting temperature 550 °C from the static roasting experiment, ore feeding rate 15 kg/h, leaching temperature, time 60 °C and 2 h, acid dosage of roasted ore mass 8%, mass ratio of liquid to solid 1:1, the relationship between air flow rate and U, Mo leaching rate is presented in Fig. 6. Under the different air flow rate, the roasted ore productivity is basically same. It can be observed that the effect of air flow rate on Mo leaching rate is bigger than U leaching rate which almost keep stable between 92 and 93.2%. With the increasing of air flow rate, ore duration in the furnace is shortened, which influence the dissociation degree of jordisite wrapper and can't sufficiently oxidize the molybdenum minerals. The Mo leaching rate difference of air flow rate 0.55 and 0.77 m/s is 4.7% from 90 to 85.3%, so the suitable duration of ore in the furnace is the requirement of U–Mo ore roasting effect. Although the lower air flow rate is favorable for U, Mo extraction, which have lower roasted ore productivity, a air flow rate of 0.64 m/s is recommended, taking the U, Mo extraction and energy utilization efficiency into consideration.

The air volume of fluidized roasting process for U–Mo ore needs to satisfy the ore fluidized status and sufficient oxidation. The theoretical air volume for U–Mo ore oxidation roasting is 180 m^3/t through the calculation, and the practical air volume is 200 m^3/t based on air excess coefficient 1.1–1.2:1. In order to maintain air flow rate 0.64 m/s, the air volume need to achieve 10 m^3/h, so the air ore ratio changes following the change of ore feeding rate. Figure 7 clearly indicates that the effect of air ore ratio on U, Mo leaching rate. The U, Mo leaching rate under different air ore ratio have little variation as 92.3–93.2% and 88.7–89.5% respectively. Low air ore ratio shorten the ore duration in the fluidized bed furnace to the average value 0.38 h which still more than optimal parameter 3 min obtained by the oxidation roasting conditions experiments via muffle furnace. This is the reason why the U, Mo leaching rate are essentially stable for different air ore ratio. Considering the roasted ore productivity, energy consumption and U, Mo leaching rate, air ore ratio 0.3 m^3/kg is chosen as optimal condition in this study.

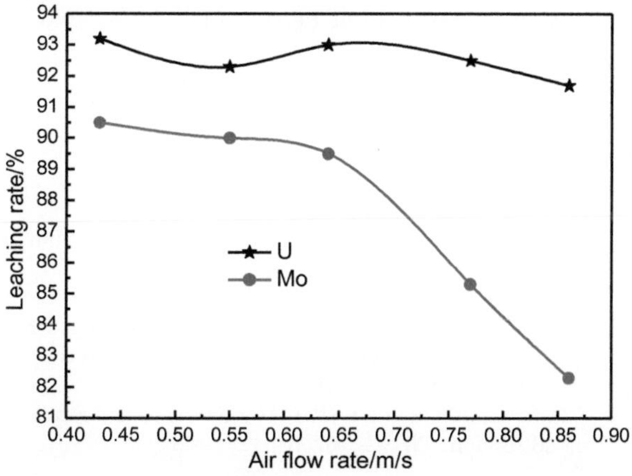

Fig. 6 Effect of air flow rate on U, Mo leaching rate

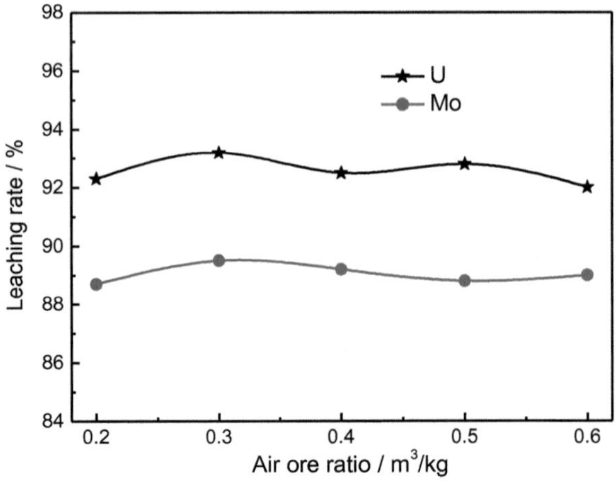

Fig. 7 Effect of air ore ratio on U, Mo leaching rate

Conclusions

1. The flotation beneficiation is not appropriate because of U, Mo loss rate 8.33 and 8.80% via water leaching. In the usual oxidation acid leaching process, the H_2SO_4 (92%), H_2O_2 (35%) dosage recommended are 10 and 6–8% of ore mass, and U, Mo leaching rate are 92.71 and 52.16% respectively.
2. For the acid pressure leaching, U, Mo leaching rate 93.75 and 76.07% are achieved under optimal temperature 130–140 °C and H_2O_2 dosage 8% combined with pressure air 0.6 MPa. The pressure air in the autoclave can efficiently impede H_2O_2 decomposition in the ore pulp and enhance oxidation effect of U, Mo.
3. Compared three approaches for the refractory U–Mo ore treatment, fluidized bed furnace roasting followed with acid leaching is the most enhanced metallurgical method, which can destroy wrapper jordisite to release U, Mo content and improve U, Mo leaching rate to 93.2 and 89.5% under parameters including air flow rate 0.64 m/s, air ore ratio 0.3 m^3/kg and roasting temperature 550 °C.

References

1. Maslov OD, Tserenpil S, Norov N (2010) Uranium recovery from coal ash dumps of Mongolia. Solid Fuel Chem 44(6):433–438
2. Lei XF, Qi GX, Sun YL (2014) Removal of uranium and gross radioactivity from coal bottom ash by $CaCl_2$ roasting followed by HNO_3 leaching. J Hazard Mater 276:346–352
3. Ritchy GM, Wong EW (1985) Influence of cations on crud formation in uranium circuits. Hydrometallurgy 15:55–61
4. Qi GX, Lei XF, Sun FY (2016) Performance and mechanisms of coal fly ash-derived C–S–H on adsorption. CIESC J 67(10):4255–4263

5. Xing HM, Liu Y, Chen G (2006) Multipurpose recovery techniques of uranium–molybdenum intergrown ores at abroad. Uranium mining Metall 25(4):186–191
6. Li Q (2012) Research on bioleaching of low-grade uranium-bearing ores and fluoride-tolerant mechanism of bioleaching microorganisms. Ph.D. thesis, Central South University
7. Shakir K, Aziz M, Beheir SG (1992) Studies on uranium recovery from a uranium-bearing phosphatic sandstone by a combined heap leaching-liquid-gel extraction process. 1-Heap leaching. Hydrometallurgy 31(1/2):29–40
8. Rao KA, Sreenivas T, Vinjamur M (2014) Continuous leaching of uranium from an Indian ore: residence time scale up and heat effects. Hydrometallurgy 146:119–127
9. Zeng L (2011) Extraction and separation of molybdenum, vanadium, nickel, cobalt from acid leach solution. Ph.D. thesis, Central South University
10. Liu L, Xue JL, Liu K (2016) Complex leaching process of scheelite in hydrochloric and phosphoric solutions. JOM 68(9):2455–2462
11. Cao ZF (2010) The research on novel hydrometallurgy for molybdenite concentrate and its mechanisms. Ph.D. thesis, Central South University
12. Guan WJ (2013) Study on solvent extraction separation of Mo and W by H_2O_2-complexation with a mixture extractant TRPO/TBP. Ph.D. thesis, Central South University
13. Wang SF, Wei C, Deng ZG (2013) Extraction of molybdenum and nickel from Ni–Mo ore by pressure acid leaching. Trans Nonferrous Met Soc China 23:3083–3088
14. Demopoulos GP (1985) Acid pressure leaching of a sulphidic uranium ore with emphasis on radium extraction. Hydrometallurgy 15:219–242
15. Mattus AJ, Torma AE (1980) A comparison of carbonate leaching of a low-grade uranium ore at atmospheric and increasing pressures. Hydrometallurgy 5:179–190
16. Wu CY, Yu HF, Zhang HF (2012) Extraction of aluminum by pressure acid-leaching method from coal fly ash. Trans Nonferrous Met Soc China 22:2282–2288
17. Yang YS, Yu Q, Hu N (2016) Effect of microwave pretreatment on leaching behavior of uranium in heap-leached uranium tailings and its mechanism. China J Nonferrous Met 26 (6):1356–1363
18. Li QG (2010) Study on production ammonium molybdate from Ni–Mo ore. Ph.D. thesis, Central South University

Thermodynamics Analysis on the Process of Decarburization and Vanadium Protection by CO_2

Zhuolin Liu, Ting-an Zhang, Liping Liu, Guozhi Lyu, Yi Luo, Zhihe Dou and Xijuan Pan

Abstract Due to the high energy consumption, long technical process, low usage rate of vanadium during the traditional smelting vanadium steel process, a new method for decarburization by CO_2 to directly smelt vanadium-containing steel in blast furnace was proposed. In this paper, the thermodynamics of CO_2 decarburization in molten iron process was calculated out by FactSage. The thermodynamic calculations indicated that in standard state, the transformation temperature of C and V was 1357 °C; the oxidation order of element was Si, C, V, Mn and S when using CO_2 as oxidant. The experiment of blowing CO_2 temperature showed that decarburization reaction temperature was 1550 °C, carbon content could be decreased to 0.676%, the decarburization rate was 88%. There was no effect on vanadium content in molten iron during the decarburization process, vanadium content was 0.323%, which meets the request of vanadium mould steel.

Keywords CO_2 · Decarburization and vanadium protection · Thermodynamics FactSage

Introduction

There is a lot of magnetite which contains vanadium and titanium in China [1], resources of vanadium element in the industry have a important strategy [2]. Its in the form of compounds, alloy and vanadium metal are widely used in metallurgy, chemical industry, aerospace and other fields [3, 4]. The role of vanadium in steel is enhancing the hardenability and carbides, high temperature resistance, a strong secondary hardening effect, improving hardness, refining grain, stable structure, and wide applications in alloy tool steel [5–7].

Z. Liu · T. Zhang (✉) · L. Liu · G. Lyu · Y. Luo · Z. Dou · X. Pan
Key Laboratory of Ecological Metallurgy of Multi-metal Intergrown Ores of Ministry of Education, Special Metallurgy and Process Engineering Institute, School of Metallurgy of Northeastern University, Shenyang 110819, China
e-mail: zta2000@163.net

H. Kim et al. (eds.), *Rare Metal Technology 2018*, The Minerals, Metals & Materials Series, https://doi.org/10.1007/978-3-319-72350-1_24

Nowadays, the production of vanadium steel are mainly in three ways, vanadium iron alloying, vanadium slag alloying and vanadium oxide direct alloying [8, 9]. The production process of vanadium ferroalloy is long. Firstly, vanadium is blown into vanadium slag by oxygen, then vanadium is extracted from vanadium slag, and vanadium iron is obtained by reduction [10]. The process not only consumes large amounts of energy and raises the cost of steel, but also brings sever problems to the environment. Vanadium content is generally low and impurity content is high in vanadium slag [11–13]. The vanadium slag is directly added to the molten iron [14], which can cause excessive inclusion in the steel and influence the purity and mechanical properties of steel. The V_2O_5 volatility is relatively large and easy to lose. Direct addition of V_2O_5 makes it easy to change the composition of the steel liquid [15–17], resulting in low yield of vanadium [18]. For the whole reaction flow chart is as follows (Fig. 1).

In view of the high energy consumption, long technical process, low usage rate of vanadium during the traditional smelting vanadium steel process, a new method for decarburization by CO_2 to directly smelt vanadium-containing steel in blast furnace was proposed. In other words, vanadium-containing molten iron in the blast furnace is blown by CO_2 gas, without being blown by O_2 gas. This procedure make use of the low oxidation of CO_2 to get rid of carbon in molten iron, and won't react with vanadium in molten iron so that achieves the goal of lowering carbon and remaining vanadium, and realize the aim that use vanadium-containing molten iron

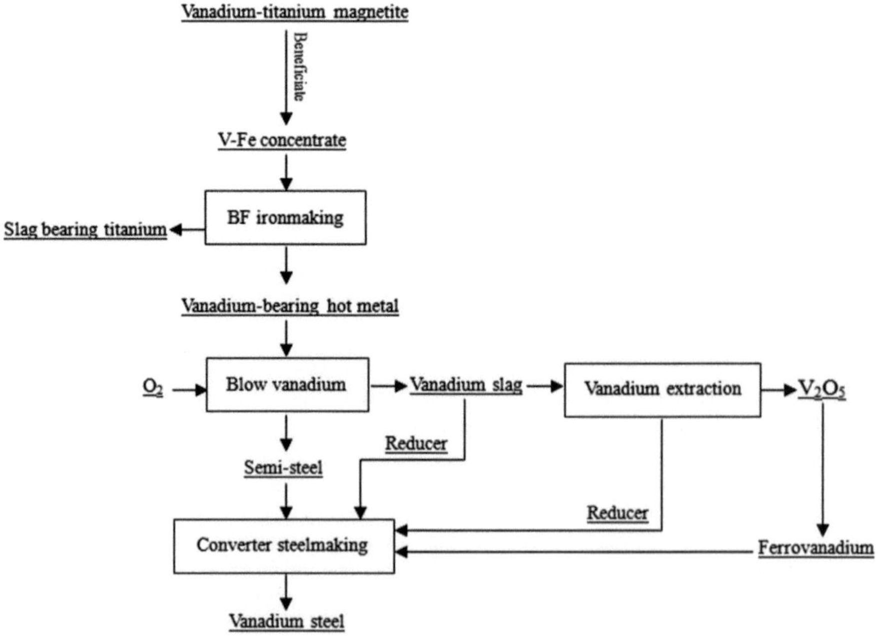

Fig. 1 Traditional smelting process of vanadium steel

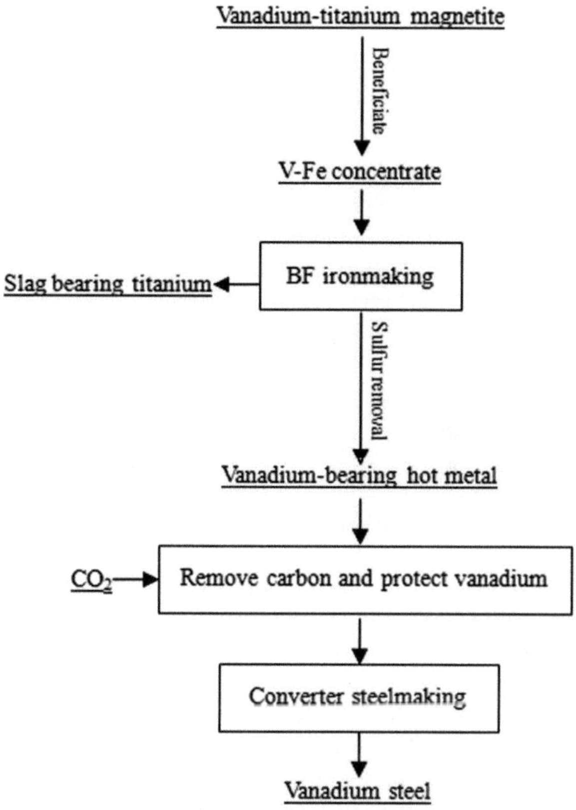

Fig. 2 The process of directly smelt vanadium-bearing steels by vanadium-bearing hot metal with the way blowing CO_2

in the blast furnace smelt vanadium steel directly. For the whole reaction flow chart is as follows.

Decarburization and vanadium protection by CO_2 is the key link of the whole process. In this paper, the thermodynamics of CO_2 decarburization in molten iron process was calculated by FactSage to explore the theoretical possibilities of this process. Experiments in induction furnace were conducted to verify the results of thermodynamic calculations (Fig. 2).

Materials and Experiments

Raw Materials

In this work, cast iron and vanadium-containing 40% of vanadium iron are used to configure vanadium-containing molten iron in induction furnace. As the preparation of molten iron is used in graphite crucible, graphite will carburize the molten iron,

Table 1 Chemical composition of vanadium bearing hot metal from lab (mass, %)

Compositions	C/%	Fe/%	V/%	P/%	S/%	Si/%	Mn/%
1#	5.68	91.85	0.355	0.086	0.035	0.986	0.075
2#	5.58	92.45	0.354	0.078	0.026	0.983	0.074
3#	5.56	92.56	0.350	0.056	0.038	0.991	0.074
Average	5.61	92.29	0.353	0.073	0.033	0.987	0.074

Table 2 Chemical composition of vanadium bearing hot metal from Panzhihua Iron and Steel Co., Ltd. (mass, %)

Compositions	C/%	V/%	P/%	S/%	Si/%	Mn/%
Atomization furnace	4.30	0.32	0.043	0.077	0.082	0.16
Ld converter	4.33	0.29	0.039	0.080	0.080	0.18

leading to high carbon content. The chemical compositions of molten iron is given in the following Tables 1 and 2.

Compared with the vanadium-containing molten iron produced by Panzhihua Iron and Steel Co., Ltd., the silicon element is seriously exceeded, and in the actual production of the steel-making, the silicon will be oxidized into the slag by oxygen and the silicate forms with other metal oxides. Slag basicity is not conducive to steel-making, so before the converter steel-making part of the steel will be pre-desilication, thereby reducing the silicon content of molten iron, but because of the experiment in the process, did not observe the obvious slag layer. Therefore, it will not cause the alkaline lowering of slag, thus affecting the blowing process.

Experiments

In this paper, we propose the method of using CO_2 decarbonization to protect vanadium, which is to use CO_2 and carbon reaction to generate CO, so as to decarbonize, and the CO_2 will not react with vanadium when higher than the selective oxidation temperature. The experimental procedure is as follows:

(1) Preparation of vanadium-containing iron: cast iron and vanadium iron are used to configure vanadium-containing molten iron by graphite crucible in induction furnace. Detection of its chemical composition, the results shown in Table 1.
(2) Cutting: vanadium-containing iron was divided into small iron bars of about 300 g with sawing and wire cutting, and the size was approximately 2.3 cm × 2.3 cm × 7 cm.
(3) Decarburization research: put the iron in the corundum crucible, coat a graphite crucible, put it in the induction furnace, gradually raise the voltage, adjust the temperature of the induction furnace, and blow into CO_2 for decarbonation

experiment. After the reaction gradually reduces the induction furnace voltage and cooling.

(4) Test: remove the corundum crucible, broken and take out the metal piece, make the sample. The content of C, V, Si, Mn, P and S in the molten iron after decarburization was determined and to study the effect of decarburization experiment.

Results and Discussion

Thermodynamics Analysis of Decarburization Process

In order to examine the oxidation reaction of various elements in vanadium containing hot metal, the thermodynamic calculation for main reactions was carried out by using FactSage. The results are shown in the following Table 3.

The standard Gibbs free energy of each reaction is varied with temperature as shown in Fig. 3. The thermodynamic calculations indicated that in standard state, the transformation temperature of C and V was 1630 K. From Fig. 1, according to the oxidation reaction standard Gibbs free energy, it can be seen that when the temperature is 1473–1630 K, the oxidation order of element was Si, V, Mn, C and S. When the temperature range is 1630–1787 K, the oxidation order of element is Si, C, V, Mn and S. When the temperature range is 1787–1973 K, the oxidation order of element is C, Si, V, Mn and S. When the temperature is higher than 1630 K, the Gibbs free energy of the oxidation reaction of C is more negative than that of V, which indicates that when the temperature is higher than 1630 K, the oxidation reaction of C takes precedence. From the thermodynamics, the starting temperature of the oxidation reaction should be higher than 1630 K. With the increase of the oxidation temperature, the Gibbs free energy of C is more negative

Table 3 The oxidation reaction of various elements in vanadium containing hot metal

Reaction equations	ΔG^{θ} J/mol
$[C] + CO_2(g) = 2CO(g)$	142095−127.9T
$[V] + CO_2(g) = VO + CO(g)$	−115910 + 35.10T
$2/3[V] + CO_2(g) = 1/3V_2O_3 + CO(g)$	−101952 + 21.75T
$1/2[V] + CO_2(g) = 1/2VO_2 + CO(g)$	−27011−66.55T
$2/5[V] + CO_2(g) = 1/5V_2O_5(g) + CO(g)$	−496−2.506T
$1/2[Si] + CO_2(g) = 1/2SiO_2 + CO(g)$	−128415.5 + 23.41T
$[Mn] + CO_2(g) = MnO + CO(g)$	−128688 + 40.08T
$2/5[P] + CO_2(g) = 1/10P_4O_{10} + CO_2(g)$	15873.2 + 22.29T
$1/2[P] + CO_2(g) = 1/2PO_2 + CO(g)$	148846−45.48T
$[P] + CO_2(g) = PO + CO(g)$	307260−75.20T
$[S] + CO_2(g) = SO_2(g) + CO(g)$	120260−74.12T

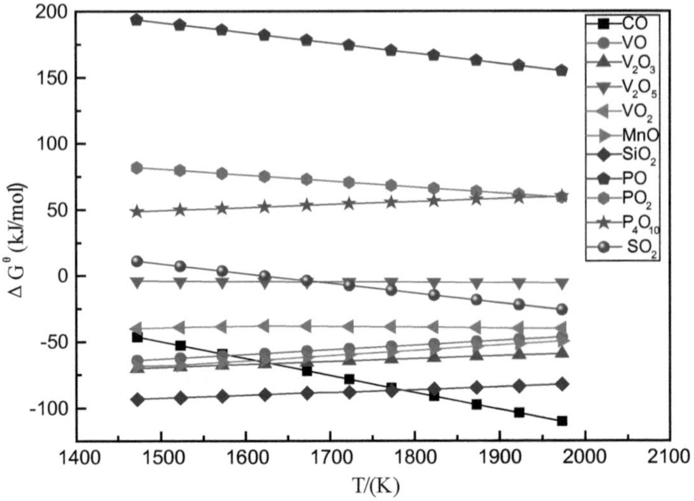

Fig. 3 The standard Gibbs free energy change for the main reaction by CO_2 decarburization

than V, indicating that the increase of the temperature is beneficial to the process of decarburization and vanadium protection.

Effect of Different Oxidation Temperature on Decarburization Process

Based on the results of thermodynamics analysis of oxidation process, this section mainly inspected the effects of temperature on oxidation process of vanadium containing hot metal as feedstock. So the blowing time (80 min) and gas flow (0.6 L/min) were certain with different temperature (1350–1650 °C) for the single factor experiments. The oxidation rate was shown in Table 4 and Fig. 4.

The experiment temperature is higher, the lower of carbon content in molten steel, the effect of decarburization is better. From the thermodynamic calculation can also get, raise the temperature, the decarburization reaction Gibbs free energy is reduced, which is more conducive to the process of decarburization reaction. When the temperature is 1550 °C, carbon content could be decreased to 0.676%, the

Table 4 Effect of reaction temperature on C and V content in hot metal

Temperature/°C	C/%	V/%
1350	2.26	0.325
1450	1.83	0.342
1550	0.676	0.323
1650	0.669	0.348

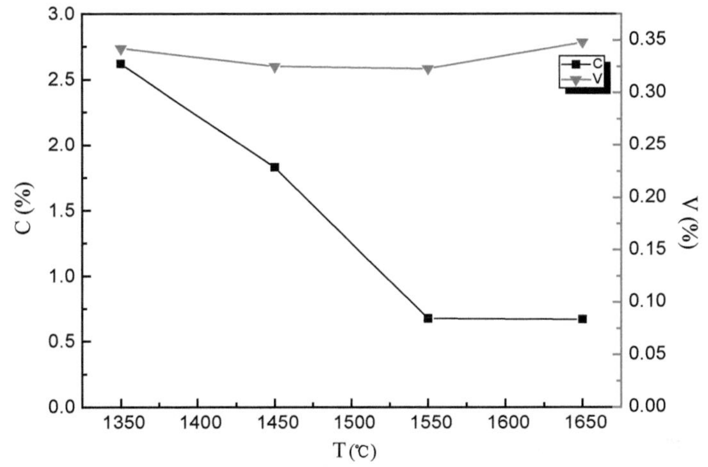

Fig. 4 Effect of reaction temperature on C and V content in hot metal

decarburization rate was 88%, and the later increase temperature has little effect on the decarburization effect. The content of vanadium is basically unchanged, and the average oxidation rate of vanadium is only 6.21%, which achieved the goal of decarburizing and vanadium protection. Due to the residual carbon content in the molten steel at 1650 and 1550 °C is litter difference, and it is difficult to heat up at high temperatures, so in this experiment, 1550 °C is selected as the appropriate reaction temperature.

Conclusions

In view of the high energy consumption, long technical process, low usage rate of vanadium during the traditional smelting vanadium steel process, a new method for directly smelt vanadium-containing steels by vanadium-bearing hot metal with the way blowing CO_2 to remove carbon and protect vanadium in blast furnace was proposed. The results of thermodynamic calculations show that, the conversion temperature of decarburization and vanadium protection is 1630 K under standard conditions. And the oxidation order of each element is Si, C, V and Mn respectively under experimental temperature. From the single factor experiments of blowing CO_2 to decarbonize, the optimum conditions can be concluded: in the condition of gas flow 0.6 L/min, decarburization reaction temperature 1550 °C, reaction time 80 min, carbon content rate is cut to 0.676%, vanadium content rate is cut to 0.323%, which meets the request of most vanadium steel.

Acknowledgements This research was supported by the National Natural Science Foundation of China (51422403, 51774078), the Fundamental Research Funds for the Central Universities (N162505002) and National Basic Research Program of China (973 program, No. 2013CB 632606).

References

1. Moskalyk RR, Alfantazi AM (2003) Processing of vanadium: a review. Mineral Eng 16 (9):793–805
2. Diesek BG, Stevens CO, Grossbeck ML (1996) Fatigue and crack growth behavior of V–Cr–Ti alloys. J Nucl Mater 237:488–491
3. Fujiwara M, Takanashi K, Satou M et al (2004) Influence of Cr, Ti, concentrations on oxidation and corrosion resistance of V–Cr–Ti type alloys. J Nucl Mater 329:452–456
4. Reichman SH, Kosin JE, Meyerink JF Titanium-aluminum-vanadium alloys and products made using such alloys: United States, US6053993
5. Lipamikov VN, Ettmayer P (1997) Effect of vacancy ordering on structure and some properties of vanadium carbide. In: 14th International plansee seminar, Austria, pp. 12–16
6. Shin DY, Wee CH, Kim MS (2007) Distribution behavior of vanadium and phosphorus between slag and molten steel. Met Mater Int 13(2):171–176
7. Wu LL, Xiao FR, Wang YC (2013) Effect of vanadium on microstructure and wear resistance of Ni–Cr alloyed cast iron. Int J Cast Met Res 26(3):176–179
8. Hui W, Zhang Y, Shao C (2016) Effect of cooling rate and vanadium content on the microstructure and hardness of medium carbon forging steel. J Mater Sci Technol 32(6): 545–551
9. Jiang T, Wang S, Guo Y (2016) Effects of basicity and MgO in slag on the behaviors of smelting vanadium titano-magnetite in the direct reduction-electric furnace process. Metals 6 (5):107
10. Chua Q, Zhang M, Li J (2017) Influence of vanadium filler on the properties of titanium and steel TIG welded joints. J Mater Process Technol 240(2):293–304
11. Wu EH, Rong ZHU, Yang SL (2016) Influences of technological parameter smelting-separation process for metallized pellets of vanadium-bearing titanomagnetite concentrates. J Iron Steel Res 23(7):655–660
12. Song SQ, Xue ZL, Zhang B (2014) Smelting vanadium steel by direct alloying with self-reduction briquette of V_2O_5. Ironmaking Steelmaking 41(1):7–11
13. Liu XH, Chen DF, Li YK (2013) Investigation on efficient vanadium extraction from vanadium-bearing molten iron. Adv Mater Res 785–786:58–62
14. Wang WX, Xue ZL, Song SQ (2012) Research on smelting vanadium steel by silicothermic reduction direct alloying with V_2O_5. Mater Res 476–478:164–169
15. Cao HT, Dong XP, Pan Z (2016) Surface alloying of high-vanadium high-speed steel on ductile iron using plasma transferred arc technique: microstructure and wear properties. Mater Des 100:223–234
16. Zhou Y, Li Z (2006) Theory and technology of making high speed steel by direct alloying of tungsten molybdenum vanadium oxide ore. China Tungsten Ind 1:4
17. Yong Z, Zhengbang L (2006) Industrial test of high speed steel making in electric arc furnace by direct reduction and alloying with tungsten molybdenum vanadium oxide ore. Spec Steel 27(1):42–44
18. Anderson SH, Foulard J, Lutgen N (1989) Inert gas technology for the production of low nitrogen steel. In: 47th electric furnace conference proceedings, pp. 365–375

Purification of a Nigerian Wolframite Ore for Improved Industrial Applications

Alafara A. Baba, Muhammed O. Muhammed, Mustapha A. Raji, Kuranga I. Ayinla, Aishat Y. Abdulkareem, Misitura Lawal, Folahan A. Adekola, Abdul G. F. Alabi and Rafiu B. Bale

Abstract Increasing demands for pure wolframite ore in the form of ammonium paratungstate (APT), tungsten oxide cannot be over emphasized. Thus, treatment of a Nigerian wolframite ore containing admixtures of *scheelite* ($CaWO_4$: 96-900-9631) and *stolzite* ($PbWO_4$: 96-900-9813) by hydrometallurgical route was investigated in hydrochloric acid chelated with phosphoric acid to extract tungsten via solvent extraction. During leaching, parameters such as leachant concentration, chelate dosage and reaction temperature on ore dissolution were examined. At optimal conditions (1.5 mol/L HCl + 2.0 mol/L H_3PO_4 solution; 75 °C), 95.0% of the initial 10 g/L ore reacted within 120 min. The derived activation energy of 56.80 kJ/mol supported the proposed dissolution mechanism. The unreacted product analyzed by XRD was found to contain siliceous impurities. Tungsten recovery from leachate was carried out by solvent extraction with Aliquat 336 in

A. A. Baba (✉) · M. O. Muhammed (✉) · M. A. Raji · K. I. Ayinla · A. Y. Abdulkareem F. A. Adekola
Department of Industrial Chemistry, University of Ilorin, P.M.B 1515, Ilorin 240003, Nigeria
e-mail: baalafara@yahoo.com
e-mail:

A. Y. Abdulkareem
National Mathematical Centre, Sheda-Kwali, P.M.B 118, Abuja, Nigeria

M. Lawal
Department of Chemistry, Kebbi State University of Science and Technology, P.M.B. 1144, Aliero, Kebbi State, Nigeria

A. G. F. Alabi
Department of Materials and Metallurgical Engineering, University of Ilorin, P.M.B 1515, Ilorin 240003, Nigeria

A. G. F. Alabi
Department of Material Science and Engineering, Kwara State University, P.M.B 1530, Malete, Nigeria

R. B. Bale
Department of Geology and Mineral Sciences, University of Ilorin, P.M.B 1515, Ilorin 240003, Nigeria

H. Kim et al. (eds.), *Rare Metal Technology 2018*, The Minerals, Metals & Materials Series, https://doi.org/10.1007/978-3-319-72350-1_25

kerosene. Pure tungsten was recovered as APT and beneficiated to produce high grade industrial tungsten oxide of industrial values.

Keywords Wolframite · Scheelite · Stolzite · Hydrochloric acid Phosphoric acid · Leaching · Solvent extraction · Tungsten · Ammonium paratungstate

Introduction

Tungsten, found to exist in nature as wolframite ((Fe, Mn)WO_4) and scheelite ($CaWO_4$) is an important metal used for wide arrays of applications [1]. It is an essential commodity whose unusual properties makes an important contribution through its use as cemented carbide and high speed steel tools, to the achievement of high productivity levels in metals, wood working construction, mining and wears protection on which the world's economic wellbeing depends. Other important applications of tungsten metal include chemical, catalysts, X-ray tubes, super alloys and radiation shielding [2–4].

At present, world's tungsten reserves are estimated to be 3×10^6 tonnes and most tungsten ores contain less than 1% WO_3. Hence, natural and synthetic scheelites are the most important raw materials used in tungsten production for defined industrial applications [5]. For instance, many Nigeria wolframite ore deposits contain admixtures of scheelite and stolzite which could be easily processed to obtain the desired products. Consequently, ammonium paratungstate (APT) is the main intermediate product for production of industrial tungsten metal and tungsten carbide. APT is the fundamental raw material for hard metal production, illumination and coating industry [5, 6].

Conventionally, alkali leaching (NaOH or Na_2CO_3) and solvent extraction with organic ammonium in acidic media is widely adopted for APT production, however high volume of acid is needed for neutralizing and acidifying the leach liquor which eventually becomes the useless inorganic material in the raffinate, poised to be harmful to the environment [5, 7, 8]. Consequently, APT is produced by acidic leaching but tungstic acid formation in the acidic solutions covers the unreacted particles. This interrupts the contact between acidic ions and particle surface and eventually the reaction cannot continue. Therefore, to address the above shortcomings, the use of complex reagents such as oxalic acid, tartaric acid, citric acid and phosphoric acid could be adopted to promote the formation of water soluble tungsten salts [5, 9]. However, due to the ease of complexation, phosphoric acid was chosen as the chelating agent for this study, aimed at establishing physico-chemical, kinetics and extraction parameters to purify a Nigerian wolframite ore for improved industrial utilization by hydrometallurgical process.

Materials and Method

The wolframite ore used for this study was sourced from Nane Didan Soro (N 12′ 03′ 46.5 E 066″ 29′77.7), Bungudu Local Government Area of Zamfara State, Nigeria. The various fractions of the sample used was produced by pulverizing with acetone-rinsed hammer and milled. The milled sample was sieved into three different particle sizes (−112 + 90, −90 + 75 and −63 + 45 μm) with the aid of ASTM standard sieve. Considering the ore high surface area, all experiments were performed only with −63 + 45 fraction, unless otherwise stated [10]. Characterization of the raw ore and selected products at optimal conditions were accordingly carried out by X-ray diffraction, X-ray fluorescence, Scanning electron microscopy, Atomic absorption spectroscopy and Fourier Transform Infra-red spectroscopic techniques.

Leaching Tests

Since tungsten can easily be complexed with phosphoric acid to form a soluble heteropoly compound [1], therefore the leaching test was carried out in a 600 mL glass reactor equipped with a mechanical stirrer. 100 mL of leaching solution ($HCl + H_3PO_4$ solution) was introduced into the reactor and heated to the desired temperature. The optimal acid concentration which gave the highest dissolution efficiency was used for further optimization studies such as reaction temperature and particle size variation. The fraction of ore dissolved in the leachant was evaluated from the difference between the mass of the dissolved and undissolved at various leaching time intervals (0–120 min) after drying. The residual leachate at 70 °C in 1.5 mol/L HCl + 2.0 mol/L H_3PO_4 solution was analyzed by X-ray diffraction and Scanning Electron microscope respectively. The appropriate dissolution kinetics parameter for improved wolframite ore dissolution conditions was determined by applying and testing the various Shrinking core models. Characterization of the original ore sample and the leach residues after leaching at optimal conditions were examined accordingly.

Solvent Extraction/Beneficiation Tests

The quantity of tungsten extracted from the ore was spectrophotometrically determined [11]. The organic and aqeous phase were kept at a ratio of 1:1. A 25 ml of the leach liquor (aqueous) and organic phases (ALIQUAT 336 dissolved in kerosene) were mixed in a stoppered bottle and shaken for 25 min. The concentration of tungsten ion in the organic phase was successfully determined after equilibration and phase separation from the difference between its concentration in the aqeous phase after extraction and prior to the extraction. Pure tungstate salt was selectively precipitated from the poly-tungstate solution by adding a chelate breaking agent ($NH_3 \cdot H_2O$) for 10 ± 2 min at 25 °C. Calcinating APT at 500 °C produces high tungsten trioxide of industrial value [12].

Results and Discussion

Characterization Studies

The wolframite ore analyzed by X-ray fluorescence (XRF) contain 36.32 wt% W, 18.31 wt% Pb, 14.80 wt% Au and 4.63 wt% Fe. The mineralogical examination by X-ray diffraction (XRD) showed the principal occurrence of scheelite ($CaWO_4$: 96-900-9631), Stolzite ($PbWO_4$: 96-900-9813) and quartz (SiO_2: 96-900-9667) in the selected ore particle size.

Leaching Results

The leaching experiments shows that the wolframite ore dissolution in HCl solution increases rapidly with increasing acid concentration, temperature with decreasing particle diameter at moderate stirring as follows:

(i) **H_3PO_4 + HCl concentration variation**: HCl concentration was varied between 0.1–2.0 mol/L and H_3PO_4 concentration between (0.1–2.5 mol/L) at 55 °C within 5–120 min using −63 ± 45 μm particle size. Increasing the acid concentration apparently increases the ore dissolution up till 1.5 mol/L HCl solution. During this period, 51.8% of the ore reacted within 120 min. Further increase in acid concentration to 2.0 mol/L lead to a decrease in ore dissolution which can be attributed to precipitation phenomenon at higher acid concentration. However, addition of H_3PO_4 as chelate to the solution not only helps to prevent the precipitation of insoluble tungstic acid and wrapping, but also helps to improve the percentage of ore dissolved. The optimal phosphoric acid concentration was found to be 2.0 mol/L. Apparently leaching at 1.5 mol/L HCl + 2.0 mol/L H_3PO_4 solution gave the highest dissolution efficiency of ≈95% during 120 min.

(ii) **Reaction Temperature variation**: The effect of reaction temperature on the extent of the ore dissolution was examined from 27 to 75 °C. Other leaching conditions are: 1.5 mol/L HCl + 2.0 mol/L H_3PO_4 solution, 5–120 min, −63 ± 45 µm. Thus, increasing the reaction temperature from 27 to 75 °C appreciably increases the ore dissolution from 28.7 to 94.47%.

(iii) **Particle size variation**: The dissolution results with particle size variations: −112 + 90, −90 + 75 and −63 + 45 µm gave 60.9, 74.8 and 95% respectively at optimal conditions (1.5 mol/L HCl + 2.0 mol/L H_3PO_4 solution) within 120 min. The unleached residue analyzed by XRD was found to contain siliceous impurities (≈5%) and could serve as a by-product in some defined industries.

The kinetics data evaluated by the shrinking core model was found to follow the proposed chemical control mechanism with activation energy of 56.80 kJ/mol. Also, the reaction order determined from the appropriate kinetic data was found to be 0.93 ≈ 1, indicating a first order dissolution process.

Solvent Extraction Studies

Solvent extraction of wolframite ore leach liquor containing 630.3 mg/L W, 23.1 mg/L Fe and 0.8 mg/L Mn was carried out by 0.15 mol/L ALIQUAT 336 in kerosene at 27 ± 2 °C for 25 min contact time. Prior to obtaining pure tungsten solution, iron and other gangues in the leach liquor were selectively precipitated at pH 3.5 to achieve improved extraction and beneficiation efficiencies [13]. About 98% purified tungsten solution was successfully extracted at pH 2–3 as APT. The produced APT was further purified by evaporation and crystallization to obtain APT of high purity (97.8%). Finally, the APT was calcined to produce a high grade tungsten trioxide which could be used for some defined industrial applications. The operational hydrometallurgical scheme for purifying a Nigerian wolframite ore to obtain high grade tungsten oxide is summarized in Fig. 1.

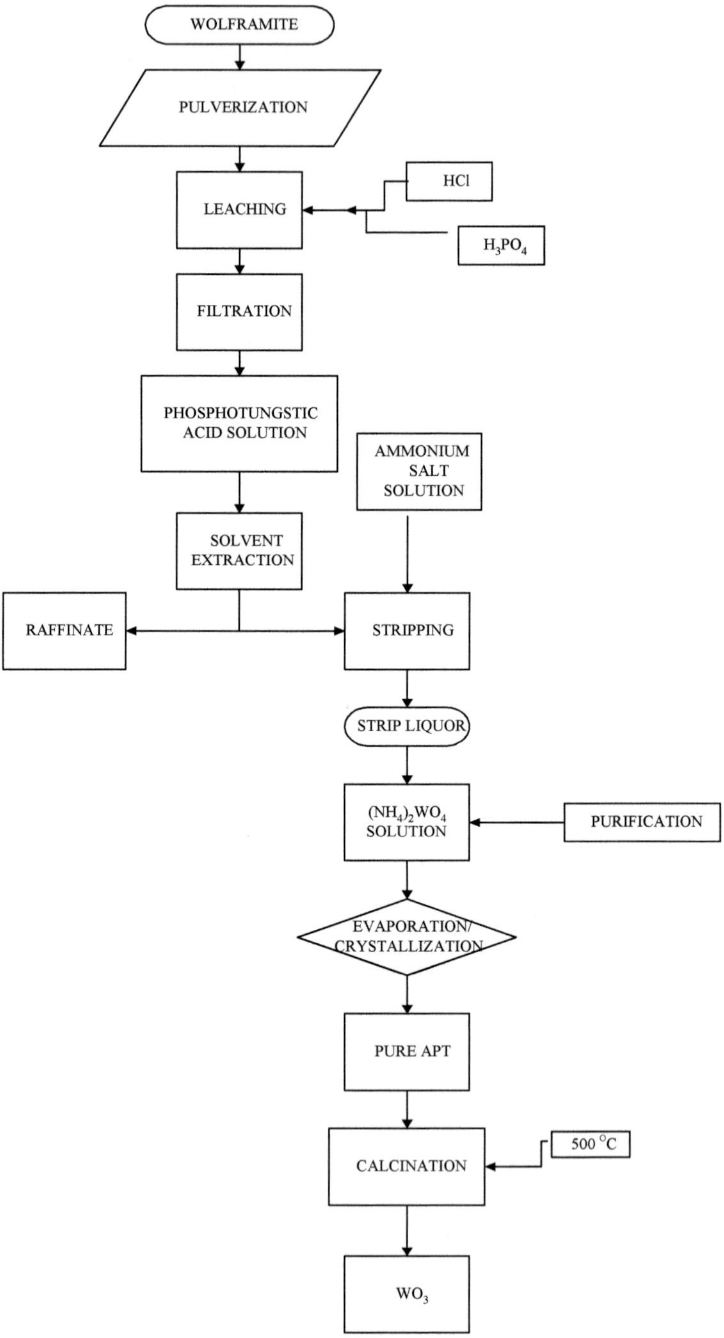

Fig. 1 Hydrometallurgical flowsheet for purification of a Nigerian wolframite ore for improved industrial applications

Conclusion

From the results of this study, it can be concluded that a Nigerian wolframite ore can be a viable source of APT which could be processed to obtain other tungsten compounds. Solvent extraction and precipitation was used in purify the resultant solution from the hydrochloric acid and phosphoric acid leaching of Nane didan Soro Bungudu wolframite ore by ALIQUAT 336 extractant. Tungsten extraction by Aliquat 336 gave an extraction efficiency of 98%. Hence, this study affirms the possibility of producing high grade and industrially applicable tungsten trioxide by hydrometallurgical route. This could contribute to the country's economic growth and development in the replacement of its current hard earned foreign exchange through petroleum explorations.

Acknowledgements The authors wish to thank:

(i) CSIR–Institute of Minerals and Materials Technology, Bhubaneswar-751013, India for their benevolence in supplying the ALIQUAT 336 extractant used for this study;

(ii) I. N. Venantius Nig. Ltd. Located at No. 39, St. Finbarrs Road, Akoka, Lagos, Nigeria: A supplier of Scientific, Hospital Equipment, Laboratory furniture & fittings, Chemicals, Reagents and General Contractor; and

(iii) KAM Industries Nigeria Limited, Ilorin, Kwara State: One of the leading indigenous Steel Industries in Nigeria for their immeasurable supports for this research and part sponsorship to attend the TMS 2018 Meeting in Arizona, U.S.A.

References

1. Zhang W, Yang J, Zhao Z, Wang W, Li J (2016) Coordination leaching of tungsten from scheelite concentrate with phosphorus in nitric acid. J Cent South Univ 23:1312–1317
2. Koribanics NM, Tuorto SJ, Lopez-Chiaffarelli N, McGuinness LR, Häggblom MM, Williams KH, Long PE, Kerkhof LJ (2015) Spatial distribution of an uranium-respiring betaproteobacterium at rifle, CO field research site. PLoS One. PMC ID: 4395306. doi:https://doi.org/10.1371/journal.pone.0123378
3. Amer AM (2000) Investigation of the direct hydrometallurgical processing of mechanically activated low-grade wolframite concentrate. Hydrometallurgy 58:251–259
4. Moby M (1995) Wolfram. Metal 49(9):S568
5. Kahruman C, Yusufoglu I (2006) Leaching kinetics of synthetic $CaWO_4$ in HCl solutions containing H_3PO_4 as chelating agent. Hydrometallurgy 81:182–189
6. Guiqing Z, Wenjuan G, Liansheng X, Qixiu Z (2016) A novel process for tungsten hydrometallurgy based on direct solvent extraction in alkaline medium. Hydrometallurgy 165:233–237
7. Hairunnisha S, Sendil GK, Rethinaraj JP, Srinivasan GN, Adaikkalam P, Kulandaisamy S (2007) Studies on the preparation of pure ammonium paratungstate from tungsten alloy scrap. Hydrometallurgy 85:65–71
8. Zhao Z, Liang Tiang T, Li H (2011) Kinetics of sodium hydroxide leaching of scheelite. Int J Refract Met Hard Mater 29:289–292
9. Sebahattin G, Servet T, Cuneyt A, Ismail D (1999) Acidic leaching of scheelite concentrate and production of hetero-poly-tungstate salt. Hydrometallurgy 51:227–238

10. Fieberg MM, Coetzee CFB (1986) The recovery of tungsten from South African low-grade scheelite concentrates. Mintek, report no: M164D, the Council for Mineral Tech., pp. 1–22
11. Joginder R, Baweja RK, Sushil K (2015) Spectrophotometric determination of tungsten (VI) after extraction of its 6-chloro-3-hydroxy-2-phenyl-4-oxo-4H-1-benzopyran complex into chloroform. Archives Appl Sci Res 5(3):81–89
12. Dimitrijevic V, Dimitrijevi M, Milanovi D (2004) Recovery of tungsten from low-grade scheelite concentrate by soda ash roast—leach method. J Mining Metall 40A(1):75–89
13. Paulino JF, Afonso JC, Mantovano JL, Vianna CA, Cunha JWSD (2012) Recovery of tungsten by liquid–liquid extraction from a concentrate after fusion with sodium hydroxide. Hydrometallurgy:127–128, 121–124

Extraction of Vanadium and Chromium from the Material Containing Chromium, Titanium and Vanadium

Sheng Huang, Shengfan Zhou, Fuhong Xie, Bianfang Chen, Biao Liu, Qi Ge, Mingyu Wang and Xuewen Wang

Abstract To recover iron, titanium, vanadium and chromium from vanadium bearing titanomagnetite, a process of direct reduction has been proposed. After magnetic separation, a material containing titanium, vanadium and chromium was obtained. In this paper, a salt-roasting process was proposed to extract vanadium and chromium from the material containing titanium, vanadium and chromium. The effects of the several parameters that included roasting temperature, roasting time and the addition of sodium carbonate were investigated. Under the most suitable conditions including a roasting temperature of 900 °C, roasting time of 2 h and sodium carbonate addition of 33 wt%, the leaching of vanadium and chromium reached 91.2 and 68.4%, respectively. After leaching, the leach residue can be used as the raw material for extraction of titanium.

Keywords Leaching · Vanadium · Chromium · Salt-roasting · Vanadium bearing titanomagnetite

Introduction

The Hongge mineral deposit in the Panzhihua-Xichang Area of China is the biggest vanadium bearing titanomagnetite, with titanomagnetite reserves of 3.55 billion tons [1]. It is also the largest chromium bearing deposit in China [2]. Vanadium, chromium and titanium are all important strategic metals and industrial raw materials, and the comprehensive recovery of vanadium, chromium and titanium from the Hongge mineral deposit is therefore required for reasonable use. At present, the use of vanadium bearing titanomagnetite is to smelt it in a blast furnace.

S. Huang · S. Zhou · B. Chen · B. Liu · Q. Ge · M. Wang (✉) · X. Wang
School of Metallurgy and Environment, Central South University, Changsha 410083, China
e-mail: wmydxx@163.com

F. Xie
Hainan Tai Xin Mining Industry CO., LTD, Haikou 570125, China

H. Kim et al. (eds.), *Rare Metal Technology 2018*, The Minerals, Metals & Materials Series, https://doi.org/10.1007/978-3-319-72350-1_26

Vanadium and chromium can be recovered in the form of chromium bearing vanadium slag by molten iron selective oxidation. However, the Ti component is concentrated into the slag which can not be utilized [3, 4].

To recover iron, titanium, vanadium and chromium from vanadium bearing titanomagnetite, a process of direct reduction and magnetic separation has been proposed [5–8]. After that, sponge iron and the concentration material containing the oxides of titanium, vanadium and chromium are obtained. The sponge iron can be used as the raw material for producing high quality steel. Therefore, how to extract titanium, vanadium and chromium from the concentration material containing the oxides of titanium, vanadium and chromium became one of the key problems.

The objective of the present work was to study the extraction of vanadium and chromium from the concentration material containing titanium, vanadium and chromium by salt-roasting followed by water leaching. The effect of experimental parameters including the roasting temperature, roasting time and the addition of sodium carbonate on the leaching of vanadium and chromium was investigated.

Experimental

The XRF analysis of the material containing titanium, vanadium and chromium is shown in Table 1. As shown in Table 1, the titanium oxide is the main component. The X-ray diffraction (XRD) pattern of the material containing titanium, vanadium and chromium is shown in Fig. 1. As can be seen, there are three main crystal mineral phases, namely, magnesium titanium oxide, iron and augite. All the other reagents used in tests were of analytical grade.

The charge was composed of a weighed amount of the material containing titanium, vanadium and chromium mixed with sodium carbonate. This charge was put in a crucible and heated gradually in an electric muffle furnace at a predetermined temperature with the door open. After roasting, the calcine was leached by water in an agitated flask, which was heated by an electric jacket. After leaching 2 h at 95 °C with the liquid-to-solid rate of 4 mL/g, the supernatant was vacuum filtered and the leach residue cake was submitted to successive rinsing with water; lasting up to 10 min, before it was dried and analyzed for vanadium and chromium contents. Vanadium and chromium were determined by inductively coupled plasma (ICP) emission spectroscopy with a PS-6 PLASMA SPECTROVAC, BAIRD (USA).

Table 1 XRF analysis of the material containing titanium, vanadium and chromium

Element	O	Ti	Fe	Mg	Si	Al	Ca	Na	Cr	V	Mn
wt%	43.03	22.33	7.37	7.30	8.30	6.27	2.47	0.90	0.65	1.10	0.46

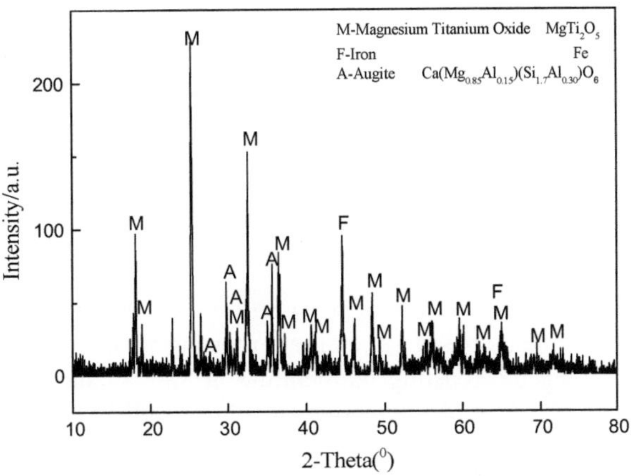

Fig. 1 XRD analysis of the material containing titanium, vanadium and chromium

Results and Discussion

Effect of Sodium Carbonate Addition

To determine the consumption of sodium carbonate in the roasting process, a predetermined weight of sodium carbonate was mixed with 50 g the material containing titanium, vanadium and chromium and then roasted at 800 °C for 2 h, and the results are shown in Fig. 2.

It is observed that there is a rapid increase in the leaching of vanadium with the addition of sodium carbonate from 17 to 33 wt%. Further addition of sodium carbonate resulted in insignificant increases in the leaching of the vanadium. For the extraction of chromium, the leaching of chromium increased with the increase of sodium carbonate addition from 17 to 45 wt%. Comparing the leaching of

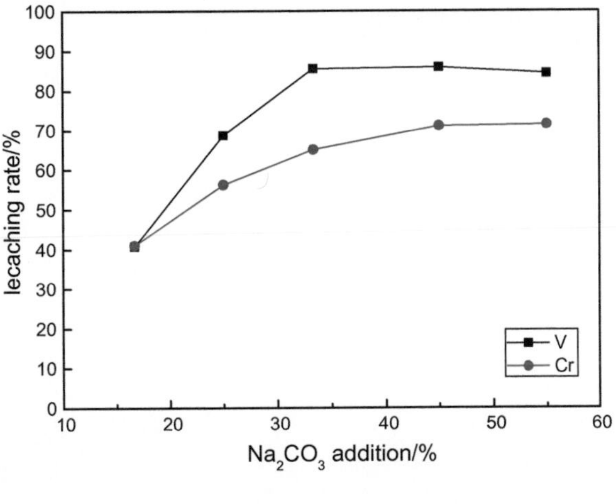

Fig. 2 Effect of sodium carbonate addition on the leaching of vanadium and chromium

vanadium to chromium, the effect of sodium carbonate addition on the leaching of chromium was not as significant. All things considered, 33 wt% sodium carbonate addition was chosen as the most suitable.

Effect of Roasting Time

To discover the role of roasting time on the leaching of vanadium and chromium, experiments based on alternate roasting times from 1 to 3 h were carried out, while keeping the roasting temperature at 800 °C and the sodium carbonate addition at 33 wt%. The results shown in Fig. 3 indicate that the leaching of vanadium and chromium increased gradually with the increase of roasting time from 1 to 2 h. However, above 2 h the leaching of vanadium and chromium did not increase anymore, and the leaching of chromium was even somewhat reduced. According to the study, the long residence time of metallurgical materials in high temperature area will hinder the transformation of low valent vanadium and chromium [9]. Therefore, for subsequent experiments the roasting time was kept to 2 h.

Effect of Roasting Temperature

The roasting temperature was varied in the range of 750–950 °C, with an interval of 50 °C. The roasting time was 2 h and the addition of sodium carbonate was 33 wt%, and the results are illustrated in Fig. 4.

It can be seen that there is an increase in the leaching of vanadium with the roasting temperature from 850 to 900 °C. After 900 °C, the increase in the leaching of vanadium is no longer obvious. For the extraction of chromium, there is a rapid

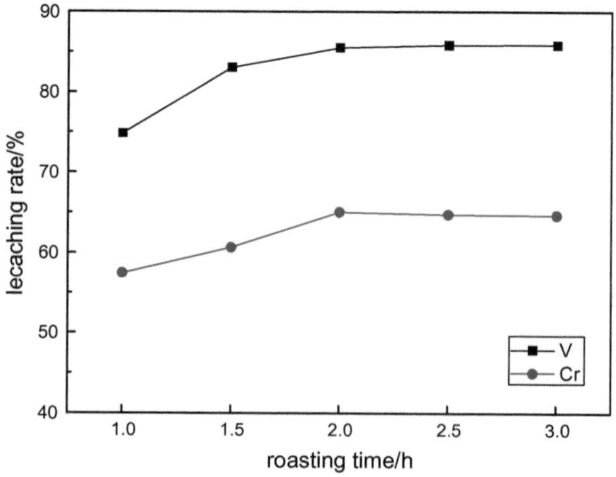

Fig. 3 Effect of roasting time on the leaching of vanadium and chromium

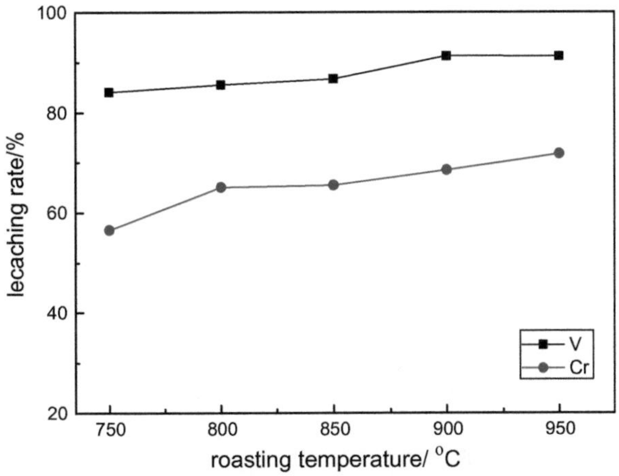

Fig. 4 Effect of roasting temperature on the leaching of vanadium and chromium

increase with the increase of temperature from 750 to 800 °C. All things considered, the roasting temperature of 900 °C was chosen for subsequent experiments.

Mechanism of Salt-Roasting

Based on the preceding experimental results, the most suitable operating conditions for roasting would be: roasting temperature 900 °C, roasting time 3 h and the sodium carbonate addition of 33 wt%. Under these conditions, leaching of vanadium and chromium from the material containing titanium, vanadium and chromium was 91.2 and 68.4% (average of the three results), respectively. After leaching, the leach residue can be used as the raw material for extraction of titanium.

Figure 5 shows the TG/DSC curves of the mixture of the material containing titanium, vanadium and chromium with sodium carbonate. According to the TG curve, the roasting process has three main stages. The first stage is a period where adhesive water on the material containing titanium, vanadium and chromium leaves the surfaces of the material below 100 °C. The second stage occurs between 200 and 500 °C and there is an increase in quality of 1.26%. In this interval, the oxidation reactions of low-valence metal or low-valence metal oxide were produced. Over 500 °C, there is a fast weightless process where V_2O_5 and Cr_2O_3 reacted with sodium carbonate to form sodium vanadate and sodium chromate. It also can be seen that weightlessness end temperature was about 900 °C which is the reaction end point of the salt roasting.

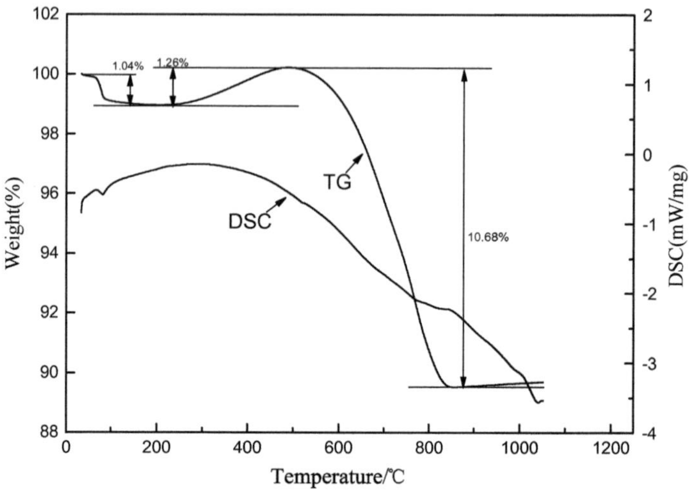

Fig. 5 TG/DSC curve of the mixture of the material containing titanium, vanadium and chromium with sodium carbonate

Conclusions

The salt roasting process can not only extract vanadium effectively from the material containing titanium, vanadium and chromium but also extract chromium. Under the most suitable conditions including a roasting temperature of 900 °C, roasting time of 2 h and sodium carbonate addition of 33 wt%, the leaching of vanadium and chromium from the material containing titanium, vanadium and chromium reached 91.2 and 68.4%, respectively. After leaching, the leach residue can be used as the raw material for extraction of titanium.

Acknowledgements This study was financially supported by the Hunan Provincial Natural Science Foundation of China (2016JJ2142) and the Open-End Fund for the Valuable and Precision Instruments of Central South University (CSUZC201711).

References

1. Luo JH, Wu EH, Li JH, Liao XJ, Tang R, Yang SL (2015) Studies on mineralogical characteristics of iron concentrates in Hongge vanadium titano-magnetite. Iron Steel Vanadium Titanium 36(2):73–77
2. Guizhen H, Xinghong D, Hewei Q, Zheng Z, Taiping L (2014) Present status and development perspective of non-blast furnace ironmaking technology. Multipurpose Utilization Mineral Resour 3:1–7
3. Zhang L, Zhang LN, Wang MY, Li GQ, Sui ZT (2007) Precipitation selectivity of perovskite phase from Ti-bearing blast furnace slag under dynamic oxidation conditions. J Non-Cryst Solids 353:2214–2220

4. Wang MY, Wang XW, He YH, Lou TP, Sui ZT (2008) Isothermal precipitation and growth process of perovskite phase in oxidized titanium bearing slag. Trans Nonferrous Metals Soc China 18:459–462
5. Wang MY, Zhou SF, Wang XW, Chen BF, Yang HX, Wang SK, Luo PF (2016) Recovery of iron from chromium vanadium-bearing titanomagnetite concentrate by direct reduction. JOM 68(10):2698–2703
6. Zhao LS, Wang LN, Qi T, Chen DS, Zhao HX, Liu YH (2014) A novel method to extract iron, titanium, vanadium, and chromium from high-chromium vanadium-bearing titanomagnetite concentrates. Hydrometallurgy 149:106–109
7. Zhao LS, Wang LN, Chen DS, Zhao HX, Liu YH, Qi T (2015) Behaviors of vanadium and chromium in coal-based direct reduction of high-chromium vanadium-bearing titanomagnetite concentrates followed by magnetic separation. Trans Nonferrous Metals Soc China 25 (4):1325–1333
8. Liu SS, Guo YF, Qiu GZ, Jiang T, Chen F (2014) Solid-state reduction kinetics and mechanism of pre-oxidized vanadium–titanium magnetite concentrate. Trans Nonferrous Metals Soc China 24(10):3372–3377
9. Mao LQ (2013) Separation and extraction of vanadium and chromium from slag containing chromium and vanadium. Northestern University, Shenyang

Extraction Separation of V and Fe in High Acid and High Iron Solution

Weiguang Zhang, Ting-an Zhang, Guozhi Lyu, Yajing Tian, Biyu Long and Xuejiao Cao

Abstract Aim to the separation problem of high concentration V–Fe in stronger acidic leaching solution of vanadium slag, a novel chelating extraction system is proposed to separate V and Fe. Influence factors including initial pH value, the extractant concentration, the phase ratio, temperature and time are investigated. The experiment results show that single extraction rate of V and Fe reach 85.57 and 0.39% respectively, when pH value is −0.4, organic phase composition is 30% Mextral 973H + 70% sulfonated kerosene oil, O:A = 1:1, stirring time is 10 min, extraction temperature is 30 °C. The industrial experiment results of three levels countercurrent extraction show that the total extraction rate of vanadium and iron can reach 97.44 and 0.23%, separation factor is 6410. The effective separation of V and Fe in high acid and high iron solution can be accomplished.

Keywords Extraction separation · V and Fe · High acid and high iron solution Chelating extraction

Introduction

As a strategic metal, vanadium is an important additive in modern industry. Now, vanadium is mainly extracted from vanadium slag and stone coal in China [1–3]. Traditional technology for extracting vanadium is roasting leaching precipitation technology, whose disadvantages are low recovery rate of vanadium, high energy

W. Zhang (✉) · T. Zhang · G. Lyu · Y. Tian · B. Long · X. Cao
Key Laboratory of Ecological Utilization of Multi-Metal Intergrown Ores of Education Ministry, School of Materials and Metallurgy of Northeastern University, Shenyang 110819, Liaoning, China
e-mail: zhangwg@smm.neu.edu.cn

T. Zhang
e-mail: zta2000@163.net

G. Lyu
e-mail: lvgz@smm.neu.edu.cn

H. Kim et al. (eds.), *Rare Metal Technology 2018*, The Minerals, Metals & Materials Series, https://doi.org/10.1007/978-3-319-72350-1_27

consumption, and serious air pollution [4, 5]. Therefore, with the increasingly strict of environmental requirement and increasingly severe market competition, developing a novel extraction technology is significant for vanadium industry.

Nowadays, acid leaching process for vanadium extraction has been widely studied due to high recovery rate of V and good working environment. However, impurities such as Fe, Al, Mn, Cr, P are leached together with V into the solution. Especially, the existence of many iron ions in leaching solution have seriously affected the purification and precipitation process of vanadium. Therefore, separation of vanadium and iron in strong acidic solution become the top priority of vanadium purification process and key technology of extracting vanadium by acid leaching method, which would be of great significance to the whole vanadium industry [6, 7].

Solvent extraction, ion exchange, and adsorption methods are widely used for the V–Fe separation in acidic leaching solution. In particular, solvent extraction has advantages of better selectivity, faster equilibrium speed and larger processing capacity. Now, common extraction agents for extracting vanadium are amine extractant, neutral extractant and acidic extractant [8–13]. Among them, acidic extractant P204 has been widely used in industrial production. However, there are some problems in extraction process of P204. Firstly, the pH value of P204 extraction equilibrium is larger than pH value of leaching solution. Therefore, feed liquid need to be neutralized before the solvent extraction of P204; Secondly, the extraction selectivity of V and Fe has a little difference because of same cation exchange mechanism. Therefore, separation effect of V and Fe through P204 solvent extraction is not very good; Thirdly, vanadium is difficult to strip completely and strip liquor also need oxidation treatment before the precipitation of vanadium [14–17].

Based on these, our group proposes that chelating extractant is used for directly extracting V from high acid and high iron solution, which can reduce working procedure including neutralization and reduction of feed liquid [18, 19]. There are some advantages as follows. Firstly, hydrogen-ion concentration has a little impact on extraction equilibrium. Therefore, vanadium can be extracted in high acid solution without neutralization; Secondly, the extraction selectivity of vanadium (V^{5+}) is much larger than iron (Fe^{3+}) through chelating extraction. Therefore, separation effect of V–Fe will be better than P204; Thirdly, vanadium is easy to strip completely and strip liquor can be directly precipitated without oxidation treatment.

Materials and Methods

Materials and Analysis

The commercial grade Mextral 973H and sulphonated-kerosene supplied by Chongqing Haokang Chemical Co., Ltd (CN) were used in the study. The extractant Mextral 973H was used without any further purification. Its structures

Scheme 1 Structures of the main components for Mextral 973H used in this study

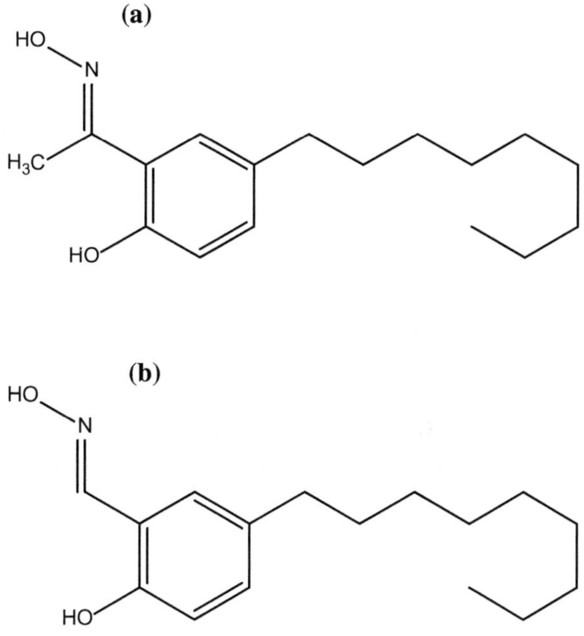

of the main components (a) 2-hydroxy-5-nonylacetophenone oxime and (b) 5-Nonylsalicylaldoxime) were shown in Scheme 1. The sulphonated-kerosene was used as the diluent.

The other chemicals used were of analytical grade, and all aqueous solutions were prepared using distilled water. The stock solutions were prepared by dissolving V_2O_5 and $Fe_2(SO_4)_3$ (Aladdin, Shanghai) in H_2SO_4 acid to required concentration, pH of the aqueous phase was adjusted by the addition of the H_2SO_4 acid or sodium hydroxide solution. The initial V(V) concentrations were maintained at 10 g/L and initial Fe(III) concentrations were maintained at 30 g/L. The pH of the solutions was measured by digital pH meter (PHS-3F) (Leici, CN). The concentration of metal ions in the aqueous phase was determined by inductively coupled plasma atomic emission spectrometry (ICP-AES) (Leeman, USA).

Experimental Procedure

Extraction studies were carried out by mechanical stirring of the aqueous and organic solutions in a glass reactor at a certain temperature, time and stirring speed. All of the reported points for the extraction were obtained by running a single batch-single element experiment. The pH was kept constant by adding 2 M sulfuric acid or 4 M sodium hydroxide. After extraction, phase separation was performed using separatory funnels at room temperature. The aqueous solutions were then filtered and analyzed by inductively coupled plasma (ICP). Only the metal concentrations were analyzed and the extraction was calculated from the initial and final aqueous solution concentrations. Organic solutions were not analyzed.

Results and Discussion

Effect of Extractant Dosage on Extraction of V and Fe

The effect of extractant dosage on the extraction of vanadium and iron is shown in Fig. 1. Mextral 973H dosage varies from 5 to 50% at room temperature, with pH value of −0.4, an O/A phase ratio of 1:1, reaction time of 10 min, stirring speed of 1100 r/min.

As seen from Fig. 1, Mextral 973H dosage greatly influences vanadium extraction as the extraction ratio of vanadium increased from 2.99 to 84.51% as the Mextral 973H dosage increased from 5 to 30%. Increasing Mextral 973H dosage to 50%, extraction ratio of vanadium increased slowly and can reach 97.16% when extractant dosage was 50%. Whereas, iron extraction remained almost constant in the dosage range, which was below of 5%. It demonstrates that Mextral 973H has high selectivity on vanadium(V) from high acidity of solution, which can be used to separate vanadium(V) from Fe. In conclusion, considering the difficulty of phase separation and economic cost, 30% of Mextral 973H dosage is chosen for the separation of V and Fe.

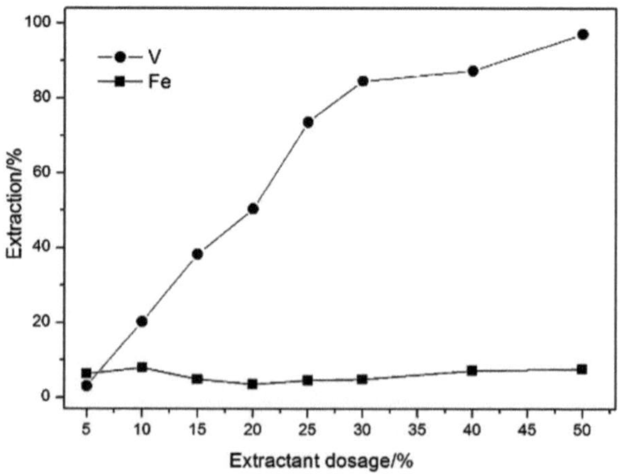

Fig. 1 Effect of extractant dosage on extraction of V and Fe

Effect of Initial Solution pH on Extraction of V and Fe

The effect of initial solution pH on the extraction of V and Fe using an organic solution of 30% (v/v) Mextral 973H is shown in Fig. 2. The pH is selected in the range of −0.8–0.2 at room temperature, with an O/A phase ratio of 1:1, reaction time of 10 min, stirring speed of 1100 r/min.

It can be seen from Fig. 2 that the feed acidity has no significant effect on the extraction of V and Fe in the pH range. Extraction ratio of V and Fe basically remained 80–88 and 0–5%. According to $M^{n+}_{aq} + nHL_{org} = MLn_{org} + nH_{aq}$, vanadium (V) has much stronger affinity to extractant than H^+ and Fe in high acidity. It is noted that a small amount of Fe was extracted with the decrease of acidity. The extraction of vanadium(V) indicated the extraction by Mextral 973H was very strong in high acidity up to −0.8 of pH value, demonstrating that Mextral 973H has high selectivity on vanadium(V) from high acidity of solution, which can be used to separate vanadium(V) from Fe.

Effect of Phase Ratio O/A on Extraction of V and Fe

The effect of phase ratio O/A on the extraction of V and Fe using an organic solution of 30% (v/v) Mextral 973H is shown in Fig. 3.

The O/A is selected in the range of 2:1–1:5 at room temperature, with an initial solution pH of −0.4, reaction time of 10 min, stirring speed of 1100 r/min. Figure 3 shows that extraction ratio of V decreased from 90.30 to 21.48% as phase ratio varied from 2:1 to 1:5. Meanwhile, extraction ratio of Fe remained between 0 and 5%. To obtain high extraction of vanadium in practice and save cost, the A/O ratio should be kept at 1:1.

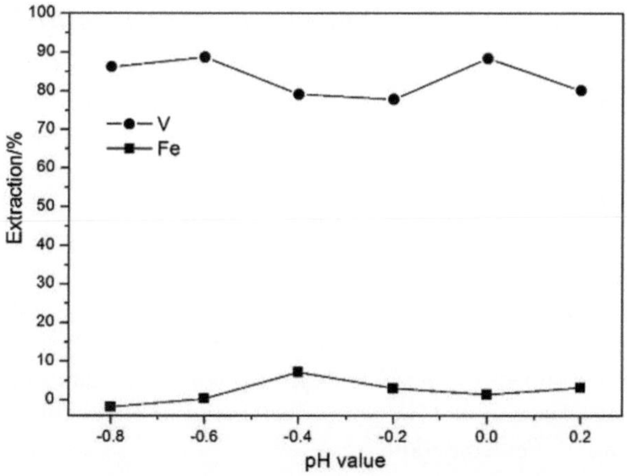

Fig. 2 Effect of initial solution pH on extraction of V and Fe

Fig. 3 Effect of phase ratio on extraction of V and Fe

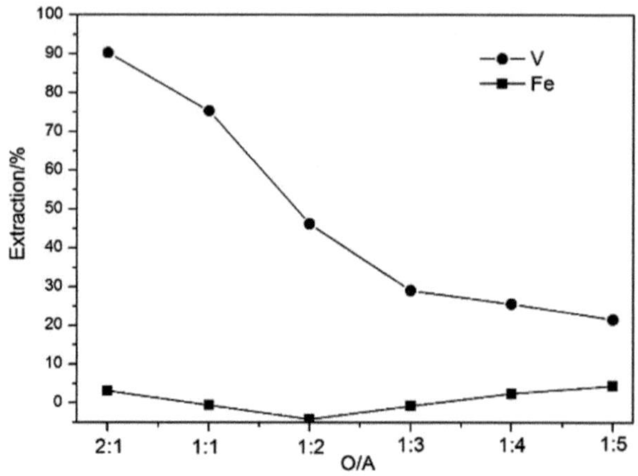

Fig. 4 Effect of stirring speed on extraction of V and Fe

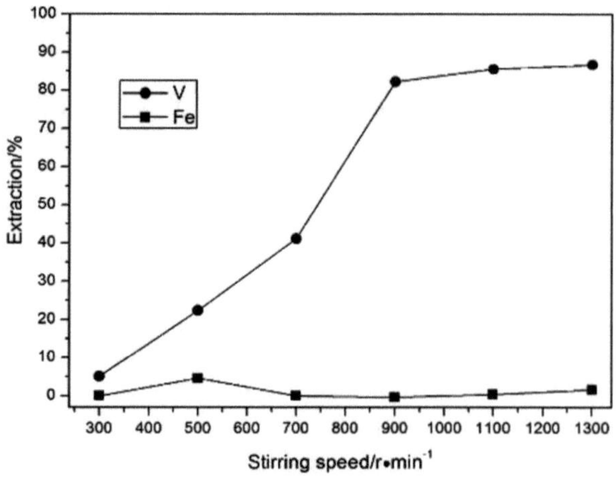

Effect of Stirring Speed on Extraction of V and Fe

The effect of stirring speed on the extraction of V and Fe using an organic solution of 30% (v/v) Mextral 973H is shown in Fig. 4. The stirring speed is selected in the range of 300–1300 r/min at room temperature, with an O/A phase ratio of 1:1, an initial solution pH of −0.4, reaction time of 10 min.

As seen from Fig. 4, stirring speed has a significant impact on the vanadium extraction. Extraction ratio of V increased from 5.09 to 85.57% as the stirring speed increased from 300 to 1100 r/min. Increasing stirring speed to 1300 r/min, extraction ratio of V slightly increased to 86.75%. Whereas, stirring speed had a little impact on the Fe extraction, whose extraction ratio basically remained about 1%. Therefore, 1100 r/min of stirring speed is relatively suitable for V extraction and V–Fe separation.

Effect of Reaction Time on Extraction of V and Fe

The effect of reaction time on the extraction of V and Fe using an organic solution of 30% (v/v) Mextral 973H at room temperature is shown in Fig. 5. The reaction time is selected in the range of 1–20 min, with an O/A phase ratio of 1:1, an initial solution pH of −0.4 and stirring speed of 1100 r/min.

As seen from Fig. 5, reaction time has a significant impact on the vanadium extraction. Extraction ratio of V increased from 30.09% at 1 min to 79.97% at 10 min. Increasing reaction time to 20 min, extraction ratio of V only slightly increased to 86.84%. Whereas, reaction time has a little impact on the Fe extraction, whose extraction ratio basically remained between 1 and 5%. Therefore, 10 min is relatively suitable for V extraction and V–Fe separation.

Effect of Reaction Temperature on Extraction of V and Fe

The effect of reaction temperature on the extraction of V and Fe using an organic solution of 30% (v/v) Mextral 973H is shown in Fig. 6. The reaction temperature is selected in the range of 25–60 °C, with an O/A phase ratio of 1:1, an initial solution pH of −0.4, stirring speed of 1100 r/min and reaction time of 10 min.

As seen from Fig. 6, reaction temperature has a little effect on extraction of V and Fe within the temperature range. Extraction ratio of vanadium decreased slightly from 85.63% at 25 °C to 79.52% at 60 °C. Meanwhile, extraction ratio of Fe basically remained between 2 and 5%.

It is well known that high temperature will promote the volatilization of organic phase and lead to the waste of extractant. Meanwhile, high temperature results in the high energy consumption and high cost. Therefore, 25–30 °C was chosen as the optimum temperature.

Fig. 5 Effect of reaction time on extraction of V and Fe

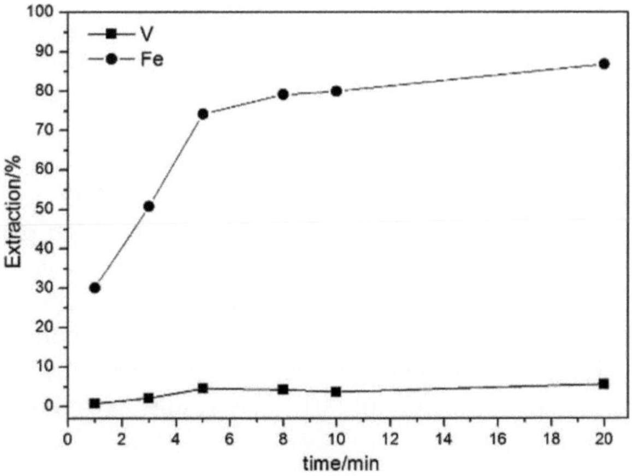

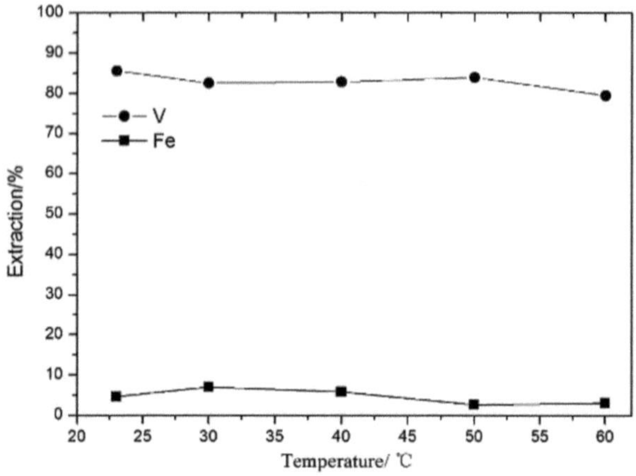

Fig. 6 Effect of reaction temperature on extraction of V and Fe

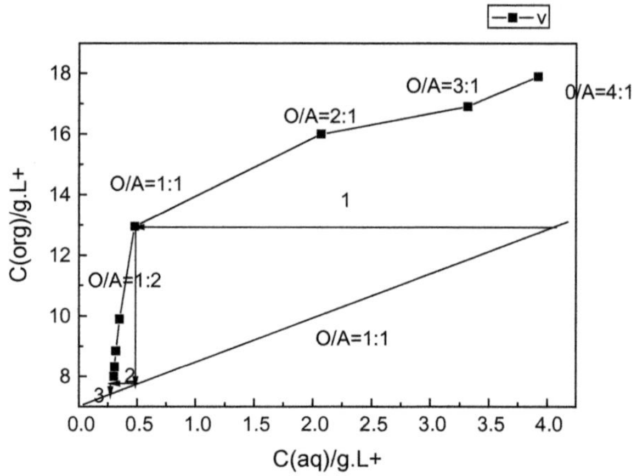

Fig. 7 Extraction isotherm of vanadium

Extraction Isotherm of Vanadium

To determine the number of stages required at a chosen volume phase ratio, extraction isotherms were obtained with 30% (v/v) Mextral 973H in sulfonated kerosene using the acidic leach solution. The feed solution and the organic phase were contacted at different phase ratios (A/O = 1:5 to 5:1) for 20 min while keeping the total volume of phases constant. After phase separation both the phases were analyzed. The McCabe-Thiele plot (Fig. 7) shows that the complete extraction of vanadium is theoretically possible in three stages using an O/A flow ratio of 1.0.

To confirm the predictions from the extraction isotherm, a three-stage counter-current simulation experiment was carried out with 30% (v/v) Mextral 973H at flow ratio of 1:1. The results are shown in Table 1. The industrial

Table 1 A three-stage counter-current simulation experiment

Counter-current level	Extraction ratio of Fe(%)	Extraction ratio of V(%)
A1	1.08	37.44
A2	1.89	97.37
A3	0.23	97.44

experiment results of three levels counter-current extraction show that the total extraction rate of vanadium and iron can reach 97.44 and 0.23%, separation factor is 6410.

Conclusion

(1) Vanadium in acidic leaching solution can be effectively extracted by 30% (v/v) Mextral 973H and 70% (v/v) sulfonated kerosene, with the advantages of good selectivity, fast extraction speed and high recovery rate.
(2) The experiment results show that single extraction rate of V and Fe reach 85.57 and 0.39% respectively, when pH value is −0.4, organic phase composition is 30% Mextral 973H + 70% sulfonated kerosene oil, O:A = 1:1, stirring time is 10 min, extraction temperature is 30°C. The industrial experiment results of three levels countercurrent extraction show that the total extraction rate of vanadium and iron can reach 97.44 and 0.23%, separation factor is 6410. The effective separation of V and Fe in high acid and high iron solution can be accomplished.

Acknowledgements This work was financially supported by the National Natural Science Foundation of China (Nos. 51504059) and the Fundamental Research Funds for the Central Universities (N162504016).

References

1. Moskalyk R, Alfantazi A (2003) Processing of vanadium: a review. Miner Eng 16:793–805
2. Habashi F (1998) Handbook of extractive metallurgy. Wiley-VCH, New York
3. Zhang Y-M, Bao S-X, Liu T, Chen T-J, Huang J (2011) The technology of extracting vanadium from stone coal in China: history, current status and future prospects. Hydrometallurgy 109:116–124
4. Yang Z, Li H-Y, Yin X-C, Yan Z-M, Yan X-M, Xie B (2014) Leaching kinetics of calcification roasted vanadium slag with high CaO content by sulfuric acid. Int J Miner Process 133:105–111
5. Ognyanova A, Ozturk A, De Michelis I, Ferella F, Taglieri G, Akcil A, Veglio F (2009) Metal extraction from spent sulfuric acid catalyst through alkaline and acidic leaching. Hydrometallurgy 100:20–28

6. Aarabi-Karasgania M, Rashchia F, Mostoufi N (2010) Leaching of vanadium from LD converter slag using sulfuric acid. Hydrometallurgy 102:14–21
7. Li M, Wei C, Qiu S (2010) Kinetics of vanadium dissolution from black shale in pressure acid leaching. Hydrometallurgy 104:193–200
8. Tavakoli MR, Dreisinger DB (2014) Separation of vanadium from iron by solvent extraction using acidic and neutral organophosporus extractants. Hydrometallurgy 141:17–23
9. Chagnes A, Rager MN, Courtaud B (2010) Speciation of vanadium (V) extracted from acidic sulfate media by trioctylamine in n-dodecane modified with 1-tridecanol. Hydrometallurgy 104:20–24
10. Zhang G, Chen D, Zhao W (2016) A novel synergistic extraction method for recovering vanadium (V) from high-acidity chloride leaching liquor. Sep Purif Technol 165:166–172
11. Zhang P, Inoue K (1996) Recovery of metal values from spent hydrodesulfurization catalysts by liquid-liquid extraction. Energy Fuels 9:231–239
12. Zeng L, Cheng CY (2009) A literature review of the recovery of molybdenum and vanadium from spent hydrodesulphurisation catalysts part II: separation and purification. Hydrometallurgy 98:10–20
13. Lozano LJ, Godinez C (2003) Comparative study of solvent extraction of vanadium from sulphate solutions by primene 81R and alamine 336. Miner Eng 16:291–294
14. Cheraghi A, Ardakani MS, Alamdari EK (2015) Thermodynamics of vanadium (V) solvent extraction by mixture of D2EHPA and TBP. Int J Miner Process 138:49–54
15. Wang L, Zhang Y, Liu T (2013) Comparison of ion exchange and solvent extraction in recovering vanadium from sulfuric acid leach solutions of stone coal. Hydrometallurgy 131–132:1–7
16. Xingbin L, Wei C, Deng Z (2011) Selective solvent extraction of vanadium over iron from a stone coal/black shale acid leach solution by D2EHPA/TBP. Hydrometallurgy 105:359–363
17. Zhang Y, Zhang T, Lv G (2016) Synergistic extraction of vanadium (IV) in sulfuric acid media using a mixture of D2EHPA and EHEHPA. Hydrometallurgy 166:87–93
18. Li QG, Xu L, Qi ZS (2013) Study on extraction of vanadium from high acid leach solution of stone coal using extractant HBL@101. Adv Mater Res 634:3248–3255
19. Nguyen TH, Lee MS (2015) Separation of molybdenum(VI) and tungsten(VI) from sulfate solutions by solvent extraction with LIX 63 and PC 88A. Hydrometallurgy 155:51–55

Batch Studies for Removing Vanadium(V) and Chromium(VI) from Aqueous Solution Using Anion Exchange Resin

Yang Yang, Hong-Yi Li, Min-Min Lin and Bing Xie

Abstract After precipitation of ammonium vanadates from vanadium-containing leaching liquor, vanadium (V(V)) and chromium (Cr(VI)) exist in the waste water in form of vanadate ions and chromate ions. In order to efficiently extract and recover V(V) and Cr(VI) from waste water, an environment-friendly method based on anion exchange was investigated in batch tests using macroporous weak base resin Dex-V. The adsorption conditions optimizes in detail, including the effect of pH, initial metal concentration, resin volume, contact time and temperature. The ion-exchange adsorption process for both V(V) and Cr(VI) showed maximum capacity at pH 3.0 for an initial concentration of 100 mg L^{-1}, and reached equilibrium after about 40 min of contact at 298 K. The maximum adsorption ratio under the optimum conditions for V(V) and Cr(VI) was 98.98 and 99.87%, respectively. Result shows that the established method is a promising alternative for simultaneous extraction of vanadium and chromium.

Keywords Vanadium · Chromium · Anion exchange · Adsorption

Introduction

Vanadium and chromium are of great importance elements in many fields such as metallurgical, material and chemical industries for its various applications and are accompanying elements of magnetite ores [1, 2]. In industry, Vanadium and chromium are the main valuable elements in the converter slag which is called vanadium slag of steels melting [3, 4]. By roasting with sodium or calcium salts and leaching with water, vanadium is extracted from slag to the V-containing liquor [5–7]. After removing the impurities and neutralizing the liquor, vanadium existed in the form of vanadate ions, which are recovered through precipitation [8].

Y. Yang · H. -Y. Li (✉) · M. -M. Lin · B. Xie
College of Materials Science and Engineering, Chongqing University,
Chongqing 400044, China
e-mail: hongyi.li@cqu.edu.cn

H. Kim et al. (eds.), *Rare Metal Technology 2018*, The Minerals, Metals & Materials Series, https://doi.org/10.1007/978-3-319-72350-1_28

Then the waste water containing vanadium (V(V)) and chromium (Cr(VI)) is formed after filtering, which is highly toxic and can affect human physiology, accumulate in food chain, and cause several ailments [4, 9]. Therefore, the removal of Cr(VI) and V(V) from waste water is one of the most essential issues in view of environmental protection and resource recovery.

Various technologies applied to remove Cr(VI)and V(V) from waste water included chemical precipitation, solvent extraction, capillary electrophoresis, microbe treatment, and ion exchange [10–13]. The methods of precipitation were restricted by the low purification of products and the high consumption of acid. The heavy use of chemical reagent during solvent extraction process is likely to cause secondary pollution. Low efficiency and small capacity limit the wide application of microbe treatments and capillary electrophoresis. The main advantages of ion exchange are recovery of metal value, selectivity, less sludge volume produced and the meeting of strict discharge specifications [14]. However, previous studies using ion exchange primarily focus on removing Cr(VI) and V(V) from aqueous solution separately [14, 15]. Rare researches propose simple technology for simultaneous removal of V(V) and Cr(VI) from a mixed solution. Fan et al. [16] used D314 resin, a weak-base anion exchange resin, to selectively adsorb vanadium from the solution containing V(V) and Cr(VI) in two resin columns, so the processing period by double-adsorption are increased accordingly. Therefore, proposing a simple and efficient method to remove Cr(VI) and V(V) from waste water require continuous effort.

In this work, weak base macropose resin Dex-V was introduced to achieve extraction of V(V) and Cr(VI) from mixed solution in a simple way for the purpose of environmental conservation and resource recycling. V(V) and Cr(VI) anions in V–Cr containing solution are adsorbed simultaneously, then desorbed selectively and recovered separately. The influences of various parameters for extraction were studied in detail in order to obtain optimal adsorption condition, which brings about a possible high-efficiency approach to recover heavy metal from waste water in an eco-friendly way.

Experimental

Chemicals and Solutions

All chemicals used in this study were analytical grade reagents, and deionized water produced by AQUELIX5 (Millipore Corp., USA) was used for preparing all solutions and standards. Stock solution of 1.0 g L^{-1} V(V) and Cr(VI) were prepared by dissolving 3.92 g $NaVO_4$ and 2.25 g Na_2CrO_4 in 500 mL deionized water separately. The mixed solution of V(V) and Cr(VI) concentration of 100 mg L^{-1} was made by blending 15 mL of each V(V) and Cr(VI) stock solution and diluting into 150 mL.

Table 1 Physical and chemical properties of Dex-V anion exchange resin

Properties	Dex-V anion exchange resin
Structure	Macro porous styrene-based chain-like copolymers
Functional groups	$-N(CH_3)_2$
Physical form	White or faint yellow spherical particle
Ionic form	Cl^-
Particle size(mm, wet)	0.50–0.60
Water retention capacity (%)	48–58
Mass total capacity (mmol/g dry resin)	≥ 4.8
Bulk density (g/ml)	0.65–0.75
Operating pH range	1–10
Maximum operating temperature	80

Ion-Exchange Resin

The macroporous weak base anion exchange resin Dex-V was selected in this study (analog to Amberlite IRA-94, Zhengzhou Qinshi Technology Co., Ltd., Hangzhou, China), since it has great adsorption capacity for both vanadium anions and chromium anions. Some properties of the resin Dex-V were listed in Table 1. First, the resin in the wet form was activated with 3% NaOH solution, then transformed to Cl^- type by using 3% HCl solution and rinsed with deionized water before use. After adsorption, the loaded resin was stripped by 3% NaOH and regenerated by 3% HCl. Finally, after rinsed adequately, the neutral resin was stored in saturated NaCl solution.

Batch Adsorption Experiments

The experiments were carried out with the synthetic mixed solution by varying the adsorption conditions, including solution pH, initial concentration, resin volume and contacting time. The volume of the solution was fixed in 150 mL in the conical flask (250 mL) on the magnetic stirring apparatus. The investigated pH range is 3.0–8.0, initial concentration is 40–140 mg L^{-1}, resin volume is 6–12 mL, and contact time is from 5 to 40 min. All experiments were carried out at same temperature 298 K. Each time point corresponds to a separate batch test. Vanadium and chromium uptakes achieved by determining the metal concentrations before and after contact with the resin. After adsorption, the concentrations of V and Cr in the resin solution were determined by inductively coupled plasma-atomic emission spectrometry (ICP-AES, Optima 4300DV, Perkin Elmer, USA).

Results and Discussion

Effect of pH Value of Solution

The effect of pH value of solution on adsorption ratio of vanadium (V(V)) and chromium (Cr(VI)) was investigated in a range of 3.0–8.0. As shown in Fig. 1, with the increase of pH value, the adsorption ratio of V(V) decreased rapidly and the ratio for Cr (VI) reduced slightly, both peaked to 98.83% (V(V)) and 99.87% (Cr (VI)) at pH 3.0. It is believed that the adsorption ratio has close relationship with charge/mass ratio (Z/N) of ions. Low Z/N ratio of ions leads to high adsorption ratio because the adsorption capacity depends on electrostatic attraction force between ions and active sites on resin surface. The ionic forms of vanadium in solution vary greatly at different pH values, which leads to the raise of Z/N ratio as well as the drop of adsorption ratio at high pH value. As shown in Table 2, charge-to-mass rations (Z/N) of V(V)-bearing ions increase with the change of ionic forms of vanadium at different pH values. At pH 1.0–2.0, vanadium exist in cations VO_2^+, and at pH = 3.0, vanadium anions exist majorly in form of $H_2V_{10}O_{28}^{4-}$ (Z/N = 0.4); then the Z/N ratio increases with the increasing pH values. Therefore, the maximum adsorption ratio of V(V) was obtained at pH 3.0, at which pH the Z/N ratio of 0.4 was the lowest. Chromium anions exist in solution in form of three states: $HCrO_4^-$, $Cr_2O_7^-$ (at pH < 5, Z/N = 1.0) and CrO_4^- (at pH > 7, Z/N = 2.0). According to the equilibrium constant of Cr(VI) shown in Table 3, chromium anions exist in form of $Cr_2O_7^-$ in aqueous solution at pH 3.0.

Therefore, the maximum adsorption ratio of V(V) and Cr(VI) was achieved at pH 3.0 because of lowest Z/N ratio of ions. The anion exchange adsorption reactions during this process were shown in Eqs. (1) and (2).

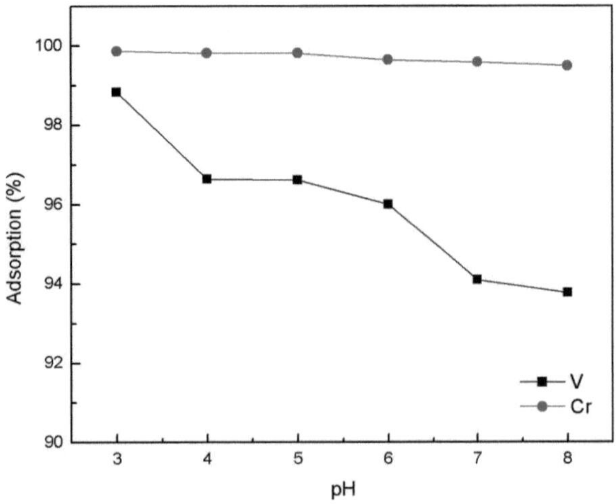

Fig. 1 Effect of pH value of solution on adsorbing of V(V) and Cr(VI)

Table 2 Charge-to-mass rations(Z/N) of V(V)-bearing ions at different pH

pH	Major existing form of V(V)-bearing ions	Z/N ratio value
1–2	VO_2^+	–
2–3	$H_2V_{10}O_{28}{}^{4-}$, $HV_{10}O_{28}{}^{5-}$ $H_2V_{10}O_{28}{}^{4-} \longrightarrow H^+ 5V_2O_5 \cdot H_2O \downarrow$	4:10 = 0.4
3–4	$H_2V_{10}O_{28}{}^{4-}$, $HV_{10}O_{28}{}^{5-}$ $HV_{10}O_{28}{}^{5-} \longrightarrow H^+ H_2V_{10}O_{28}{}^{4-}$	4:10 = 0.4
4–5	$V_{10}O_{28}{}^{6-}$, $HV_{10}O_{28}{}^{5-}$ $V_{10}O_{28}{}^{6-} \longrightarrow H^+ HV_{10}O_{28}{}^{5-}$	5:10 = 0.5
5–7	$V_4O_{12}{}^{4-}$, $V_{10}O_{28}{}^{6-}$ $V_4O_{12}{}^{4-} \longrightarrow H^+ V_{10}O_{28}{}^{6-}$	6:10 = 0.6
7–10	$V_2O_7{}^{4-}$, $V_4O_{12}{}^{4-}$ $V_2O_7{}^{4-} \longrightarrow H^+ V_4O_{12}{}^{4-}$	4:4 = 1

Table 3 The ionization reaction and equilibrium constant of Cr(VI) in water

Reaction	lgK (298 K)
$H_2CrO_4 \Leftrightarrow H^+ + HCrO_4^-$	−0.8
$HCrO_4^- \Leftrightarrow H^+ + CrO_4{}^{2-}$	−6.5
$2HCrO_4^- \Leftrightarrow H_2O + Cr_2O_7{}^{2-}$	1.52

$$\text{Adsorption of V(V)} \quad 4\overline{R - Cl} + H_2V_{10}O_{28}{}^{4-} \rightarrow \overline{R_4 - H_2V_{10}O_{28}} + 4Cl^- \quad (1)$$

$$\text{Adsorption of Cr (VI)} \quad 2\overline{R - Cl} + Cr_2O_7{}^{2-} \rightarrow \overline{R_2 - Cr_2O_7} + 2Cl^- \quad (2)$$

Effect of Initial Concentration of Solution

The initial concentration of vanadium (V) and chromium (VI) plays a role in anion exchange adsorption with resin, so the effect was investigated in a range of 40–140 mg/L according to actual ingredient content of industry wastewater. The results of the study were shown in Fig. 2, the adsorption ratio of V(V) and Cr(VI) both declined mildly with the initial concentration increased in solution, within the scope of 98.22–95.81% for V(V) and of 99.84–99.69% for Cr(VI). Based on the existing experiment result and the fact that the leaching liquor of sodium roasted vanadium slag with high chromium content contains high concentration of vanadium (V) and chromium (VI), the initial concentration of V(V) and Cr(VI) in solution at batch experiments was chosen at 100 mg L^{-1}.

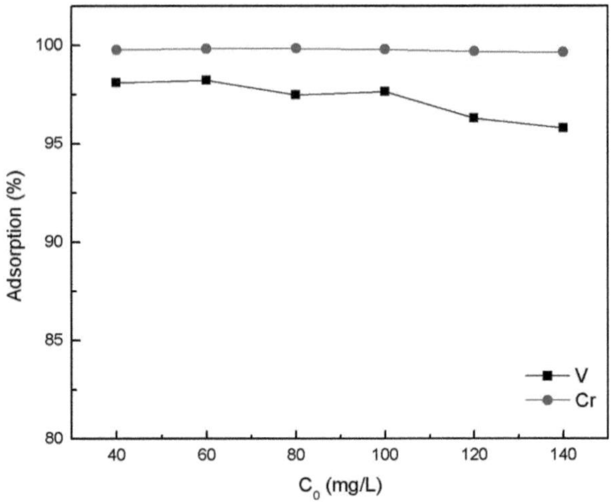

Fig. 2 Effect of initial concentration on adsorbing of V(V) and Cr(VI)

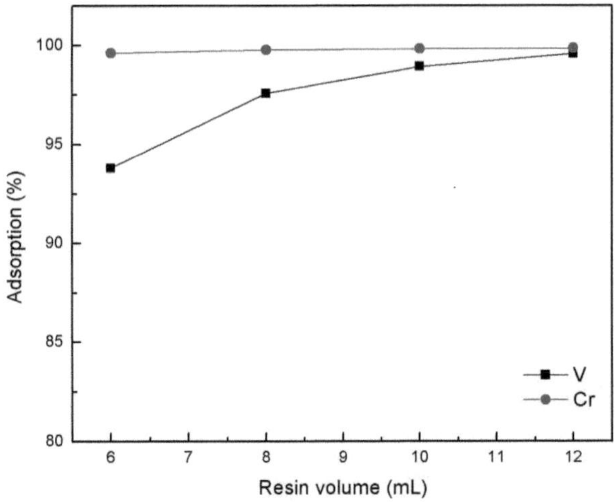

Fig. 3 Effect of resin volume on adsorbing of V(V) and Cr(VI)

Effect of Resin Volume

In order to achieve abundant saturated adsorption of V(V) and Cr(VI) in solution according to the initial concentration of 100 mg/L, the effect of resin volume was studied by adding 6, 8, 10, 12 mL Dex-V resin respectively into 150 mL mixed solution. It is apparent that, the amount of absorbed metal ion per unit mass increases by increasing the resin amount. Therefore, it can be seen from Fig. 3, the adsorption ratio of V(V) increases massively with the addition of resin volume, while the ratio of Cr(VI) remained over 99.61% and the maximum reached to 99.84%. Since with a fixed initial metal concentration, the fraction of metal ions

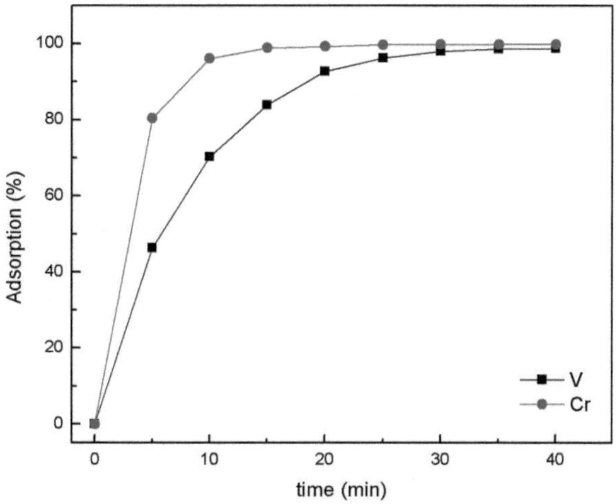

Fig. 4 Effect of contact time on adsorbing of V(V) and Cr(VI)

removed from aqueous phase increases as the sorbent dosage increased, adding 12 mL resin to 150 mL mixed solution that concentrated on 100 mg L^{-1} at pH 3.0 was chosen as optimum.

Effect of Contact Time

Adsorption time is an important index to study adsorption ratio of V(V) and Cr(VI), and the result of this experiment shown in Fig. 4. In the early stage of adsorption (10 min), there not only were many active adsorption sites on the resins surface, but also the big concentration difference of V(V) and Cr(VI) anions between resin surface and solution. This leads to the enhancement of mass transfer, so the adsorption ratio was relatively fast. As time went on, active sites were occupied by V(V) and Cr(VI) anions. The number of active sites decreased as well as the concentration gradient descent, which resulted in the decrease of adsorbing speed. After 30 min, the adsorption rate decreased then reached a balance after 40 min, the adsorption ratio of V(V) and Cr(VI) achieved 98.76 and 99.80% respectively.

Conclusions

Experiments confirmed that by ion-exchange resin Dex-V, vanadium and chromium can be efficiently extracted from Cr(VI)-containing vanadate(V(V)) solution. The V(V) and Cr(VI) anions can be effectively absorbed onto resin surface, leading to the removal of V(V) and Cr(VI) from aqueous solution. Under the optimum condition, including the initial pH value at 3.0, resin volume of 12 ml and the contact time of 40 min at fixed initial concentration of 100 mg L^{-1}, the maximum adsorption ratio of V(V) and Cr(VI) anions by the Dex-V resin were 98.83 and 99.87% separately.

Acknowledgements National Natural Science Foundation of China (No. 51474041 and No. 51674051), National Basic Research Program of China (No. 2013CB632604), National Key Technology Research and Development Program of China (2015BAB19B02) and Fundamental Research Funds for the Central Universities of China (No. 106112015CDJZR465505) financially supported this work.

References

1. Moskalyk RR et al (2003) Processing of vanadium: a review. Miner Eng 16(9):793–805
2. Lee YS et al (2012) Role of alloying elements in vanadium-based binary alloy membrane for hydrogen separation. J Membr Sci 423:332–341
3. Li HY et al (2015) Asynchronous extraction of vanadium and chromium from vanadium slag by stepwise sodium roasting–water leaching. Hydrometallurgy 156:124–135
4. Jardine PM et al (1999) Fate and transport of hexavalent chromium in undisturbed heterogeneous soil. Environ Sci Technol 33:2939–2944
5. Li XS et al (2012) Extraction of vanadium from high calcium vanadium slag using direct roasting and soda leaching. Int J Mineral Metall Mater 19(7):595–601
6. Sadykhov GB (2008) Oxidation of titanium–vanadium slags with the participation of Na_2O and its effect on the behavior of vanadium. Russ Metall (Metally) 6:449–458
7. Li HY et al (2016) Selective leaching of vanadium in calcification-roasted vanadium slag by ammonium carbonate. Hydrometallurgy 160:18–25
8. Li HY et al (2016) Removal of V(V) from aqueous Cr(VI)-bearing solution using anion exchange resin: equilibrium and kinetics in batch studies. Hydrometallurgy 165:381–389
9. Park SJ et al (2001) Adsorption behaviors of chromium(III) and (VI) on electroless Cu-plated activated carbon fibers. J Colloid Interface Sci 243:316–320
10. Ning PG et al (2014) Selective extraction and deep separation of V(V) and Cr(VI) in the leaching solution of chromium-bearing vanadium slag with primary amine LK-N21. Sep Purif Technol 137:109–115
11. Song YC et al (1998) Influence of electron donor and toxic materials on the activity of sulfate reducing bacteria for the treatment of electroplating wastewater. Water Sci Technol 38: 187–194
12. Yeh CF et al (2004) Speciation of V, Cr and Fe by capillary electrophoresis-bandpass reaction cell inductively coupled plasma mass spectrometry. Chromatography A 1029(1):255–261
13. Liao XP et al (2008) Adsorption of metal anions of vanadium and chromium on Zr(IV)-impregnated collagen fiber. Adsorption 14(1):55–64
14. Godea F et al (2005) Removal of Cr(VI) from aqueous solution by two Lewatit-anion exchange resins. J Hazard Mater B 119:175–182
15. Rengaraj S et al (2001) Removal of chromium from water and wastewater by ion exchange resins. J Hazard Mater 87(1):273–287
16. Fan YY et al (2013) Separation and recovery of chromium and vanadium from vanadium-containing chromate solution by ion exchange. Hydrometallurgy 136:31–35

Part V
Poster Session

Effects of Pre-oxidation on the Kinetics of Iron Leaching from Ilmenite in Hydrochloric Acid Solution

Junyi Xiang, Qingyun Huang, Wei Lv, Xuewei Lv and Chenguang Bai

Abstract The kinetics of iron leaching from ilmenite in hydrochloric acid solutions were investigated. The effects of pre-oxidation roasting, leaching temperature, and holding time on iron extraction rate were determined. The results obtained show that the extraction of iron is significantly enhanced by pre-oxidation roasting and increasing leaching temperature from 75 to 108 °C. The results of the kinetic analysis of the leaching data indicate that the reaction is controlled by diffusion through the product layer. The apparent activation energy for the extraction of iron has been calculated to be 62.5 and 55.1 kJ/mol for untreated and oxidized ilmenite, respectively.

Keywords Ilmenite · Leaching · Hydrochloric acid · Kinetics Pre-oxidation

Introduction

Ilmenite is one of the primary sources of titanium dioxide which is the most important intermediate in the manufacture of pigments, paints, welding rod, ceramics, and in other miscellaneous applications such as papers and inks [1]. Chloride process is an economical and environment friendly way to produce titanium dioxide [2]. However, it needs high grades raw materials, such as mineral rutile, upgraded titanium slag (UGS), or synthetic rutile [3, 4]. Acid leaching is a common way to upgrade ilmenite into synthetic rutile by removing the acid soluble impurities, such as iron, magnesium, and calcium. As a leachant, hydrochloric acid

J. Xiang · W. Lv · X. Lv (✉) · C. Bai
School of Materials Science and Engineering, Chongqing University, Chongqing 400044, China
e-mail: lvxuewei@163.com

Q. Huang
School of Metallurgical and Materials Engineering, Chongqing University of Science and Technology, Chongqing 401331, China

H. Kim et al. (eds.), *Rare Metal Technology 2018*, The Minerals, Metals & Materials Series, https://doi.org/10.1007/978-3-319-72350-1_29

is preferred to other acids because it allows comparatively easier recovery of the acid from its waste solution [5].

Most leaching studies of ilmenite by hydrochloric acid were performed in order to obtain the best parameters and reaction mechanisms for upgrading ilmenite to synthetic rutile. Hydrochloric acid concentration and leaching temperature have significant influences on the dissolution of iron and titanium in the ilmenite. Leaching efficiency improved with the increase of hydrochloric acid concentration and temperature [6–8]. The initial acid to solid ratio also influence the dissolution of iron and titanium. And in general, the higher acid to solid ratio, the higher dissolution rate of iron and titanium. However, high acid concentration and acid to solid ratio will delay the precipitation of $TiOCl_2$ [9]. The dissolution rates are inversely proportional to the average initial diameter of the particle [5, 10].

The pretreatment of the ilmenite also has significantly influence on the upgrading of ilmenite. It was reported that pre-oxidation roasting resulting a rough and porous surface of the ilmenite, which increased the specific area [11, 12]. In our previous study, both reduction and oxidation roasting pretreatment can enhance the dissolution of iron and facilitate the hydrolysis of the dissolved titanium. However, oxidation roasting pretreatment can significantly reduce the formation of fine products which cannot fulfil the requirement of particle size for the fluidizing chlorination process [13, 14]. The effect of peroxidation roasting on the dissolution behavior of iron in hydrochloric acid was investigated this paper. In addition, the kinetics of the leaching process have also been investigated and discussed.

Experimental

Materials

Ilmenite for this study was taken from the milling area in Yunnan province, southwest China. Table 1 presents the main chemical compositions of the received sample. All other reagents used in this experiments were of analytical grade and of purity no less than 99%.

Mineralogical analyses of the samples were done using X-ray diffraction (XRD) analyzer (Rigaku D/max 2500 PC, Japan). As shown in Fig. 1a, the major minerals present in the received sample is ilmenite ($FeTiO_3$).

Table 1 Chemical analysis of the samples, wt%

Samples	TiO_2	Fe_2O_3	FeO	CaO	MgO	SiO_2
Untreated	45.92	17.15	31.67	0.15	1.21	2.84
Oxidized	45.55	41.44	7.83	0.15	1.20	2.83

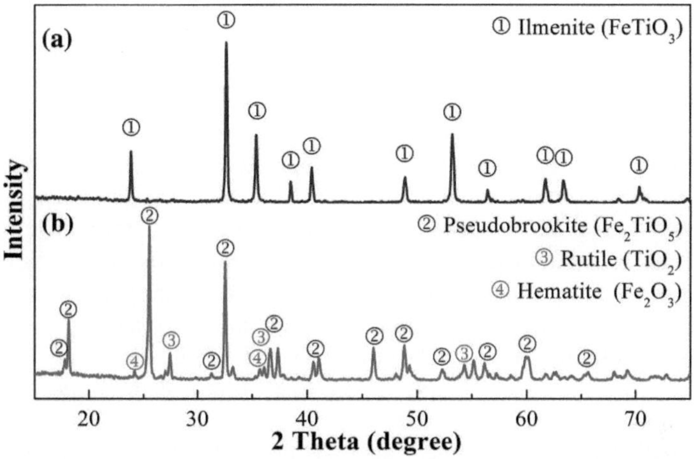

Fig. 1 XRD patterns of the samples: **a** untreated ilmenite; **b** oxidized ilmenite

Oxidation Roasting

Oxidation roasting was performed in a laboratory open muffle furnace. The ilmenite was first placed in an alumina crucible and then introduced into the furnace once the furnace reached a certain temperature. The roasting was conducted at temperature of 900 °C and lasting 1 h. After roasting, the crucible was removed from the furnace and cooled to room temperature in the air. After oxidation roasting, the content of ferrous iron dramatically decreased from 31.67 to 7.83% while the content of ferric iron increased from 17.15 to 41.44% (as shown in Table 1). As shown in Fig. 1b, the major minerals present in the oxidized ilmenite are pseudobrookite (Fe_2TiO_5), rutile (TiO_2) and hematite (Fe_2O_3).

Acid Leaching

Both of the untreated and roasted ilmenites were crushed to a particle size smaller than 0.125 mm. The solid samples were leached in hydrochloric acid at certain temperature, L/S ratio and time. The leaching experiments were performed at an atmospheric pressure in a four-necked round-bottomed flask. The details of apparatus can be found elsewhere [13].

For each leaching test, 50 g of solid sample was mixed with 500 mL of hydrochloric acid (26 wt% HCl). During leaching, 5 mL of suspension was siphoned from the flask at regular time intervals, centrifuging, and chemically analyzing the supernatant solution.

Results and Discussion

Iron Leaching

Figure 2 shows the effect of leaching temperature and time on the leaching efficiency of iron from untreated and oxidized ilmenites.

As shown in Fig. 2a, leaching temperature and holding time have significant effects on the dissolution of iron and maximum extraction was achieved at the highest temperature of 108 °C and longest time of 420 min. It's also can be found that the leaching efficiency of iron from untreated ilmenite gradually increased with the extending holding time to 420 min at 75 and 90 °C. However, the leaching efficiency of iron increased sharply at the first 60 min then increased slowly at longer leaching time at 108 °C.

The oxidation roasting treatment also has a significant influence on the dissolution of iron. As shown in Fig. 2b, the leaching efficiency of iron for oxidized ilmenite are much higher than untreated ilmenite. The maximum leaching efficiencies of iron for untreated ilmenite leached at temperature of 75, 90 and 108 °C are 31.7, 42.1, and 71.8%, respectively. While the maximum leaching efficiencies for oxidized ilmenite leached at temperature of 75, 90 and 108 °C are 50.6, 74.6, and 93.4%, respectively.

Kinetic Analysis

In order to identify the reaction mechanism of the leaching process, the experimental results obtained in this study were examined utilizing the well-known shrinking core model under the assumption that the particles are homogeneous and

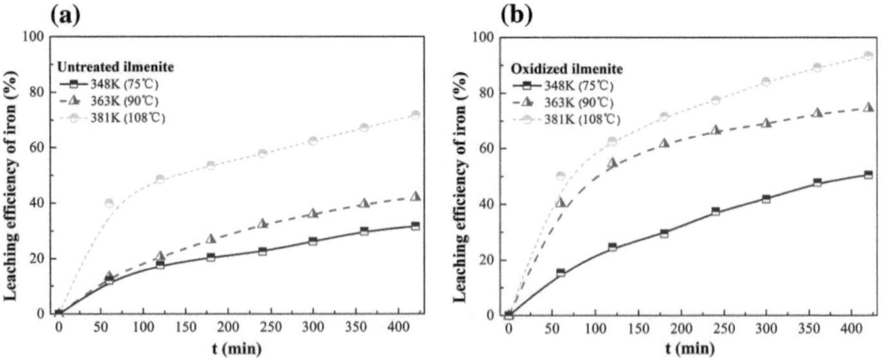

Fig. 2 Effect of leaching temperature on leaching efficiency of iron from samples: **a** untreated ilmenite; **b** oxidized ilmenite. (26% HCl, leaching time 420 min, L/S = 10:1, particle size smaller than 0.125 mm)

spherical. Two established kinetic models were used, as expressed by the following equations:

$$1-(1-a)^{\frac{1}{3}}=\frac{Mk_cC_At}{dr}=k_1t \quad (1)$$

$$1+2(1-a)-3(1-a)^{\frac{2}{3}}=\frac{6uMDC_At}{dr^2}=k_3t \quad (2)$$

Where a is fraction of iron dissolved at time t (min); M is the molecular weight of the solid reactant (kg/mol); k_c is the chemical rate constant (cm/min); C_A is the concentration of acid (mol/m^3); r is the initial radius of the solid reactant (r); k_1, and k_2 (m/min) are the rate constants of surface chemical reactions and internal diffusion, respectively.

Using the date shown in Fig. 2, the right hand side of Eqs. 1 and 2 are plotted versus time for untreated and oxidized ilmenites, respectively. The fitting results are shown in Figs. 3 and 4. Larger regression coefficients were obtained for the type of product layer diffusion control. Low regression coefficients were obtained for the type of chemical reaction control. Therefore, the dissolution of iron was controlled by the diffusion through product layer.

From the results in Fig. 4, the rate constants for different reaction temperatures were calculated and Arrhenius plot were constructed in Fig. 5. According to Arrhenius equation, apparent activation energy E_a for the leaching of iron were calculated to be 62.5 and 55.1 kJ/mol for untreated and oxidized ilmenite, respectively. Thus, it is confirmed that the apparent activation energy decreased after the oxidation roasting pretreatment. It's also indicate that the pre-oxidation treatment enhances the reactivity of the ilmenite, and facilitates the leaching of iron.

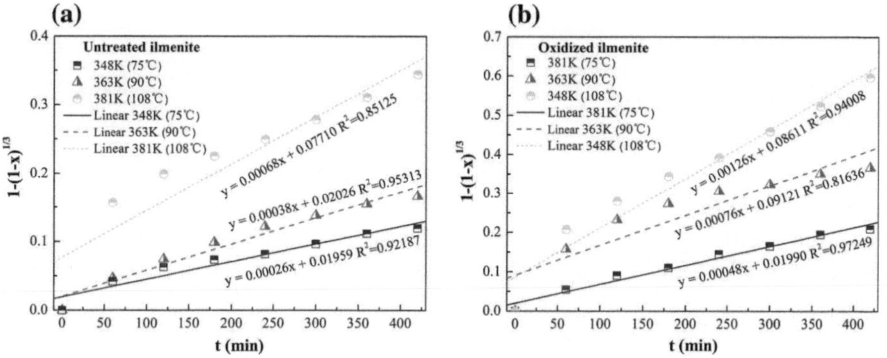

Fig. 3 Plots of 1 − (1− a)$^{1/3}$ versus time at various reaction temperature: **a** untreated ilmenite; **b** oxidized ilmenite

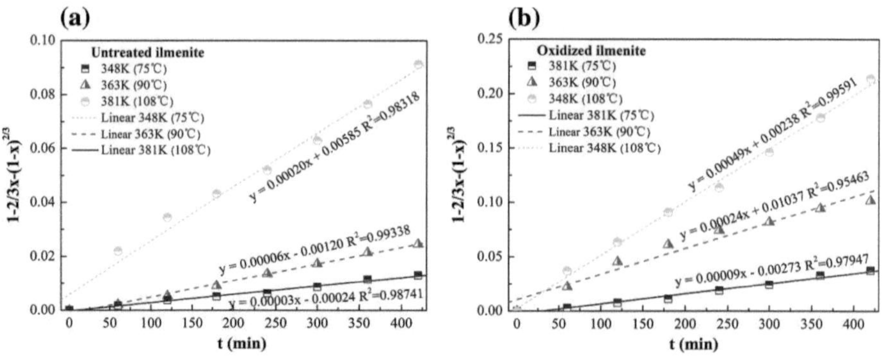

Fig. 4 Plots of $1 + 2(1 - a) - 3(1 - a)^{2/3}$ versus time at various reaction temperature: **a** untreated ilmenite; **b** oxidized ilmenite

Fig. 5 Arrhenius plot for leaching of untreated and oxidized ilmenite

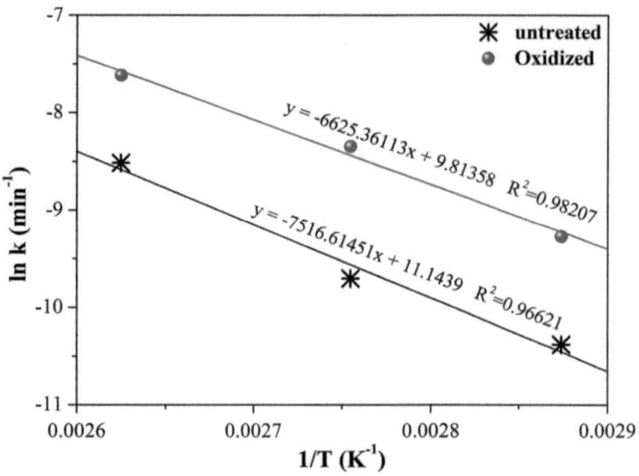

Conclusions

In this study, peroxidation roasting treatment was found to significantly improve the leaching efficiency of iron from ilmenite. The maximum leaching efficiency of iron increased from 71.8 to 93.4% with oxidation roasting pretreatment. Leaching temperature also found to obviously affect the leaching of iron. The leaching efficiency of iron nearly doubled when raising the temperature from 75 to 108 °C. The kinetic analysis indicated that the leaching of iron from both of untreated and oxidized ilmenites were controlled by the diffusion through the product layer. Furthermore, pre-oxidation roasting decreased the apparent energy from 62.5 and 55.1 kJ/mol.

Acknowledgements This work was supported by the Natural Science Foundation of China [grant numbers 51404047 and 51234010] and the Basic and Frontier Research Program of Chongqing [grant numbers cstc2014jcyjA50011]

References

1. Braun JH, Baidins A, Marganski RE (1992) TiO_2 pigment technology: a review. Prog Org Coat 20(2):105–138
2. Mackey TS (1974) Acid leaching of ilmenite into synthetic rutile. Ind Eng Chem Product Res Dev 13(1):9–18
3. Zhang W, Zhu Z, Cheng CY (2011) A literature review of titanium metallurgical processes. Hydrometallurgy 108(3):177–188
4. Mahmoud MHH, Afifi AAI, Ibrahim IA (2004) Reductive leaching of ilmenite ore in hydrochloric acid for preparation of synthetic rutile. Hydrometallurgy 73(1):99–109
5. Olanipekun E (1999) A kinetic study of the leaching of a Nigerian ilmenite ore by hydrochloric acid. Hydrometallurgy 53(1):1–10
6. Sarker MK, Rashid A, Kurny ASW (2006) Kinetics of leaching of oxidized and reduced ilmenite in dilute hydrochloric acid solutions. Int J Miner Process 80(2):223–228
7. Tsuchida H, Narita E, Takeuchi H et al (1982) Manufacture of high pure titanium (IV) oxide by the chloride process: kinetic study on leaching of ilmenite ore in concentrated hydrochloric acid solution. Bull Chem Soc Jpn 55(6):1934–1938
8. Vásquez R, Molina A (2012) Effects of thermal preoxidation on reductive leaching of ilmenite. Miner Eng 39:99–105
9. Van Dyk JP, Vegter NM, Pistorius PC (2002) Kinetics of ilmenite dissolution in hydrochloric acid. Hydrometallurgy 65(1):31–36
10. Li C, Liang B, Wang HY (2008) Preparation of synthetic rutile by hydrochloric acid leaching of mechanically activated Panzhihua ilmenite. Hydrometallurgy 91(1):121–129
11. Lv W, Lv XW, Zhang YY et al (2017) Isothermal oxidation kinetics of ilmenite concentrate powder from Panzhihua in air. Powder Technol. https://doi.org/10.1016/j.powtec.2017.07.058
12. Zhang YY, Lv W, Lv XW et al (2017) Oxidation kinetics of ilmenite concentrate by non-isothermal thermogravimetric analysis. J Iron Steel Res 24:678–684
13. Liu SL, Xiang JY (2016) The effects of thermal pretreatment on leaching of yunnan ilmenite with hydrochloric acid. Metall Mater Trans B 47(2):1334–1339
14. Xiang JY, Liu SL, Lv XW et al (2016) Preparation of rutile from ilmenite concentrate through pressure leaching with hydrochloric acid. Metall Mater Trans B 48(2):1–9

Extraction of Vanadium from Vanadium-Containing APV-Precipitated Wastewater by W/O Microemulsion System

Yun Guo, Hong-Yi Li, Min-Min Lin and Bing Xie

Abstract The extraction of vanadium from vanadium-containing APV-precipitated wastewater by W/O microemulsion system was studied. In the N263/n-heptane/isoamyl alcohol/NaCl microemulsion system, N263 played double functions of cationic surfactant and extractant to form microemulsion in n-heptane, and isoamyl alcohol was added in the microemulsion as a cosurfactant. The extraction of vanadium by W/O microemulsion systems effectively accelerated the extraction and improved the extractatability because of the enormous increase in the micro-interfacial surface area. Due to ion association with extractant, vanadium was extracted into the microemulsion phase. The influence of different parameters on the extraction ratio (E%) were investigated, including the volume ratio of aqueous phase to microemulsion, surfactant concentration, pH value of the feed solutions, cosurfactant concentration, extraction time as well as temperature. In optimum conditions, the extraction ratio (E%) can reach 97.2%.

Keywords Microemulsion · Extraction · N263 · Vanadium

In modern times, vanadium is mainly used in iron and steel industry, battery industry, pharmaceutical industry, chemical industry and other fields. Among the above, the application of vanadium in iron and steel industry is one of the most mature and widespread [1]. Titanomagnetite resources in China is rich and they have better development prospects [2]. Vanadium slag obtained from smelted Titanomagnetite is the main raw material to produce vanadium product. The vanadium is converted into water-soluble sodium vanadate and water-insoluble calcium vanadate in roasting process [3]. Then the leaching treatment is generally carried out using the corresponding leaching solution. The obtained vanadium-containing leaching liquor is treated to get ammonium polyvanadate precipitations

Y. Guo · H. -Y. Li (✉) · M. -M. Lin · B. Xie
College of Materials Science and Engineering, Chongqing University, Chongqing 400044, China
e-mail: hongyi.li@cqu.edu.cn

H. Kim et al. (eds.), *Rare Metal Technology 2018*, The Minerals, Metals & Materials Series, https://doi.org/10.1007/978-3-319-72350-1_30

(APV) and finally the product of V_2O_5. Meanwhile, the vanadium-containing APV-precipitated wastewater after vanadate precipitation was also obtained, from which vanadium must be extracted for resource utilization and environmental safety considerations [4].

The process of extracting vanadium from vanadium-containing APV-precipitated wastewater by microemulsion was studied in this paper. Microemulsions are thermodynamically stable, homogeneous and optically isotropic solutions [5]. There are four types of microemulsion systems called Winsor I, Winsor II, Winsor III, and Winsor IV [6]. Microemulsions are made from water, a surfactant, an organic solvent and sometimes an alcohol as a cosurfactant, which have been widely used in extraction of metallic ions, such as europium, [7] germanium, [8] gold, [9] cobalt and other metals [10–12]. In a Winsor II system, the tremendous rise of the micro-interfacial surface area in the microemulsion phase and the participation of the microemulsion globules in transporting metal ions from the aqueous phase to the organic phase are the main motive factors to accelerate extraction of metal ions [13–15]. Viewed from the driving force of microemulsion extraction, the ionic W/O microemulsion is mainly utilized to extracte metal ions by the electrostatic attraction between the headgroups of surfactant and metal ions. For nonionic microemulsion, extractants needs to be added, which react with metal ions to form complex in the oil phase [16–18]. In this paper, N263 played double functions of cationic surfactant and extractant.

This article aimed to apply W/O microemulsion to the treatment of vanadium-containing APV-precipitated wastewater. Extraction of vanadium from the simulated vanadium-containing wastewater was studied with W/O microemulsion. The influence of different parameters on the extraction ratio (E%) were investigated, including the volume ratio of aqueous phase to microemulsion, surfactant concentration, pH value of the feed solutions, cosurfactant concentration, extraction time as well as temperature. In this paper, we provided both theoretical basis and corresponding extraction conditions for extraction of vanadium from vanadium-containing wastewater by W/O microemulsion.

Experimental

Instruments and Reagents

Inductively Coupled Plasma Optical Emission Spectrometer (ICP-OES; Optima 800, PerkinElmer), Electronic Balance (FA2004 N, Mettler Toledo), Constant Temperature Heating Magnetic Stirrer (DF-101S, GongYi Electronic Equipment Limited Liability Company), Aquelix5 Pure Water Meter (ZTLC00031, Millipore Corporation).

Sodium orthovanadate dodecahydrate (CP, Sinopharm Chemical Reagent Co., Ltd), sodium hydroxide (AR, Chengdu Kelong Chemical Reagent Factory), hydrogen chloride (AR, Chongqing Chuandong Chemical Co., Ltd), sodium chloride (AR, Chengdu Kelong Chemical Reagent Factory), tricaprylylmethylammonium chloride(N263, $R_3CH_3N^+Cl^-$, $R=C_8H_{17}$ (Fig 1); AR, Chengdu Micxy Chemical Co., Ltd), n-heptane (AR, Chengdu Kelong Chemical Reagent Factory), isoamyl alcohol (AR, Chengdu Kelong Chemical Reagent Factory).

All aqueous solutions were prepared using distilled water.

Fig. 1 Structure of N263

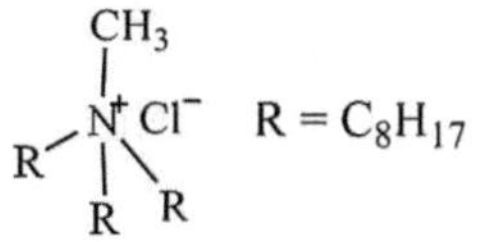

Preparation of Microemulsion and Feed Solutions

The organic phase was prepared by injecting N263, 0.5 mol/L sodium chloride and 0.5 mol/L sodium hydroxide solutions, isoamyl alcohol in n-heptane. The resulting mixtures form microemulsions spontaneously whenever the composition is adequate.

The feed solution was prepared by adding $Na_3VO_4{\cdot}12H_2O$ to deionized water. The pH of the feed solution was adjusted by HCl and NaOH. In the following studies, the initial vanadium concentration is 1 g/L.

Experimental Methods

In the following studies, unless stated specially, the temperature was generally maintained at 300 ± 1 K, and $R = 1$. The volume ratio of aqueous phase to microemulsion was expressed as R. The extraction ratio was expressed as E%, which was calculated using Eq. 1:

$$E\% = \frac{c_0 - c_t}{c_0} \times 100\% \qquad (1)$$

where c_0 and c_t were the total amount of vanadium in the feed solutions at t = 0 and t = t, respectively. Vanadium extraction ratio were measured after stirring the two phases in Constant Temperature Heating Magnetic Stirrer for 5 min. The mixture was kept still until the two phase separated completely. The vanadium concentration in aqueous solution were measured by Inductively Coupled Plasma Optical Emission Spectrometer and the vanadium concentration in organic solution were calculated by mass balance.

Extraction Mechanism

The existing forms of vanadium in the solution is related to the pH and vanadium concentration of the solution. This article examined the weak acid feed solution. If the feed solution was too much acidity, vanadium ions are precipitated from the feed solution which was detrimental to extract. Vanadium can form a variety of companions with different numbers of oxygen atoms; for example, $V_{10}O_{28}{}^{6-}$, $HV_{10}O_{28}{}^{5-}$, $V_4O_{12}{}^{4-}$, $V_3O_9{}^{3-}$, $HV_2O_7{}^{3-}$, $V_2O_7{}^{4-}$, $VO_3{}^{-}$, $HVO_4{}^{2-}$ and $VO_4{}^{3-}$ [19]. The pH of the feed solution was controlled in the 6–7. The reaction between N263 ($(C_8H_{17})_3CH_3NCl$) and vanadate anions, can be expressed by the following equations:

$$V_{10}O_{28}^{6-} + 6(C_8H_{17})_3CH_3NCl \rightarrow \left((C_8H_{17})_3CH_3N\right)_6V_{10}O_{28} + 6Cl^- \quad (2)$$

The vanadate anions combined with N263 on the external aqueous/organic interface to form the complex, then the association diffuses to the inner organic/aqueous interface.

Results and Discussion

Effect of the pH of the Feed Solution on Vanadium Extraction

With the purpose of investigating the influence of the pH of the feed solution on the vanadium extraction, experiments were conducted in the pH range between 4 and 11. The obtained results were shown in Fig. 2. The result indicates that with the

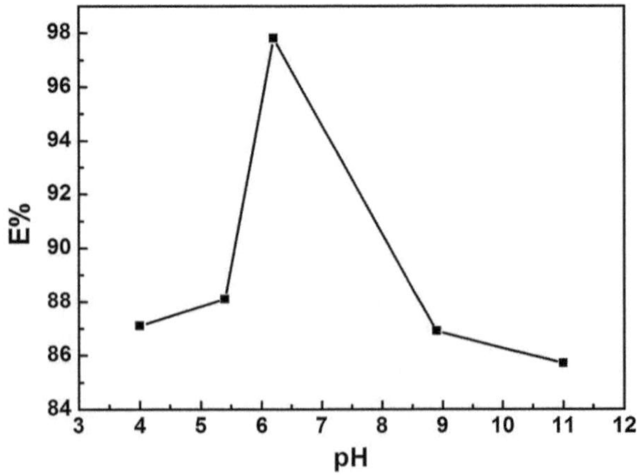

Fig. 2 Effect of the pH of the feed solution on the extraction ratio (E%) of vanadium. Oil phase: *w*(N263) = 10%; *w* (isoamyl alcohol) = 20%; inner aqueous is 0.5 mol/L NaCl and 0.5 mol/L NaOH solution. Feed solution: vanadium 1 g/L

increase in pH value, the vanadium extraction ratio presented a first increasing and then decreasing trend. This is mainly due to that vanadium in solutions with various pH values have different existing forms. The combination of vanadate anions and N263 was thus different. When pH was 6.2, the extraction ratio reached maximum. Correspondingly, the vanadate anion complexed with N263 to form $((C_8H_{17})_3CH_3N)_6V_{10}O_{28}$. On the basis of these results, an external phase pH value of 6.2 was chosen for the following experiments.

Effect of the Extraction Time on Vanadium Extraction

The influence of contact time of microemulsion and feed solution on the vanadium extraction ratio was investigated so as to obtain the optimum time required to attain equilibrium. The extraction time was varied from 2 to 6 min. It is found in Fig. 3 that the extraction ratio of vanadium can reach up to 96% within 3 min and the extraction equilibrium was achieved rapidly within 5 min. Thus, 5 min could be chosen as an appropriate time in the following experiment.

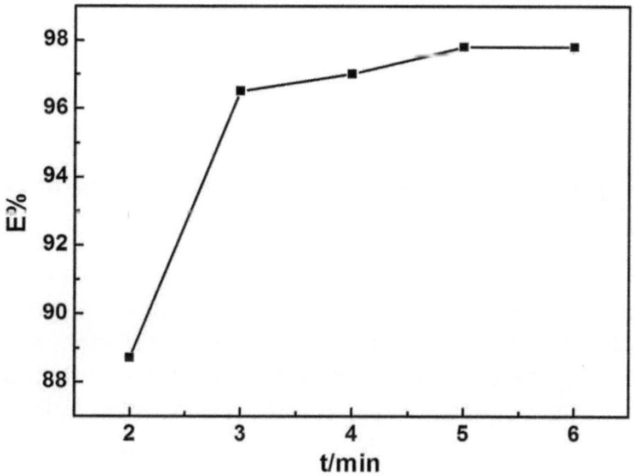

Fig. 3 Effect of the extraction time on the extraction ratio (E%) of vanadium. Oil phase: *w* (N263) = 10%; *w*(isoamyl alcohol) = 20%; inner aqueous is 0.5 mol/L NaCl and 0.5 mol/L NaOH solution. Feed solution: vanadium 1 g/L; pH 6.2

Effect of the N263 Concentration on Vanadium Extraction

The effect of N263 concentration on the extraction ratio of vanadium is shown in Fig. 4. The amount of microemulsion produced increased with increasing surfactant and extractant (N263) concentrations which could accelerate the extraction and improve the extractability of vanadium. It was possible that the absorption quantity

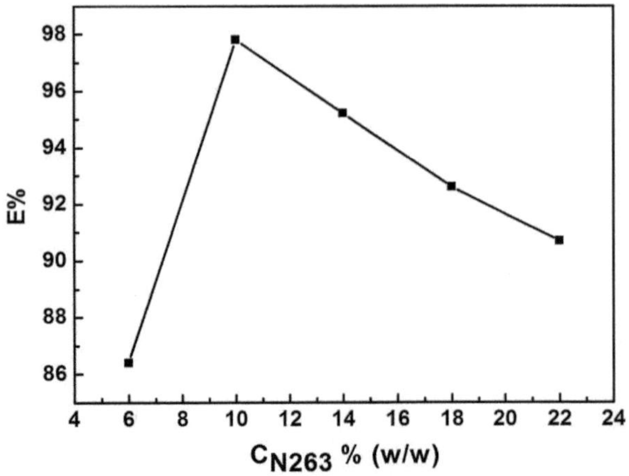

Fig. 4 Effect of N263 concentration on the extraction ratio (E%) of vanadium. Oil phase: *w* (N263) = 6–22%; *w*(isoamyl alcohol) = 20%; inner aqueous is 0.5 mol/L NaCl and 0.5 mol/L NaOH solution. Feed solution: vanadium 1 g/L; pH 6.2

of surfactant increases with the increasing amount of micelles formed by the surfactant and the free surfactant molecules. As the surfactant concentration increased, the extraction equilibrium moved forward and the extraction ratio increased. When the concentration of surfactant was too high, the viscosity of microemulsion increased, which was not conducive to the extraction process. As the surfactant concentration was 10%, the microemulsion system exhibited a good extraction performance.

Effect of Isoamyl Alcohol Concentration on Vanadium Extraction

Most surfactants need cosurfactants to form microemulsion. The cosurfactant is a nonionic molecule that combines with surfactant to neutralize the repulsive effect among the polar heads of surfactant. In the system used for the vanadium extraction, an amount of isoamyl alcohol was added in the organic, otherwise the system was not stable enough to extract metal ions. The effect of the concentration of isoamyl alcohol on the vanadium extraction ratio (E%) was investigated. As shown in Fig. 5, with the increase in isoamyl alcohol concentration, the extraction ratio by the microemulsion first increased to a maximum and then decreased. There was a maximum of E% when *w*% (isoamyl alcohol) was around 20%. Cosurfactants frequently have short or medium chains which can screen the repulsive forces between the charged surfactant head groups, and thus enhance the aggregation of the surfactant molecules. But too much isoamyl alcohol will reduce extraction ratio of vanadium. This result may be explained by the combination between isoamyl alcohol and N263, which made some of N263 losing their activity.

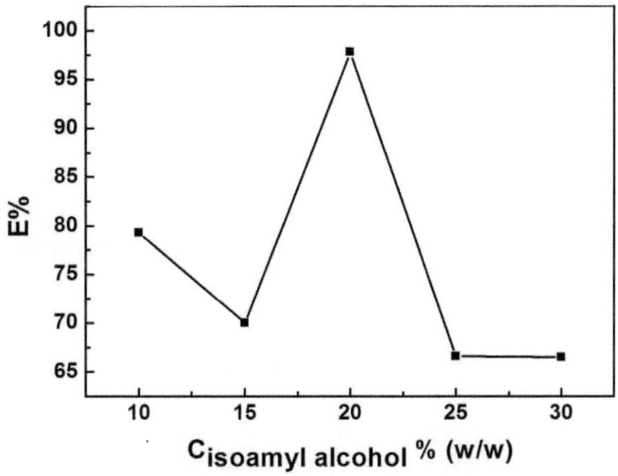

Fig. 5 Effect of isoamyl alcohol concentration on the extraction ratio (E%) of vanadium. Oil phase: *w* (N263) = 10%; *w*(isoamyl alcohol) = 10–30%; inner aqueous is 0.5 mol/L NaCl and 0.5 mol/L NaOH solution. Feed solution: vanadium 1 g/L; pH 6.2

Effect of Inner Aqueous on Vanadium Extraction

The effect of salt and alkali on the vanadium extraction ratio (E%) were investigated. Inner aqueous was produced by sodium chloride and sodium hydroxide. As seen in Fig. 6, the results indicated that with the increase in NaCl and NaOH concentration, the vanadium extraction ratio presents a first increasing and then decreasing trend. There was a maximum of E% when inner aqueous was 0.5 mol/L NaCl and 0.5 mol/L NaOH solution. This was due to that chlorine ion and hydroxide have positive effects on vanadate anion diffusion. The inner aqueous of chloride ion and hydroxide in feed solution formed a concentration difference. Chlorine ion and hydroxide migrated along the concentration gradient to provide a driving force for the migration of vanadate anion. However, the water solubilizing

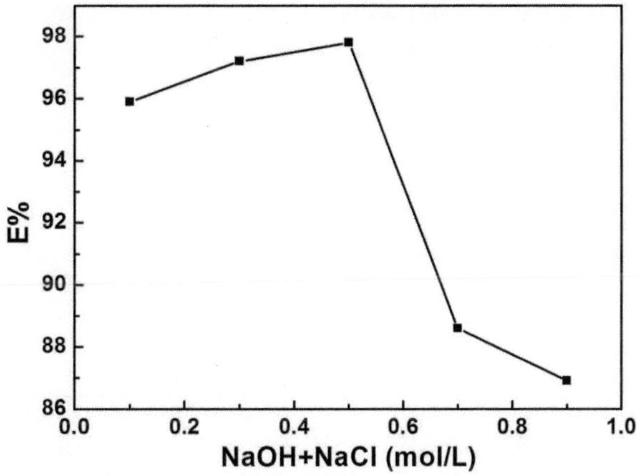

Fig. 6 Effect of salt and alkali on the extraction ratio (E%) of vanadium. Oil phase: *w*(N263) = 10%; *w*(isoamyl alcohol) = 20%; Feed solution: vanadium 1 g/L; pH 6.2

capacity of microemulsion decreased gradually with the increase in NaCl and NaOH concentrations. Due to the common ion effect, the extraction ratio of vanadium was reduced when NaCl and NaOH concentrations were higher than 0.5 mol/L.

Effect of Temperature on Vanadium Extraction

From the viewpoint of thermodynamics, extraction of vanadium can be regarded as a transfer process of vanadium from the aqueous phase to the microemulsion phase. The temperature was an important influencing factor in the extraction procedure. In the Fig. 7, it could be noted that the partition coefficient of the vanadium between the organic and aqueous phases (*D*) decreased with the temperature increasing from 300 to 320 K. This phenomenon can be explained that the extraction is exothermic and lower temperature favored the metal extraction.

The enthalpic change ($\Delta_r H^o_{\mathrm{m}}$) was calculated from the slope of the linear equation between ln*D* and 1/*T* (Eq. 3) [20].

$$\left[\frac{\partial \ln D}{\partial\left(\frac{1}{T}\right)}\right]_P = \frac{-\Delta_r H^o_{\mathrm{m}}}{R} \tag{3}$$

where *D* is the partition coefficient of vanadium between the organic and aqueous phases; *T* (K) is the temperature; R is the gas constant. The value of $\Delta_r H^o_{\mathrm{m}}$ was calculated to be −58.5 kJ/mol. It can be known that the extraction process is a spontaneously exothermic process.

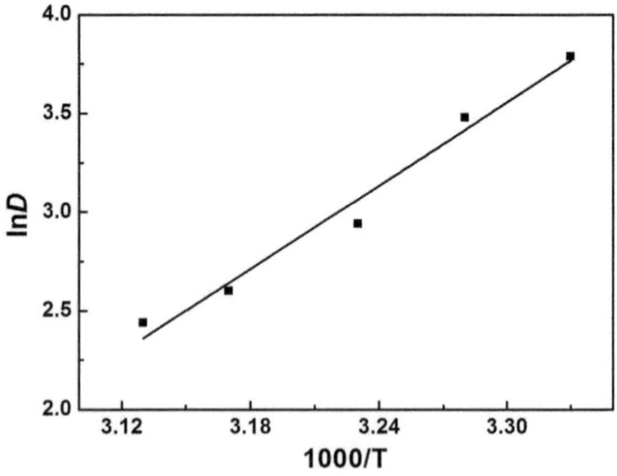

Fig. 7 Effect of temperature to the partition coefficients of the vanadium between the organic and aqueous phases. Oil phase: *w*(N263) = 10%; *w* (isoamyl alcohol) = 20%; inner aqueous is 0.5 mol/L NaCl and 0.5 mol/L NaOH solution. Feed solution: vanadium 1 g/L; pH 6.2

Effect of the R on Vanadium Extraction

The effect of the volume ratio of aqueous to microemulsion (*R*) on the vanadium extraction ratio (E%) was investigated. Figure 8 shows that the extraction ratio (E%) decreased slowly with the increase in *R* ranging from 1 to 3, and decreased drastically when *R* increasing from 3 to 5. With economic factors taken into account, *R* = 3 is the most suitable condition for vanadium extraction.

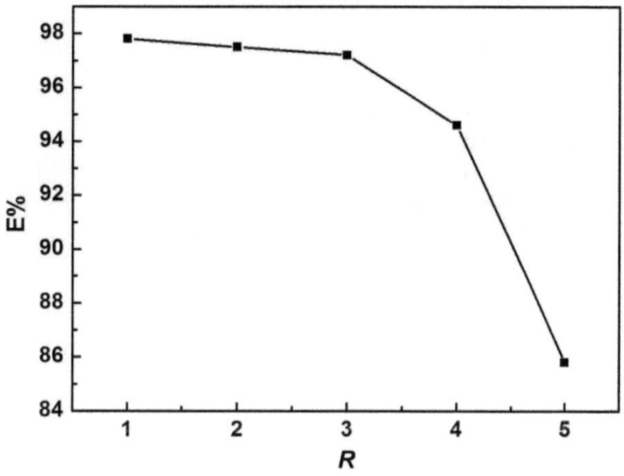

Fig. 8 Effect of *R* on the extraction ratio (E%) of vanadium. Oil phase: *w* (N263) = 10%; *w*(isoamyl alcohol) = 20%; inner aqueous is 0.5 mol/L NaCl and 0.5 mol/L NaOH solution. Feed solution: vanadium 1 g/L; pH 6.2

Conclusions

N263/isoamyl alcohol/n-heptane/NaCl was chosen for the formation of the microemulsion to extract vanadium from vanadium-containing APV-precipitated wastewater, which is highly efficient for vanadium extraction. The results obtained in the present study showed that the extraction equilibrium was immediately achieved by this W/O microemulsion and the extracted compound was probably $((C_8H_{17})_3CH_3N)_6V_{10}O_{28}$ at pH 6.2. The enthalpy of vanadium extraction into microemulsion were calculated to be −58.5 kJ/mol. Several factors which impact the vanadium extraction ratio (E%) have been investigated in this article. The extraction ratio (E%) reached above 97.2% in optimum condition of *R* = 3, *w*(N263) = 10%, *w*(isoamyl alcohol) = 20%, inner aqueous composed of 0.5 mol/L NaCl and 0.5 mol/L NaOH.

Acknowledgements The authors would like to acknowledge the financial supports by National Natural Science Foundation of China (No. 51474041 and No. 51674051), National Key Technology Research and Development Program of China (2015BAB19B02) and Fundamental Research Funds for the Central Universities of China (No. 106112015CDJZR465505).

References

1. Zhang YM et al (2011) The technology of extracting vanadium from stone coal in China: history, current status and future prospects. Hydrometallurgy 109(1–2):116–124
2. Moskalyk RR et al (2003) Processing of vanadium: a review. Miner Eng 16(9):793–805
3. Li HY et al (2015) Asynchronous extraction of vanadium and chromium from vanadium slag by stepwise sodium roasting–water leaching. Hydrometallurgy 156:124–135
4. Zhang JH et al (2015) Mechanism of vanadium slag roasting with calcium oxide. Int J Miner Process 138:20–29
5. Langevin D (1988) Microemulsions. Acc Chem Res 21(7):720–728
6. Schwuger MJ et al (1995) Microemulsions in technical processes. Chem Rev 95(4):849–864
7. Wang W et al (2012) Extraction of europium by sodium oleate/pentanol/heptane/NaCl microemulsion system. J Radioanal Nucl Chem 292(3):1093–1098
8. Liu F et al (2010) Extraction of germanium by the AOT microemulsion with N235 system. Ind Eng Chem Res 49(20):10005–10008
9. Lu WJ et al (2011) Extraction of gold(III) from hydrochloric acid solutions by CTAB/n-heptane/iso-amyl alcohol/Na_2SO_3 microemulsion. J Hazard Mater, 186(2–3):2166
10. Shang K et al (2012) Extraction of cobalt by the AOT microemulsion system. J Radioanal Nucl Chem 291(3):629–633
11. Gao S et al (2012) Solvent extraction of thorium(IV) using W/O microemulsion. Sci China Chem 55(9):1712–1718
12. Lou ZN et al (2016) Extraction of Re(VII) from hydrochloric acid medium by N263/TBP/n-heptane/NaCl microemulsion. Hydrometallurgy 165:329–335
13. Watarai H (1997) Microemulsions in separation sciences. J Chromatogr A 780(1–2):93–102
14. Fanun M (2012) Microemulsions as delivery systems. Curr Opin Colloid Interface Sci 17(5):306–313
15. Letts K et al (1976) Reactions in microemulsions. I. Metal ion incorporation by tetraphenylporphine. Chemischer Informationsdienst, 7(8)
16. Chevalier Y et al (1999) The structure of micelles and microemulsions. Rep Prog Phys 53(3):279
17. Chandra S et al (1990) Equilibrium extraction and concentration of multivalent metal ion solutions by using Winsor II microemulsions. Sep Sci Technol 25(6):711–727
18. Moulik SP et al (1998) Structure, dynamics and transport properties of microemulsions. Adv Coll Interface Sci 78(2):99–195
19. Greenwood NN et al (1998) Chemistry of the elements. Butterworth-Heinemann 10–15
20. Pei YC et al (2009) Ionic liquid-based aqueous two-phase extraction of selected proteins. Sep Purif Technol 64(3):288–295

Experimental Modeling of Nodulation in Copper Electrorefining

Yuya Nakai, Ken Adachi, Atsushi Kitada, Kazuhiro Fukami and Kuniaki Murase

Abstract Concerning nodulation in copper electrorefining, it has been reported that short circuits caused by nodulation have the largest impact on the current efficiency. Therefore, to improve the current efficiency, it is important to prevent nodulation. In this research, the nodulation growth was investigated by replicating nodulation in the laboratory. The experiments were conducted by cathodic electrodeposition on copper plates attached to copper sticks of various heights. The nodulation was analyzed in terms of growth rates versus the initial heights.

Keywords Copper · Electrorefining · Current efficiency · Nodulation

Introduction

Copper is a widely used base metal. The final process of copper production is electrorefining, where crude copper with the grade of 99.4–99.5% is used as anodes. By the electrolytic process crude copper dissolves as Cu^{2+} ions and electrodeposits at the cathode as electrolytic copper with purity of more than 99.99%.

In the electrorefining process, the ideal current efficiency for copper deposition ($Cu^{2+} + 2e^{-} = Cu$) is 100% since the deposition potential is nobler than hydrogen evolution. The actual current efficiency is, however, about 95% due to unexpected short-circuiting. The short-circuiting is caused by nodules formed on the cathodes in touch with adjacent anodes. Therefore, to improve the current efficiency, it is important to prevent nodulation.

Some impurities in crude copper do not dissolve and stick on the surface of the anode or settle in the bottom of the tank; they are called anode slime. Other impurities dissolve into the electrolyte, but in many case they are prevented from

---

Y. Nakai (✉) · K. Adachi · A. Kitada · K. Fukami · K. Murase
Department of Materials Science and Engineering, Kyoto University, Yoshida-Hommachi 36-1, Kyoto 606-8501, Japan
e-mail: nakai.yuya.35x@st.kyoto-u.ac.jp

H. Kim et al. (eds.), *Rare Metal Technology 2018*, The Minerals, Metals & Materials Series, https://doi.org/10.1007/978-3-319-72350-1_31

depositing on the cathode by appropriate electrode potential and electrolyte purification. Thus, we can obtain high-purity copper deposits and separate some impurities as anode slime. A small part of anode slime, however, floats in the electrolyte by convection and slime granules tens of microns in diameter can stick on the cathode surfaces. It is believed that this inclusion of slime particles causes nodulation, and indeed clusters of slime particles were found at the bases of nodules [1]. However, the mechanism of nodular growth of copper is not understood well.

Our ultimate goal is to reveal the mechanism of nodulation, and to elucidate how nodules are affected by anode slime and how they grow to cause short-circuiting.

Experimental

Effect of Impurities on Stuck to Cathode Nodulation

Particles of various impurities were purposefully embedded at the surface of copper cathodes, and the cathodes were used for electrodeposition. The electrolyte consisted of 0.76 mol L^{-1} $CuSO_4$, 1.94 mol L^{-1} H_2SO_4, 60 mg L^{-1} HCl, 5.0 mg L^{-1} thiourea, and 0.5 mg L^{-1} gelatin. The bath temperature was 65 °C. As the impurities for embedding, an Ag particle and polyester resin were selected as conductive and nonconductive substances, respectively, and the resulting copper cathodes were polished and buffed before use so that the impurities were evenly embedded into the copper surface. Electrodeposition of copper on the impurity-embedded copper cathodes was carried out at a current density of 35 mA cm^{-2} and the deposition behaviors were compared to that of a copper cathode without impurity.

Observation of Nodule Growth

In these experiments, we prepared copper bumps of various heights on flat and smooth copper plates, and the resulting plates were used for copper electrodeposition. The electrolyte used was the same as above. Each electrodeposition was carried out at a current density of 35 mA cm^{-2} for 40 h. The initial area of the copper plate was 5 cm × 6 cm, where the back side of the plate was covered with a fluorine resin adhesive tape. In this way, the relationship between the initial height of the bump and the growth behavior of the copper originating from the bump was examined by comparing the height of the bump before and after the electrodeposition. The bumps were made by two different methods:

Bumps Made by Hammering

Hammered using a pointed pick from the reverse side of the copper plate at a position 3 cm from the bottom. The diameter of the resulting bump was about 2 mm and the height was about 1 mm. These bumps were smooth.

Bumps Made by the Installation of Copper Sticks

On each copper plate, a hole of diameter 2 or 4 mm was vertically drilled at a position of 3 cm from the bottom. A short piece of square-section copper stick 2 or 4 mm in diameter was inserted and fixed in place with polyester resin. These bumps were square.

Results and Discussion

Effect of Impurities Stuck to Cathode on Nodulation

Figure 1 shows a picture of a cathode copper plate without impurity before and after electrodeposition for 20 h. Without impurity on the cathode, smoothly deposited copper was obtained. Figure 2 shows a picture of a copper plate with an Ag particle as the impurity before and after electrodeposition for 20 h. Copper was deposited smoothly in the same way as the plate without impurity. In contrast, Fig. 3 shows a picture of a copper plate with nonconductive polyester resin before electrodeposition, after electrodeposition for 20 h, and after electrodeposition for 40 h. After 20 h, a depression was observed at the place where polyester resin was embedded. However, after 40 h, copper was deposited smoothly. This indicates that evenly-embedded impurities have no effect on nodulous growth. For nodulation to occur, the unintended formation of bumps with a certain size appears to be necessary.

Fig. 1 Pictures of a copper plate without impurity before and after electrodeposition for 20 h

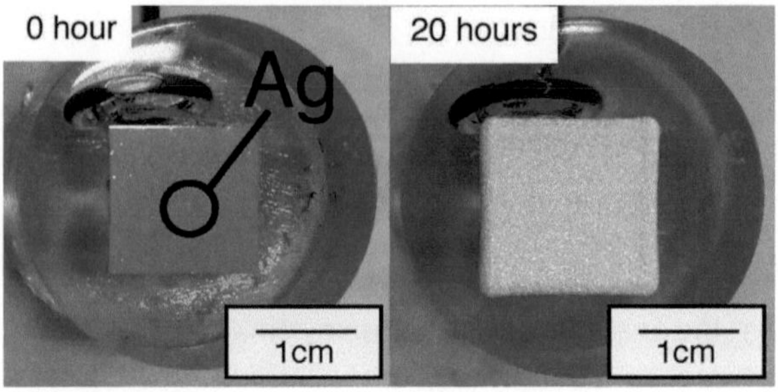

Fig. 2 Pictures of a copper plate with an evenly-embedded Ag particle before and after electrodeposition for 20 h

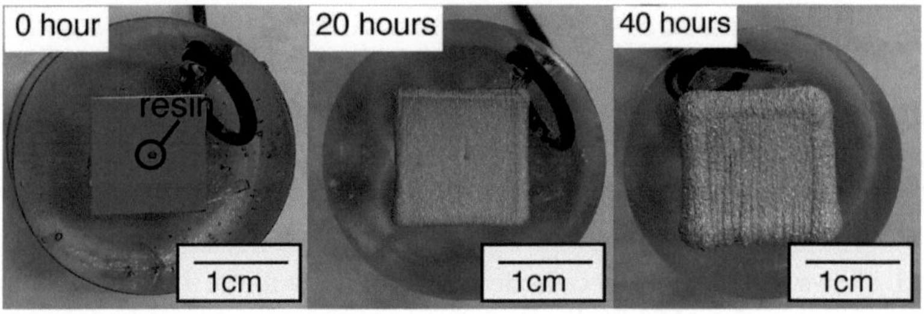

Fig. 3 Pictures of a copper plates with evenly-embedded polyester resin before electrodeposition, after electrodeposition for 20 h, and after electrodeposition for 40 h

Observation of Nodule Growth

Figure 4 shows the relationship between the initial height of the bump before electrodeposition and the change in height of the bump before and after electrodeposition. When the height of the projection before the electrodeposition was 1–10 mm, there was an almost linear relationship between the initial height and the change in height. When the initial height exceeded 10 mm, the bump grew more rapidly because of dendritic growth. In contrast, if the height was less than 1 mm, there was a characteristic relationship: when the diameter of the copper stick was 2 mm, there was little change in the height of the bump before and after electrodeposition. However, when the diameter of the copper stick was 4 mm, more rapid growth occurred compared to the linear relationship. Furthermore, in the case of the bump made by hammering, despite the diameter being about 2 mm, the projection became smaller. As a result, it is supposed that, if the initial height of the bump is less than 1 mm, the growth is influenced more by the shape of the bump than by its height.

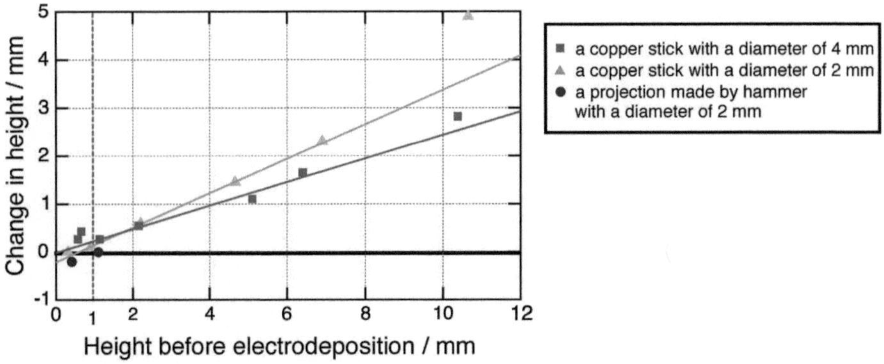

Fig. 4 Relationship between the initial height of the bump before electrodeposition and the change in height of the bump before and after electrodeposition

Conclusion

In this study, we examined the deposition behavior of copper at impurities and bumps. The results indicated that an accidental bump of a certain size is necessary to cause nodulation, and if the height of the bump is less than 1 mm the growth is more influenced by the shape of the projection than by its height. Therefore, we can infer that it is necessary for an aggregate of conductive anode slime with a certain shape and size to attach on the cathode in order to cause nodulation.

Reference

1. Dutrizac JE, Chen TT (2013) Proceedings of Copper 99–Cobre 99 international conference III, 383

Conclusion

Microfluidic Solvent Extraction of Zinc from Low Concentration Sulfate Solution Using D2EHPA

Feng Jiang, Libo Zhang, Jian Jian, Hongying Xia, Shaohua Ju and Jinhui Peng

Abstract Microfluidic solvent extraction of zinc from sulfate solution was studied using di-2-ethyl hexyl phosphoric acid, D2EHPA, in sulfonated kerosene as diluent. The influence of different parameters on the extraction of zinc, such as flow rate, temperature, flow rate ratio, extraction concentration, and saponification ratio of extractant were investigated and the conventional and microfluidic extraction were compared. Under the optimized conditions, microfluidic extraction efficiency of zinc is up to 97.1%. It is better than that conventional extraction. Therefore, microfluidic solvent extraction can achieve high efficiency and low consumption of extraction process.

Keywords Microfluidic · Solvent extraction · Zinc · Microreactor

Introduction

At present, the purification of zinc hydrometallurgy lixivium is based on the chemical precipitation and replacement methods, which not only consume a large amount of chemical reagents, but also are difficult to achieve high degree of purification [1, 2]. However, the extractive purification of zinc from zinc oxide ore leaching solution is a better selection [3–5].

F. Jiang · L. Zhang · J. Jian · H. Xia (✉) · S. Ju · J. Peng
State Key Laboratory of Complex Nonferrous Metal Resources Clean Utilization, Kunming University of Science and Technology, Kunming 650093, Yunnan, China
e-mail: hyxia81@hotmail.com

F. Jiang · L. Zhang · J. Jian · H. Xia · S. Ju · J. Peng
Key Laboratory of Unconventional Metallurgy, Ministry of Education, Kunming 650093, Yunnan, China

F. Jiang · L. Zhang · J. Jian · H. Xia · S. Ju · J. Peng
Faculty of Metallurgical and Energy Engineering, Kunming University of Science and Technology, Kunming 650093, Yunnan, China

H. Kim et al. (eds.), *Rare Metal Technology 2018*, The Minerals, Metals & Materials Series, https://doi.org/10.1007/978-3-319-72350-1_32

In recent years, microfluidic technology for the extraction process have emerged in the past few years. Hotokezaka et al. [6] studied extraction of uranium from an aqueous phase into tributyl phosphate (TBP) containing organic phase using a microchannel etched in a glass chip. Priest et al. [7] studied solvent extraction of metal ions and complexes from leach solutions containing nanoparticles by microfluidic technology. Li et al. [8] studied separation of In^{3+} and Fe^{3+} with Di-(2-ethylhexyl) phosphoric acid as an extractant using an interdigital micromixer. Yin et al. [9] intensified the extraction process in a microreactor for extraction of La(III) with 2-ethylhexyl phosphoric acid-2-ethylhexyl ester (P507). Zhang et al. [10] carried out the extraction and separation of Co from a Ni sulfate solution using a coiled flow inverter.

The above results reported in the literature indicate that microfluidic technology is applicable for the solvent extraction process and can be used for the construction of an efficient separation process. In this work, microfluidic solvent extraction of zinc from sulfate solution was studied. The influence of different parameters on the extraction of zinc, such as flow rate, temperature, flow rate ratio, extraction concentration, and saponification ratio of extractant were investigated and the conventional and microfluidic extraction were compared. Under the optimized conditions, microfluidic extraction efficiency of zinc is up to 97.1%. It is better than that conventional extraction. Therefore, microfluidic solvent extraction can achieve high efficiency and low consumption of extraction process.

Experimental

Materials and Analysis

The commercial extractant P204 (Di-(2-ethylhexyl)phosphoric acid, $C_{16}H_{35}O_4P$) was provided by Shanghai Laiyashi Chemical Co., Ltd. The diluent, 260# solvent naphtha (C11 ~ C17 alkanes mixtures primarily, and containing 4wt% aromatic hydrocarbons) were simultaneously supplied by Luoyang Aoda Chemical Co., Ltd, China. The organic phase was obtained by diluting the extractant using the diluent to specified concentration. $ZnSO_4 \cdot 7H_2O$ was analytical grade and purchased from Xilong Scientific Co., Ltd. The aqueous phase was prepared by dissolving $ZnSO_4 \cdot 7H_2O$ in deionized water. The concentration of Zn^{2+} in aqueous phase is 15 g/L at all the experiments.

Apparatus and Extraction Produce

The microfluidic experiments were performed in an interdigital micromixer with 40 μm width channels fabricated by IMM, Germany It consists of a mixing element, two opposite inlets and one upwards outlet, as shown in Fig. 1a. The constant

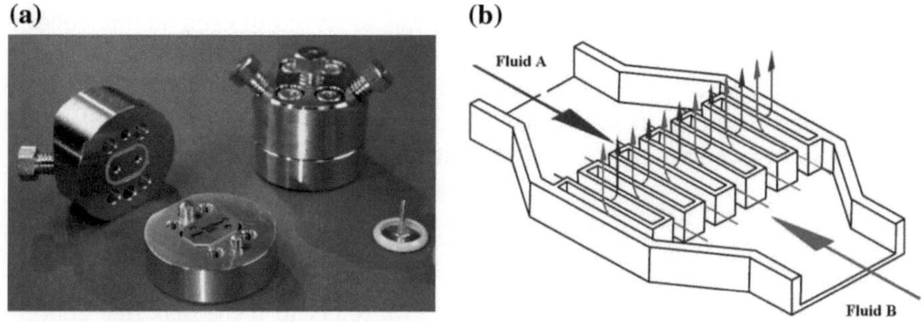

Fig. 1 Schematic diagram of micromixer **a** interdigital micromixer, **b** principle of mixing

flow pumps (HLB-4015) provided by Yanshang Instrument Factory, China, were used to pump the two fluids into the microchannels and the flow velocity was in the range of 0.1–40 mL min^{-1}. The two fluids flowed through the interdigital channels and then flowed upwards into a slit, which is perpendicular to the interdigital structure (shown in Fig. 1b [11]) where the mixing and extraction reaction took place and then fluids were clarified and separated in the separatory funnel. A schematic photo of the experimental setup is shown in Fig. 2.

The conventional solvent extraction experiment by contacting equal volumes of organic and aqueous phase (A/O volume ratio = 1) in a separatory funnel of 60 mL. The separatory funnel was immersed in a roundtrip thermostatic water bath oscillator (WHY-2), and both phases were mechanically mixed at a constant stirring speed of 350 rpm. After mixing for constant time, the mixture was allowed to stand for 10 min to obtain two individual phases.

The upper phase was the Zn^{2+} loaded organic phase and the bottom phase was the phase raffinate in both microfluidic extraction and the conventional extraction

Fig. 2 Schematic photo of microfluidic experimental setup

experiment. In all experiments, the raffinate was analyzed to estimate the concentrations of Zn^{2+} using inductively coupled plasma-atomic emission spectroscopy (AA240FS ICP-AES spectrophotometer). The concentration in the organic phase was estimated based on mass balance. The extraction efficiency, E, was calculated based on the concentration of metal present as shown in the following equation:

$$E = \frac{C_0 - C_{out}}{C_0} \times 100\% \quad (1)$$

where E represents the extraction efficiency, C_0 refers to the initial zinc concentration in the aqueous phase, and C_{out} is the zinc concentration in the aqueous phase after extraction.

Results and Discussion

Effect of Flow Rate

The effect of flow rate on the extraction of zinc in microfluidic solvent extraction was studied. The volumetric flow rate of the aqueous phase was equal to that of the organic phase, the flow rate of 5, 10, 15, 20, 25 and 30 mL/min was adapted to investigate the influence on the extraction of zinc. Figure 3 shows the effect of flow rate on the extraction of zinc in the experiment of microfluidic extraction.

The extraction efficiency of zinc increased with an increase flow rate from 5 to 30 mL/min, with the extraction efficiency of increasing to 45.27% at a flow rate of 25 mL/min. The extraction efficiency of zinc decreased to 44.63% at the flow rate of 30 mL/min.

Fig. 3 Effect of flow rate on the extraction of zinc (conditions: $[Zn^{2+}]$ = 15 g/L, room temperature, A/O = 1, extractant concentration = 30%, saponification ratio = 0)

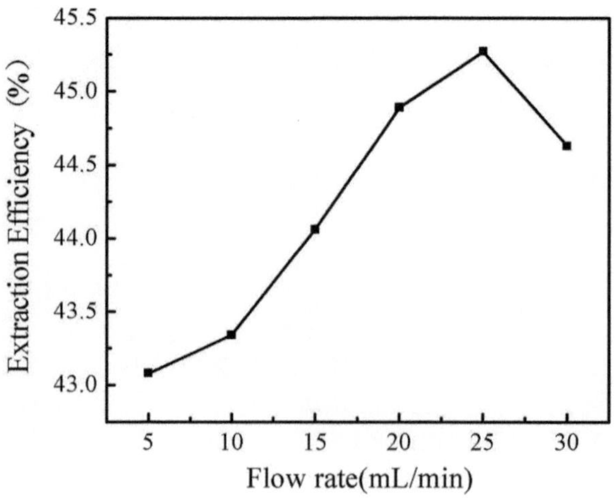

Since the internal volume of the microreactor is on the order of 8 μL (volume of the chamber), residence times were less than 0.019 s at the flow rate of 25 mL min^{-1}. Despite these extremely short residence time, the thermodynamic equilibrium could also be reached.

Effect of Temperature

It is very easy to control the temperature of extraction process in interdigital micromixer. In this experiment, the fluid tube and micromixer were placed in a constant temperature water bath to alter the process temperature. Experiments were conducted by varying the temperature in the range of 40 to 90 °C, with the extractant concentration of 30% and the flow rate of 25 mL/min. The extraction efficiency of zinc is shown in Fig. 4.

The extraction reaction is known to be exothermic and the increase in temperature is expected to shift the equilibrium reaction to reverse reducing the equilibrium conversion. However, as shown in Fig. 4, the extraction efficiency of Zn^{2+} was not significantly affected within the temperature range covered in the present study. As per equilibrium as well supported by the experimental results, room temperature was chosen as the desirable condition for other experiments.

Effect of Flow Rate Ratio

The flow rate ratio of microfluidic solvent extraction is equivalent to the phase ratio of batch solvent extraction. The effect of flow rate ratio on extraction of zinc was studied.

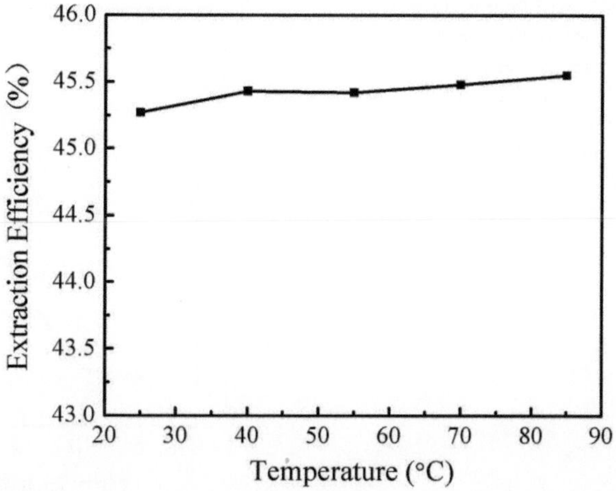

Fig. 4 Effect of temperature on the extraction of zinc (conditions: $[Zn^{2+}]$ = 15 g/L, flow rate = 25 mL/min, A/O = 1, extractant concentration = 30%, saponification ratio = 0)

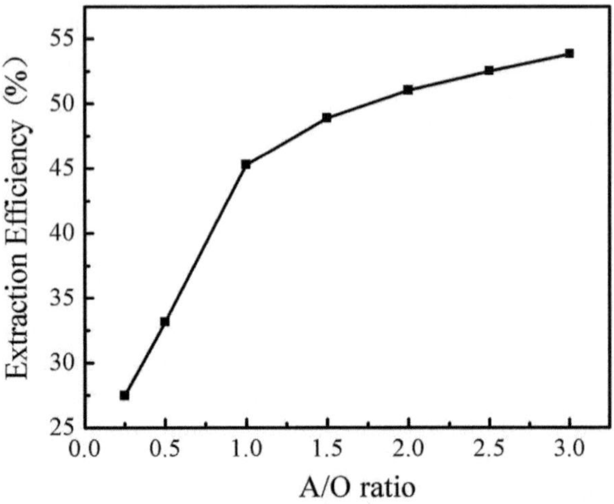

Fig. 5 Effect of phase ratio on the extraction of zinc (conditions: flow rate = 25 mL/min, room temperature, extractant concentration = 30%, saponification ratio = 0)

Figure 5 shows an increase in the extraction efficiency of Zn^{2+} with increasing of the phase ratio. The increasing of phase ratio is equivalent to the increasing of extractant concentration which improved extraction efficiency of Zn^{2+}. However, at large phase ratio the Zn^{2+} concentration in the organic phase is relatively low, which would increase the stripping cost as well as higher consumption of acid. Considering the economic costs, a phase ratio of 1:1 is recommended.

Effect of Extractant Concentration

The effect of P204 concentration in the range of 5% (V/V) to 40% (V/V) on the extraction of zinc was studied. Figure 6 shows that the extraction efficiency

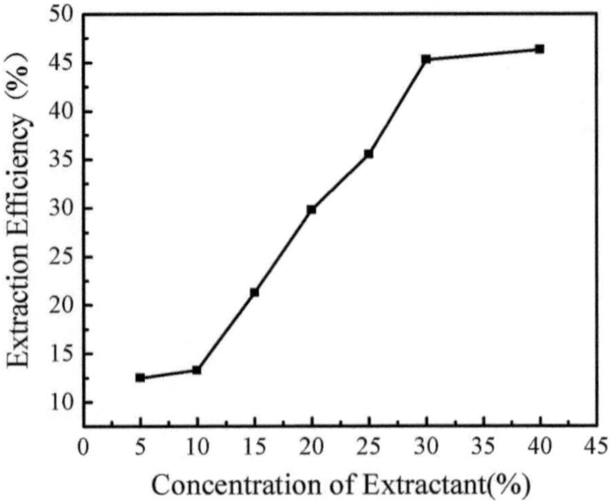

Fig. 6 Effect of D2EHPA concentration on the extraction of zinc (conditions: flow rate = 25 mL/min, room temperature, A/O = 1, saponification ratio = 0)

increased linearly with an increase in extractant concentration from 5 to 30%. The extraction efficiency of zinc was up to 45.27% at an extraction concentration of 30% and there was no significant increase when the extractant concentration was above 30%. Therefore, based on the above discussion, it can be seen that 30% (V/V) of P204 concentration is appropriate to extract zinc from the solution containing zinc ion.

Effect of Saponification Ratio

Saponification is a process in which the metal ions replace H^+ ions through acid-alkali neutralization reaction. Common saponifiers of D2EHPA are NaOH, NH_4OH and Na_2CO_3, and the organic phase is the oil-water emulsion after saponification [12]. And, the extraction reaction is the exchange process of sodium ions with metal ions, which accelerates the extraction reaction in the alkaline environment [13].

The effect of the organic phase saponified by NaOH solution on the extraction of zinc was studied. It can be seen from Fig. 7 that the extraction ability of extractant D2EHPA saponified is enhanced. The extraction efficiency of zinc increased linearly with an increase in saponification ratio from 0 to 60%. And the extraction efficiency of zinc reaches maximum, 97.1% at the saponification ratio of 60%. Therefore, the saponification ratio of 60% was selected as the preferred saponification ratio.

Fig. 7 Effect of saponification ratio on the extraction of zinc (conditions: flow rate = 25 mL/min, room temperature, A/O = 1, D2EHPA concentration = 30%)

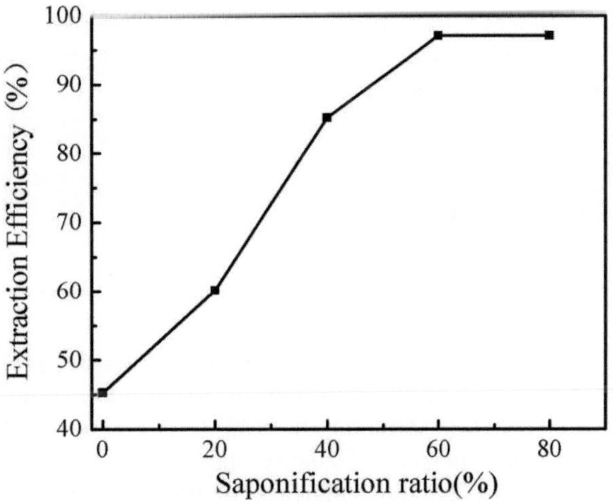

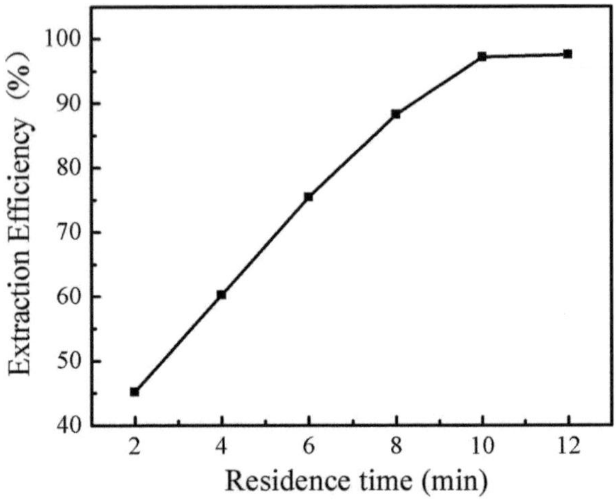

Fig. 8 Effect of residence time on the extraction of zinc in the conventional solvent extraction

Comparison Microfluidic and Conventional Solvent Extraction

The experiments of conventional solvent extraction of zinc were carried out under the preferred parameters of microfluidic solvent extraction: room temperature, phase ratio of 1:1, the D2EHPA concentration of 30%, and saponification ratio of 60%. The residence time was adapted from 2 to 12 min to investigate the effect on the extraction of zinc.

Figure 8 shows that the extraction efficiency increased linearly with an increase in residence time from 2 to 10 min. The residence time was 10 min for the extraction efficiency of zinc reaching to 97%, however, only 0.019 s was required for microfluidic solvent extraction.

Conclusion

In this work, an interdigital micromixer was used for extraction of zinc from zinc sulfate solution.

According to single factor experiments, the preferred microfluidic extraction parameters were the flow rate of 15 mL/min, room temperature, A/O ratio of 1, extractant concentration of 30% (V/V) and saponification ratio of 60%. The extraction efficiency of zinc reaches 97.1% under the preferred conditions.

A zinc extraction efficiency of 97% was achieved by microfluidic extraction technology needing 0.019 s, but batch extraction technology took 10 min.

Acknowledgements We gratefully acknowledge the support of the National Natural Science Foundation of China (Grant No. U1302271 and 51504116) and China Scholarship Council (Grant No. 201608740004).

References

1. Xing-hua L (2007) Solvent extraction of multiple nonferrous metals in the South Africa. World Nonferrous Metals 1:20–21
2. Zhang X, Li X, Cao H, Zhang Y (2010) Separation of copper, iron (III), zinc and nickel from nitrate solution by solvent extraction using LK-C2. Sep Purif Technol 70(3):306–313
3. Liu G, Zhang J (2005) Recent advance for separation of zinc from cadmium. Hydrometallurgy China 4:009
4. Fu W, Chen Q, Wu Q, Hu H, Bai L (2010) Solvent extraction of zinc from ammoniacal/ammonium chloride solutions by a sterically hindered β-diketone and its mixture with tri-n-octylphosphine oxide. Hydrometallurgy 100(3):116–121
5. Vahidi E, Rashchi F, Moradkhani D (2009) Recovery of zinc from an industrial zinc leach residue by solvent extraction using D2EHPA. Miner Eng 22(2):204–206
6. Hotokezaka H (2005) Development of the innovative nuclide separation system for high-level radioactive waste using microchannel chip-extraction behavior of metal ions from aqueous phase to organic phase in microchannel. Prog Nucl Energy 47:439–447
7. Priest C, Zhou J, Klink S, Sedev R, Ralston J (2012) Microfluidic solvent extraction of metal ions and complexes from leach solutions containing nanoparticles. Chem Eng Technol 35(7): 1312–1319
8. Li C, Jiang F, Ju S, Peng J, Wei Y, Zhang L (2017) Using an interdigital micromixer for separation of In3+ from zinc hydrometallurgical process with D2EHPA as an extractant. Arch Metall Mater 62(2):873–878
9. Yin S, Zhang L, Peng J, Li S, Ju S, Zhang L (2015) Microfluidic solvent extraction of La (III) with 2-ethylhexyl phosphoric acid-2-ethylhexyl ester (P507) by a microreactor. Chem Eng Process 91:1–6
10. Zhang L, Hessel V, Peng J, Wang Q, Zhang L (2017) Co and Ni extraction and separation in segmented micro-flow using a coiled flow inverter. Chem Eng J 307:1–8
11. Zhang L, Peng J, Ju S, Zhang L, Dai L, Liu N (2014) Microfluidic solvent extraction and separation of cobalt and nickel. RSC Adv 4(31):16081–16086
12. Qin W, Li W, Lan Z, Qiu G (2007) Simulated small-scale pilot plant heap leaching of low-grade oxide zinc ore with integrated selective extraction of zinc. Miner Eng 20(7): 694–700
13. Long H, Chai L, Qin W, Tang S (2010) Solvent extraction of zinc from zinc sulfate solution. J Cent South Univ T 17(4):760–764

Author Index

H. Kim et al. (eds.), *Rare Metal Technology 2018*, The Minerals, Metals & Materials Series, https://doi.org/10.1007/978-3-319-72350-1

Subject Index

H. Kim et al. (eds.), *Rare Metal Technology 2018*, The Minerals, Metals & Materials Series, https://doi.org/10.1007/978-3-319-72350-1

MIX
Papier aus verantwortungsvollen Quellen
Paper from responsible sources
FSC® C105338

If you have any concerns about our products,
you can contact us on
ProductSafety@springernature.com

In case Publisher is established outside the EU,
the EU authorized representative is:
Springer Nature Customer Service Center GmbH
Europaplatz 3, 69115 Heidelberg, Germany

Printed by Libri Plureos GmbH
in Hamburg, Germany